An Introduction to Computational Thermodynamics

Brian R. Greenhalgh

An Introduction to Computational Thermodynamics

Phase Equilibrium and Equations of State

Springer

Brian R. Greenhalgh
Oegstgeest, The Netherlands

ISBN 978-3-032-30316-5 ISBN 978-3-032-30317-2 (eBook)
https://doi.org/10.1007/978-3-032-30317-2

This Springer imprint is published by the registered company Springer Nature Switzerland AG
The registered company address is: Gewerbestrasse 11, 6330 Cham, Switzerland

To Claudia, Alexander and Victoria: follow your dreams.

The Purpose of This Book

This book was written for students and working professionals seeking an understanding of Thermodynamics from a computational perspective. This emphasis implies that this book is not yet another text on the main thermodynamic concepts themselves, but rather a guide on performing thermodynamic calculations. While this book assumes that the reader already has some familiarity with Thermodynamics (i.e., the four laws, and the typical language used in the subject), the most important concepts will be referenced and explained where this helps to illustrate the ideas being presented.

Essentially this book seeks to demystify and make absolute the execution of key thermodynamic calculations, principally properties from an equation of state, phase boundaries, and flash calculations. To do so, this book emphasizes the following:

- The role played by Helmholtz (vs. Gibbs) free energy as the machinery of computational thermodynamics.
- The different criteria for equilibrium (beyond $\Delta G = 0$).
- The subjectivity of terms such as "liquid," "vapor," and "solvent" applied to fluid behavior.
- The ideal gas equation of state ($PV = nRT$) is of little use in phase equilibrium modeling, as it does not describe dense phase properties. This text demonstrates how to use more capable models such as Soave Redlich Kwong (SRK). The methods outlined are readily extended to other cubic equations of state such as Peng-Robinson. Though this text focuses on equations of state, and does not formally cover activity coefficient modeling, the methods outlined can be extended accordingly.
- The more capability you build in thermodynamics, the easier it will be for you to progress. This statement applies to your general understanding, where yesterday's complexity becomes tomorrow's familiar tool. It also applies to programming; the more time you invest in correctly coding the various derivatives and routines described in this text, the faster your toolkit will expand, allowing you to easily draw on complex methods and routines as you develop further.

Acknowledgments This book was inspired by the work of the late professor Michael L. Michelson. Though I never had the chance to meet this singular contributor to Thermodynamics in person, Professor Michelsen's published work and kind advice through written correspondence had a very beneficial effect on my efforts to understand computational thermodynamics. I wrote this book not as an attempt to supersede his text (co-authored by J.M. Mollerup) "Thermodynamic Models: Fundamentals and Computational Aspects," but rather my work is intended as a companion to his, or more accurately, a primer for students seeking to better understand his work and the unique mastery of Thermodynamics he possessed. If I have succeeded, there will be more students who are able to follow in his footsteps and a wider realization that there may not come another of his stature in the field for some time, if ever. This book also derives from my collaboration and friendship with Torben Laursen and Joseph Moebus. Torben and his commercial software helped me understand a number of thermodynamic concepts, and many years after our first conversation I continue to find wisdom in his statements and anecdotes. Two examples are reproduced here for the reader's benefit: "Start simple, and add complexity as you go," and "A robust volume solver is the heart of a thermodynamics engine." If Torben taught me the concepts, Joe taught me how to implement them. For the basic lessons on coding through his patient explanation of more advanced concepts, I am deeply grateful.

Competing Interests The author has no competing interests to declare that are relevant to the content of this manuscript.

Contents

Chapter 1
Introduction

1.1 Conventions and Units

The author's intent was to write a text that would be helpful to those wishing to implement thermodynamic calculations in computer code and to demonstrate such methods without particular commitment to one language over another. For this reason, the author's source code was not copied into this text as a substitute for explaining the ideas behind it. This proved more challenging than expected, as key concepts like variable initialization, loops, and assignment operators are handled differently among coding languages. In an attempt to simplify the use of pseudocode, a number of conventions are adopted and explained in Appendix 8.3. Also note that the author used a zero-index programming language, meaning that locations within arrays start at 0, not 1. As mentioned above, every effort was made to avoid writing this text in such terms. However, there are places in the text where formally declaring this was deemed to be necessary.

The choice of units when performing thermodynamic calculations is somewhat arbitrary, aside from the need for absolute temperatures and pressures. The utility of moles vs mass can be debated, particularly when dealing with polymers and/or reactions. For various reasons, the author elected to use the following units in this text:

- Temperature in Kelvin
- Volume in cubic centimeters
- Pressure in Pascals
- Material quantity in moles (and composition in mole fraction)
- Energy (and energy terms) in Joules

These choices yield familiar results, such as the gas constant taking the value:

$$R_g = 8.3145 \frac{J}{\text{K*mol}}$$

B. R. Greenhalgh, *An Introduction to Computational Thermodynamics*,
https://doi.org/10.1007/978-3-032-30317-2_1

However, one can also see that these units do not necessarily form a consistent set. For example, energy is the product of pressure and volume from a dimensional perspective, but:

$$1J \neq (1\text{Pa})x(1\text{cm}^3)$$

As a result, some of the equations in this text have factors to reconcile the units. For example, according to the definition of compressibility factor,

$$Z \equiv \frac{PV}{nR_gT}$$

Yet the choice of units requires this relation to be written as:

$$Z = \frac{1}{1E6}\frac{PV}{nR_gT}$$

This is of course regrettable, as the former expresses its underlying concept far more elegantly than the latter. However, to correspond with pressures in Pascals, one would have to use volumes in cubic meters, which is an inconvenient unit. The author suggests that fundamental analysis be done without undue regard for units (i.e., writing $Z = \frac{PV}{nR_gT}$) and to resolve units during implementation. True, this poses a risk of numerical errors, but the importance (and reward) of seeing thermodynamics for its elegance is also to be considered. To remedy this issue, the author has tried to be explicit in terms of units where confusion might otherwise arise.

1.2 The Laws of Thermodynamics

It must be recognized that the four Laws of Thermodynamics are in fact *laws*; they are not definitions, they cannot be proved, nor does one imply any of the others. They result from experimentation and experience that suggest their truth, in that no repeatable, properly executed experiment or process so far contradicts them. This vast body of evidence has been carefully collected into four compact statements; the $0th$, $1st$, $2nd$, and $3rd$ Laws of Thermodynamics. The scope of this text requires little discussion of the $0th$ or $3rd$ Law, while limited discussions of the 1st and 2nd are offered below.

The $1st$ Law of Thermodynamics is often referred to as the conservation of energy. This law can be stated in many ways, but in the author's opinion, a very useful formalism is that "The Energy of the Universe ***cannot change***." To make use of this statement practically, one must of course be familiar with the typical language of Thermodynamics, including terms[1] like "system," "surroundings," and "boundary." If a system and its surroundings exchange energy through a boundary, the $1st$ Law presents no restriction on the energy change of either the system or the surroundings, other than that the change in the total energy (system plus surroundings) must be zero. There is also the more common statement of the First Law; "Energy cannot be created or destroyed, but simply changed in form." The author prefers the earlier statement with its implication of *total* energy, rather than the energy of a frying pan or a unit operation in a chemical plant. Such phrasing naturally raises the question of what we mean by *energy*, and here we will also assume previous experience of the reader with terms like kinetic and potential energy. Those working in the chemical industries will recognize that, beyond these "mechanical" concepts, matter also possesses attributes due to its "chemical identity." That is to say, a baseball and a hand grenade, which could be argued possess a similar mass, differ significantly even at the same velocity. The same can be said for two pure substances; water and benzene have very different properties at the same temperature and pressure. Recognition of this naturally leads to the concept of internal energy: the energy of the chemical state. This is typically given the symbol U. When combined with the "mechanical energies," the First Law of Thermodynamics can be applied to Closed Systems:

$$\Delta(E_{kin} + E_{pot} + U)_{Sys} + \Delta(E_{kin} + E_{pot} + U)_{Surr} = 0 \tag{1.1}$$

This text is not concerned with the calculation of or changes in the kinetic or potential energies (E_{kin} and E_{pot}, respectively). Also, given the broad possibilities for definition and complexity, one rarely attempts to model the properties of the surroundings. Instead, changes in the energy of the surroundings are described using the concepts of heat and work. This results in a familiar restatement of the $1st$ Law of Thermodynamics for Steady-State, Closed Systems:

$$\Delta(U)_{Sys} = -\Delta(E_k + E_p + U)_{Surr} = Q + W \tag{1.2}$$

With the formalism that properties are those of the system, the subscripts can be dropped, and a more familiar form is:

$$dU = \delta Q + \delta W \tag{1.3}$$

In keeping with standard nomenclature, we have also adopted the δ symbol to represent changes in path functions, while d is used for state functions (see Glossary). Many excellent references are available on the concepts of and distinctions

[1] Should a review be necessary, the reader is referred to the Glossary in Chap. 7 of this text or to the many introductory texts available on Thermodynamics.

between heat and work. The scope of this text is limited to boundary work, which can be expressed as:

$$\delta W_{Bound} = -PdV \tag{1.4}$$

The combination of Eq. 1.3 and Eq. 1.4 gives:

$$dU = \delta Q - PdV \tag{1.5}$$

A more general statement would include the effect of composition on the internal energy:

$$dU = \delta Q - PdV + \sum_r \mu_r dn_r \tag{1.6}$$

In Eq. 1.6, the symbol μ_i is the chemical potential of species i (See Appendix 8.8). It is important to understand that Eq. 1.6 is written for a closed system. The summation of μ_i terms does not describe the addition of material across the boundary into the system, but rather the difference in internal energy that results from the existence of molecules of one species vs. another, through chemical reactions for example. As stated above, Eq. 1.6 is *not* a general statement of the First Law, but a simplification based on the given assumptions.

The $2nd$ Law causes even more grief to students of Thermodynamics than the $1st$, and in this text only a limited attempt at clarifying this concept will be made. As with the $1st$, a compact statement of the $2nd$ Law is "The Entropy of the Universe ***cannot decrease***."

$$\Delta S_{Sys} + \Delta S_{Surr} \geq 0 \tag{1.7}$$

Though both the $1st$ and $2nd$ Laws introduce the important and surprisingly hard to define concepts of energy and entropy, the $2nd$ Law suggests something deeper than the $1st$. Strict application of the $1st$ Law alone does not tell us that liquid water cannot form ice above its melting temperature; it simply tells you the amount of heat and/or work that would be required if this were to happen. The $2nd$ Law forbids certain processes through its inequality and therefore provides a sense of direction and possibility. As with the $1st$ Law, this statement of the $2nd$ Law places no prohibition on changes in the system's or the surrounding's entropy, but rather on their total. Also, as with energy and the $1st$ Law, the $2nd$ Law is of little use without a means of calculating entropy or at least changes therein. Here the concept of a reversible process is applied, wherein the change in total entropy (system plus surroundings) is zero, leading to an equality in Eq. 1.7, and the following *definition* of entropy applies:

$$dS \equiv (\frac{\delta Q}{T})_{rev} \tag{1.8}$$

It is essential to recognize that Eq. 1.8 only holds for reversible heat transfer (the subscript rev must not be ignored). That is to say, it is a misconception that "entropy equals heat divided by temperature." Equation 1.8 is useful because entropy is a state function, providing the critical link between the $1st$ and $2nd$ Laws, leading to the concept of Thermodynamic Potentials.

1.3 Thermodynamic Potentials

Combining Eqs. 1.6 and 1.8, and considering only reversible paths between states of interest (the "rev" subscript has not been *ignored*), one has:

$$dU = TdS - PdV + \sum_{i} \mu_r dn_r \tag{1.9}$$

Note that only state functions remain in Eq. 1.9, the development of which demonstrates a major approach of this text; the terminology and assumptions have been invoked, and we are left with a purely mathematical relationship between variables with which we are (or should be) familiar. This is not to say that how we arrived at Eq. 1.9 is unimportant; a comfortable understanding on the part of the reader is assumed. Equation 1.9 is also our first look at U as a thermodynamic potential, and it suggests that U is a function of entropy, volume, and mole numbers. This may come as a surprise to some readers, who might be tempted to represent internal energy in terms of temperature and volume (or pressure). As will be shown in this text, this is of course possible, but the resulting expression is not nearly as compact as Eq. 1.9. This is an important idea that you should associate with the term *canonical* (see Glossary). Nevertheless, use of Eq. 1.9 requires information on how T, P, and μ_i depend on S, V, and $\overline{n}$. Such models are not common,[2] and even if they were, they would produce information of the form $U = f(S, V, \overline{n})$; of little general use in chemistry or chemical engineering settings. One can mathematically eliminate entropy as an independent variable from Eq. 1.9 through a *Legendre transformation*:

$$dU - d(TS) = TdS - PdV + \sum_{r} \mu_r dn_r - d(TS)$$

$$d(U - TS) = TdS - Pdn + \sum_{r} \mu_r dn_r - TdS - SdT$$

$$d(U - TS) = -SdT - PdV + \sum_{r} \mu_r dn_r \tag{1.10}$$

[2] Think about models such as the ideal-gas equation of state, or others that you know; what are the independent variables?

Some readers will be familiar with the left side of Eq. 1.10, the Helmholtz free energy:

$$A \equiv U - TS \tag{1.11}$$

For those unfamiliar with Helmholtz free energy, do not despair; what is important at this stage of the text is that you recognize that something (not the internal energy) is a canonical function of T, V, and $\overline{n}$;

$$dA = -SdT - PdV + \sum_r \mu_r dn_r \tag{1.12}$$

and, from a calculus perspective, we can conclude

$$\left(\frac{\partial A}{\partial T}\right)_{V,\overline{n}} = -S \tag{1.13}$$

$$\left(\frac{\partial A}{\partial V}\right)_{T,\overline{n}} = -P \tag{1.14}$$

$$\left(\frac{\partial A}{\partial n_i}\right)_{T,V,n_{k\neq i}} = \mu_i \tag{1.15}$$

At this early stage, Eq. 1.14 is the most important to understand. If we know the pressure as a function of T, V and $\overline{n}$, (i.e., we use a pressure explicit equation of state) we can compute differences in the Helmholtz free energy:

$$A(T, V, \overline{n}) - A(T, V_R, \overline{n}) = \int_{V_R}^{V} \left(\frac{\partial A}{\partial V}\right)_{T,\overline{n}} dV = -\int_{V_R}^{V} PdV \tag{1.16}$$

A more familiar potential is the Gibbs free energy, which we can develop via a Legendre transformation of Eq. 1.12, replacing volume as an independent variable with pressure:

$$dA + d(PV) = -SdT - PdV + \sum_r \mu_r dn_r + d(PV)$$

$$d(U - TS + PV) \equiv d(H - TS) \equiv d(G) = -SdT + VdP + \sum_r \mu_r dn_r \tag{1.17}$$

Despite the fact that most students of Thermodynamics have an idea of Gibbs free energy (and its utility in addressing problems at fixed temperature and pressure), direct use of the Gibbs free energy as a framework for thermodynamic computations

Table 1.1 Summary of potentials for simple systems

Name (Legendre transform of A)	Differential form	Derivatives
Helmholtz free energy (A)	$dA = -SdT - PdV + \sum_r \mu_r dn_r$	$\left(\frac{\partial A}{\partial T}\right)_{V,\overline{n}} = -S$
		$\left(\frac{\partial A}{\partial V}\right)_{T,\overline{n}} = -P$
		$\left(\frac{\partial A}{\partial n_i}\right)_{T,V,n_{k\neq i}} = \mu_i$
Gibbs free energy (A+PV)	$dG = -SdT + VdP + \sum_r \mu_r dn_r$	$\left(\frac{\partial G}{\partial T}\right)_{P,\overline{n}} = -S$
		$\left(\frac{\partial G}{\partial P}\right)_{T,\overline{n}} = V$
		$\left(\frac{\partial G}{\partial n_i}\right)_{T,P,n_{k\neq i}} = \mu_i$
Enthalpy (A+PV+TS)	$dH = TdS + VdP + \sum_r \mu_r dn_r$	$\left(\frac{\partial H}{\partial S}\right)_{P,\overline{n}} = T$
		$\left(\frac{\partial H}{\partial P}\right)_{S,\overline{n}} = V$
		$\left(\frac{\partial H}{\partial n_i}\right)_{S,P,n_{k\neq i}} = \mu_i$
Internal energy (A+TS)	$dU = TdS - PdV + \sum_r \mu_r dn_r$	$\left(\frac{\partial U}{\partial S}\right)_{V,\overline{n}} = T$
		$\left(\frac{\partial U}{\partial V}\right)_{S,\overline{n}} = -P$
		$\left(\frac{\partial U}{\partial n_i}\right)_{S,V,n_{k\neq i}} = \mu_i$

is not practical. The reason for this is that equations of state[3] do not provide V as a function of P, because strictly speaking, volume is not a function[4] of pressure; one can have multiple volumes for the same pressure. A good example is water at 100 °C and 1 atm; do we want the liquid or the vapor volume? Therefore knowledge of other potentials (and the relationships between them) is very useful in computational thermodynamics. Table 1.1 summarizes the potentials for simple systems.

Table 1.1 need not be committed to memory, as it can largely be generated by what is often called the "Guggenheim Scheme":

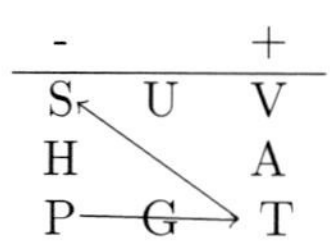

This scheme indicates that "changing G through T, leaving P alone, gives -S": $\left(\frac{\partial G}{\partial T}\right)_{P,\overline{n}} = -S$. The derivatives in Table 1.1 (except μ_i) can be generated

[3] Other than simple ones such as the ideal gas EOS, which is of limited use.

[4] The mathematical definition of a function includes the requirement that a unique output results from a unique input.

analogously. To remember the scheme, consider the sentence "**P**ressure **H**olds **S**teady **U**ntil **V**olume **A**nd **T**emperature **G**row'.'

Also, as mentioned above, the chemical potential can be written in terms of any of the potentials:

$$\mu_i = \left(\frac{\partial A}{\partial n_i}\right)_{T,V,n_{k\neq i}} = \left(\frac{\partial G}{\partial n_i}\right)_{T,P,n_{k\neq i}} = \left(\frac{\partial H}{\partial n_i}\right)_{S,P,n_{k\neq i}} = \left(\frac{\partial U}{\partial n_i}\right)_{S,V,n_{k\neq i}} \tag{1.18}$$

Finally, application of the arbitrary order (See the Appendix for a discussion of this and other mathematical concepts) of mixed second derivatives results in the useful *Maxwell Relations*. For example, from the equivalence of

$$\left(\frac{\partial^2 G}{\partial P \partial n_i}\right) = \left(\frac{\partial^2 G}{\partial n_i \partial P}\right) \tag{1.19}$$

Table 1.1 implies:

$$\left(\frac{\partial \mu_i}{\partial P}\right)_{T,\overline{n}} = \left(\frac{\partial V}{\partial n_i}\right)_{T,P,n_{k\neq i}} \tag{1.20}$$

This concept is used to great effect in Thermodynamics.

1.4 Equations of State (EOS)

The V_R in Eq. 1.16 is an arbitrary reference volume corresponding to a reference value $A(T, V_R, \overline{n}) = A_R$; we do not know the value of A_R. To illustrate this concept, consider the simplest case; the ideal-gas equation of state:

$$P^{IG}(T, V, \overline{n}) = \frac{nR_gT}{V} \tag{1.21}$$

Integration of Eq. 1.14 gives:

$$\begin{aligned} A^{IG}(T, V, \overline{n}) - A^{IG}(T, V_R, \overline{n}) &= -\int_{V_R}^{V} P^{IG} dV \\ &= -\int_{V_R}^{V} \frac{nRT}{V} dV = -nR_gTln(\frac{V}{V_R}) \end{aligned} \tag{1.22}$$

This result allows us to compute changes in Helmholtz free energy, but we do not know the values of A or A_R, only their difference. If we differentiate Eq. 1.22 with respect to temperature, and make use of Eq. 1.13, we derive the familiar result:

$$S^{IG}(T, V, \overline{n}) - S^{IG}(T, V_R, \overline{n}) = nR_g ln(\frac{V}{V_R}) \tag{1.23}$$

Note that in Eq. 1.23, the temperature is the same for the state of interest and the reference state. Differentiating 1.22 with respect to volume simply results in $P = nR_gT/V$, while differentiating 1.22 with respect to mole number gives:

$$\mu_i^{IG}(T, V, \overline{n}) - \mu_i^{IG}(T, V_R, \overline{n}) = -R_gTln(\frac{V}{V_R}) \tag{1.24}$$

These equations provide an example of how an EOS is not particularly useful for *absolute* property calculations. For example, Eq. 1.23 does not tell us the exact functional form of the entropy, only how differences in this quantity depend on volume and mole numbers. Similarly, Eq. 1.24 does not provide the exact functional form for the chemical potential of an ideal gas, only how this quantity changes as a function of temperature and volume. For the time being, put this issue aside. In the next section, we apply the same methods used in the ideal gas example on more powerful equations of state. We will then make use of a concept called a *residual property*, which is the difference between the property from our model and that from an ideal gas.

1.5 Residual Properties

Let us perform the integration specified in Eq. 1.22 for another EOS; the Soave–Redlich–Kwong Equation of State (SRK EOS) [SRK paper]:

$$P = \frac{nR_gT}{(V - nb)} - \frac{n^2a}{V(V + nb)} \tag{1.25}$$

where a and b are typically written as functions of temperature and composition. Here we have written the SRK EOS as a function of T, V and $\overline{n}$; not in terms of mole fractions. We have also written it without particular regard to unit choice, as discussed in Sect. 1.1. Finally, as will be required later, the *minimum* volume of the SRK EOS is by inspection:

$$V_{min}^{SRK} = nb \tag{1.26}$$

Before proceeding, let's rewrite the SRK EOS into a more easily integrated form:[5]

$$P = \frac{nR_gT}{(V - nb)} - \frac{n^2a}{nb}(\frac{1}{V} - \frac{1}{V + nb}) \quad (1.27)$$

The integration specified in Eq. 1.22 yields:

$$\begin{aligned} A(T, V, \overline{n}) - A(T, V_R, \overline{n}) &= \left[-nR_gTln(V - nb) + \frac{n^2a}{nb}ln\frac{V + nb}{V} \right]_{V_R}^{V} \\ &= -nR_gTln\frac{(V - nb)}{(V_R - nb)} + \frac{n^2a}{nb}ln\frac{(V + nb)V_R}{V(V_R + nb)} \quad (1.28) \end{aligned}$$

We could proceed further, but we will encounter the same challenge as with Eq. 1.22; we do not know the value of $A(T, V_R, \overline{n})$ for an SRK fluid at an arbitrary V_R. We can, however, make significant progress if we compare calculations with the SRK EOS to those of an ideal gas. If we subtract Eq. 1.22 from Eq. 1.28, we have:

$$\begin{aligned} &A(T, V, \overline{n}) - A(T, V_R, \overline{n}) - A^{IG}(T, V, \overline{n}) + A^{IG}(T, V_R, \overline{n}) \\ &= -nR_gTln\frac{(V - nb)V_R}{(V_R - nb)V} + \frac{n^2a}{nb}ln\frac{(V + nb)V_R}{V(V_R + nb)} \quad (1.29) \end{aligned}$$

While it might seem that we've simply combined two (underdefined) equations, we must recognize that the conditions $(T, V, V_R, \overline{n})$ apply to both the SRK fluid and the Ideal Gas. Further, we are entirely free to select V_R at our convenience. If we choose to let $V_R \to \infty$ (an infinitely expanded gas), and we verify that SRK trends to ideal gas behavior at infinite volume (this is an important property of a useful EOS), we can conclude that $A(T, V_R, \overline{n}) \to A^{IG}(T, V_R, \overline{n})$. We can also conclude that terms like $\frac{V_R}{V_R \pm nb} \to 1$. With this selection Eq. 1.29 can be simplified to:

$$A(T, V, \overline{n}) - A^{IG}(T, V, \overline{n}) = -nR_gTln\left[\frac{V - nb}{V}\right] + \frac{n^2a}{nb}ln\left[\frac{V + nb}{V}\right] \quad (1.30)$$

The left side of Eq. 1.30 is the definition of *Residual* Helmholtz Free Energy:

$$A^{Res} \equiv A(T, V, \overline{n}) - A^{IG}(T, V, \overline{n}) \quad (1.31)$$

[5] Note that we retain the factors of n on a and b rather than canceling. This is to retain mole *numbers* rather than mole *fractions*. The importance of this will be clarified in subsequent sections.

It is important to understand how we arrived at Eq. 1.30, especially how certain terms cancel as $V_R \to \infty$. Also, the definition of a Residual Property relies on both the fluid and the IG states having the same T, V, and $\overline{n}$. When written in terms of residual properties, Eqs. 1.13 through 1.15 become:

$$\left(\frac{\partial A^{Res}}{\partial T}\right)_{V,\overline{n}} = -S^{Res} = -S(T, V, \overline{n}) + S^{IG}(T, V, \overline{n}) \tag{1.32}$$

$$\left(\frac{\partial A^{Res}}{\partial V}\right)_{T,\overline{n}} = -P^{Res} = -P(T, V, \overline{n}) + \frac{nR_gT}{V} \tag{1.33}$$

$$\left(\frac{\partial A^{Res}}{\partial n_i}\right)_{T,V,n_{k\neq i}} = \mu_i^{Res} = \mu_i(T, V, \overline{n}) - \mu_i^{IG}(T, V, \overline{n}) \tag{1.34}$$

Equations 1.32 through 1.34 do not provide something for nothing. We can differentiate Eq. 1.30 with respect to T,V, or $\overline{n}$, and we will simply produce residual properties. Without knowing the property of the ideal gas, we will not know the property of the SRK fluid. We do know the pressure of the ideal gas and the SRK fluid, which allows us to fully formulate $\left(\frac{\partial A^{Res}}{\partial V}\right)_{T,\overline{n}}$. With regard to $\overline{n}$, the next section will demonstrate that if we can formulate μ_i^{IG} (the chemical potential of the ideal gas), we can derive a useful form of μ_i for the SRK and another EOS. With regard to T, we will see that for a large class of thermodynamic calculations, we do not require knowledge of the ideal gas entropy (or other properties arising from other derivatives); these terms often cancel algebraically in the relevant equations.[6]

1.6 The Chemical Potential of an Ideal Gas (μ_i^{IG})

Having an expression for the chemical potential of an ideal gas is essential in the use of an EOS based on residual Helmholtz free energy. We now derive this expression. We begin by integrating Eq. 1.14 while formally assuming $P^{IG} = \frac{nR_gT}{V}$:

$$A^{IG} = -nR_gTln(V) + F(T, \overline{n}) \tag{1.35}$$

[6] The ideal-gas terms cancel algebraically in many "phase" equilibrium calculations. In others (those at a specified entropy or enthalpy, and calculations for reacting systems), the ideal-gas terms are indeed necessary.

Note carefully that we must allow the unknown function F to be a function of T and $\overline{n}$, but it is not a function of V. Referring to Appendix 8.6, and particularly Eq. 8.49,[7] we choose to rewrite Eq. 1.35 as:

$$A^{IG} = -(PV)^{IG} + \sum_r \mu_r^{IG} n_r$$

$$= -nR_gT + R_gT \sum_r \big[-ln(V) + f_r(T, \overline{n}) \big] n_r \tag{1.36}$$

We again allow the unknown functions f_i to be functions of T and $\overline{n}$, but not V. For the sake of compact notation, we will omit this statement of variables. In order for the given form of A^{IG} in Eq. 1.36 to be $1st$ order homogeneous in V and $\overline{n}$ (as detailed in Appendix 8.6), we require that $\big[-ln(V) + f_i \big]$ be $0th$ order in these variables. This requires the following (see Eq. 8.47):

$$-\frac{1}{V}V + \sum_r \frac{\partial f_i}{\partial n_r} n_r = 0$$

$$\sum_r \frac{\partial f_i}{\partial n_r} n_r = 1 \tag{1.37}$$

Two points should be emphasized here. The first is that f_i is not homogeneous in $\overline{n}$ (refer to Eq. 8.47). The second point is that, regardless of the values for n_j on the left side, the summation in the final form of Eq. 1.37 always produces 1 on the right side. This suggests:

$$f_i = ln(n_i) + k_i \tag{1.38}$$

with $k_i = k_i(T)$ (see the discussion below Eq. 1.36).
Combination of Eqs. 1.36 and 1.38 yields:

$$\mu_i^{IG} = -R_gTln(V) + R_gTln(n_i) + R_gTk_i \tag{1.39}$$

In order to determine k_i, we consider that the expression for μ_i^{IG} (including k_i) needs to reproduce the pure component value of g_i^{IG}. We can use *formation* data, measured calorimetrically, and standard state element references for this purpose. Although these data are often not required in phase equilibrium calculations,[8] we will invoke them here for clarity. The Gibbs free energies of pure substances are

[7] After reviewing Appendix 8.6, you should recognize that its content does not depend on the functional form of μ^{IG}. Also, in order to understand Thermodynamics, one occasionally has to make such forward and backward steps, understanding the past in terms of the future and vice versa.

[8] They are needed in solid–fluid equilibria calculations and for reacting systems.

typically tabulated at 298.15K and 1 bar. Using an expression for ideal-gas heat capacity, the value of g_i is then determined at the desired T and 1 bar (The details, not shown here, can be found in chemistry or reaction engineering texts.). Via Table 1.1, we have:

$$g_i(T, P)^{IG} = g_i(T, Po) + R_g T ln(\frac{P}{Po}) = g_i^o(T) + R_g T ln(P) - R_g T ln(Po) \tag{1.40}$$

The Po is often omitted in expressions for Gibbs free energy, as many reference values are tabulated at $Po = 1$bar. However, serious numerical errors associated with reference states will result from failing to properly account for this reference pressure. For this reason, it is best to retain it. Rewriting Eq. 1.39 in terms of pressure, we have:

$$\begin{aligned} \mu_i^{IG} &= -R_g T ln(\frac{nR_g T}{P}) + R_g T ln(n_i) + R_g T k_i \\ &= R_g T\left[k_i - ln(R_g T) + ln(P) + ln(c_i)\right] \end{aligned} \tag{1.41}$$

Here we have written the mole fraction as $c_i = \frac{n_i}{n}$, which has naturally arisen from the derivation. Comparing Eq. 1.40 with Eq. 1.41 written for a pure component, we have:

$$k_i(T) = \frac{g_i^o(T)}{R_g T} + ln(R_g T) - ln(Po) \tag{1.42}$$

The chemical potential of an ideal gas is then:

$$\mu_i^{IG}(T, P, \overline{n}) = g_i^o + R_g T ln(\frac{P}{Po}) + R_g T ln c_i \tag{1.43}$$

This is an important result, as it allows one to write expressions for chemical potential of real fluids (i.e. not ideal gases). As will be seen shortly, it also leads to the concept of fugacity, which forms a computational basis in phase equilibrium.

1.7 Criteria for Equilibrium

Many students of thermodynamics are familiar with statements of equilibrium in terms of Gibbs free energy. In this section, we will formalize the conditions for equilibrium. To do so, we revisit Eq. 1.8 and write it more generally in terms of the Clausius Inequality (which, itself is a statement of the $2nd$ Law):

$$dS \geq \frac{\delta Q}{T} \tag{1.44}$$

Note that in Eq. 1.44, reversibility has not been formally assumed, and the statement is an inequality. It is this inequality that gives rise to the approach to equilibrium states, as when Eq. 1.44 is substituted into Eq. 1.6, one has:

$$dU \leq TdS - PdV + \sum_r \mu_r dn_r \tag{1.45}$$

In words Eq. 1.45 states that, at fixed entropy (S), Volume (V), and mole numbers (n_i), the internal energy (U) proceeds toward a minimum value, and then ceases to change. This is a statement of equilibrium, but admittedly not a very useful one.[9] The remedy, Legendre Transformation, is already known to us from Sect. 1.3. Let us eliminate entropy as a variable by subtracting $d(TS)$ from both sides of Eq. 1.45, which produces:

$$d(U - TS) = dA \leq -TdS - PdV + \sum_r \mu_r dn_r \tag{1.46}$$

Equation 1.46 states that a system held at fixed temperature (T), volume (V), and mole numbers (n_i) will minimize its Helmholtz free energy to reach equilibrium. While such constraints on a system (a rigid container immersed in an isothermal bath without reactions) are easier to visualize than those of Eq. 1.45, the result is perhaps still new to most readers, who are likely familiar with statements of equilibrium in terms of Gibbs free energy. Of course, by rewriting the statement of equilibrium in terms of pressure via Table 1.1, we can generate:

$$dG \leq -TdS + VdP + \sum_r \mu_r dn_r \tag{1.47}$$

This is a familiar result of Thermodynamics; the Gibbs free energy is minimized at equilibrium for a system at fixed temperature, pressure, and mole numbers. Many students often forget the qualifying conditions of this statement and incorrectly "generalize" this result; there is no requirement of minimum Gibbs free energy at fixed temperature, *volume*, and mole numbers. Mathematically, one can write the equilibrium condition in terms of enthalpy, developed via relationships such as $H = U + PV$ or $G = H - TS$:

$$dH \leq TdS + VdP + \sum_r \mu_r dn_r \tag{1.48}$$

One can rearrange Eq. 1.48 to represent a very familiar chemical process, throttling:

$$dS \geq \frac{dH - VdP - \sum_r \mu_r dn_r}{T} \tag{1.49}$$

[9] How many processes can you imagine conducted at fixed entropy?

Table 1.2 Summary of potentials for simple systems

Constraint (fixed variables)	Equilibrium Conditions
$(T, V, \overline{n})$	Minimized A
$(T, P, \overline{n})$	Minimized G
$(S, V, \overline{n})$	Minimized U
$(S, P, \overline{n})$	Minimized H
$(U, V, \overline{n})$ or $(H, P, \overline{n})$ (isolated)	Maximized S

At fixed enthalpy, pressure, and mole numbers, the entropy of the system is maximized.[10] One can also write Eq. 1.49 in terms of internal energy:

$$dS \geq \frac{dU + PdV - \sum_r \mu_r dn_r}{T} \tag{1.50}$$

An important interpretation of Eq. 1.50 is that the entropy of an isolated ($dU = dV = dn_i = 0$) system tends toward a maximum. This is also a statement of the $2nd$ Law.

As in the work of Michelsen and Mollerup [Michelsen book], the abovementioned results are summarized in Table 1.2.

An important conclusion from Table 1.2 is that the correct potential representing equilibrium depends on the fixed variables; minimized Gibbs free energy is not the only criteria for equilibrium, and it only applies at fixed $T, P, \overline{n}$.[11] So far, we have considered these conclusions applied to a single phase. In the next section, we will look at equilibrium conditions for systems consisting of multiple phases.

1.7.1 Multiphase Equilibrium

Consider an isolated, non-reacting system consisting of an ensemble of phases. As material and energy pass between these phases (but not out of the ensemble), the entropy of each phase j, considered as a subsystem, obeys Eq. 1.50:

$$dS^j \geq \frac{dU^j + P^j dV^j - \sum_r \mu_r^j dn_r^j}{T^j} \tag{1.51}$$

Equation 1.51 has been written in general form, allowing each phase to have its own properties (T, P, etc.). However, because the overall system is isolated, we expect the entropy of the entire system to maximize on the approach to equilibrium (Table 1.2):

$$dS = \sum_q dS^q \geq \sum_q \frac{dU^q}{T^q} + \sum_q \frac{P^q dV^q}{T^q} - \sum_q \frac{\sum_r \mu_r^q dn_r^q}{T^q} \to 0 \tag{1.52}$$

[10] Of course $T \geq 0$ for absolute temperature scales such as Kelvin.

[11] In the case of reactions, minimized G also applies at fixed T, P, and feed mole numbers.

The variables U^j, V^j and n_i^j are not independent. They are constrained by the fact that the overall system has fixed U, V, and n_i, which we express by the following constraints:

$$dU = \sum_q dU^q = 0 \tag{1.53}$$

$$dV = \sum_q dV^q = 0 \tag{1.54}$$

$$dn_i = \sum_q dn_i^q = 0 \tag{1.55}$$

If we select an arbitrary reference phase (R), we can write each of these constraints as:

$$dU^R = -\sum_{q \neq R} dU^q \tag{1.56}$$

$$dV^R = -\sum_{q \neq R} dV^q \tag{1.57}$$

$$dn_i^R = -\sum_{q \neq R} dn_i^q \tag{1.58}$$

When Eqs. 1.56 through 1.58 are substituted into Eq. 1.52, the result (at equilibrium) is:

$$\sum_{q \neq R} \left(\frac{1}{T^q} - \frac{1}{T^R} \right) dU^q + \sum_{q \neq R} \left(\frac{P^q}{T^q} - \frac{P^R}{T^R} \right) dV^q - \sum_{q \neq R} \sum_r \left(\frac{\mu_r^q}{T^q} - \frac{\mu_r^R}{T^R} \right) dn_r^q \rightarrow 0 \tag{1.59}$$

Equation 1.59 differs from Eq. 1.52, as it is written in terms of independent variables $U^{j \neq R}$, $V^{j \neq R}$, and $n_i^{j \neq R}$. Therefore, the only way to satisfy the zero on right-hand side is for each of the terms in parentheses to be independently zero. This results in:

$$T^j = T^R \tag{1.60}$$

$$P^j = P^R \tag{1.61}$$

$$\mu_i^j = \mu_i^R \tag{1.62}$$

A few clarifying remarks are in order. The temperature and pressure of a given phase may change as the equilibrium state is reached. Equations 1.60 and 1.61 simply say that once the system has reached equilibrium, all phases must have the same temperature and pressure (otherwise, energy exchange through heat transfer or boundary work would be possible). Second, Eq. 1.62 does not say that *all* chemical potentials are equal to each other at equilibrium, but rather that the chemical potential of *each* substance is the same in all phases (otherwise, energy changes through mass transfer would be possible).

1.7.2 The Gibbs Phase Rule

Phase equilibrium introduces a seemingly large number of variables. Yet, as shown in Sect. 1.7.1, coexisting phases have the same temperature, pressure, and component chemical potentials (Eqs. 1.60, 1.61 and 1.62). This significantly reduces the number of *independent* variables, or *degrees of freedom*, in a multicomponent, multiphase mixture. The Gibbs phase rule formalizes this.

Consider a mixture consisting of C-components in Π-phases. We might say that the variables that we would need defined in order to characterize (i.e., to be able to calculate properties of) the system are the temperature, pressure, and the *independent* mole fractions in each phase.[12] There are therefore $C-1$ independent mole fractions in each phase, as the last mole fraction is available by normalization. The number of such variables (or *unknowns*, UK) are:

$$UK = 2 + \Pi(C - 1)$$

The *equations* (EQ) at our disposal are the statements of equilibrium (Eq. 1.62). In a two-phase mixture, we can apply this *once*[13] (i.e., between liquid and vapor phases) for *each* component. In the general case, each component is in equilibrium across each phase compared to the reference phase, meaning there are $\Pi - 1$ independent equilibria for each species.

$$EQ = C(\Pi - 1)$$

Stating additional equilibria beyond this adds no new information, as such equilibria follow algebraically from the others. The difference between the number of equations and unknowns is the *degree of freedom* (F), which is the number of variables that must be given to fully define the system:

$$F = UK - EQ = 2 + C - \Pi \tag{1.63}$$

[12] This is one form of the *state postulate*.

[13] You are free to select the reference phase in Sect. 1.7.1, and for fluid-phase equilibrium, this is often taken to be the phase with the highest density.

This simple equation is called the Gibbs phase rule and is very useful in analyzing multicomponent, multiphase mixtures. We will now clarify its utility through a few examples.

Example 1 (Application of the Gibbs Phase Rule) Let's apply the Gibbs phase rule to mixtures containing different numbers of components and phases:

- The triple point ($\Pi = 3$) of pure water ($C = 1$) has no ($F = 0$) degrees of freedom. No specification can be made; it occurs at approximately ≈ 273.16K and 611.7Pa.
- Pure water ($C = 1$) in a single phase state ($\Pi = 1$) has two ($F = 2$) degrees of freedom; both the temperature and pressure of liquid water are required to know the density.
- Pure water ($C = 1$) in a two-phase state ($\Pi = 2$) has one ($F = 1$) degree of freedom; the temperature and pressure of saturated steam are not independent, and only one needs to be specified.
- A binary mixture ($C = 2$) in vapor liquid equilibrium (VLE) ($\Pi = 2$) has two ($F = 2$) degrees of freedom; given the temperature and pressure (assuming two phases coexist), the compositions of the phases for a binary mixture are determined.
- A three-component mixture ($C = 3$) can at most ($F = 0$) contain five coexisting phases ($\Pi = 5$); though mixtures rarely exhibit the maximum number of possible phases, it is still useful to determine such limits.

1.8 Thermodynamic Quantities

In this section, we will take a conceptual look at some important thermodynamic quantities.

1.8.1 Compressibility Factor

The compressibility factor of a mixture is defined as:

$$Z \equiv \frac{PV}{nR_gT} \tag{1.64}$$

The compressibility factor is sometimes considered as a test for ideal-gas behavior ($Z^{IG} = 1$), but for us, it will simply be a convenient symbol for the multiple instances of PV that occur in thermodynamic expressions. Also we will use Z as the basis for calculating pressure from temperature, volume, and mole numbers.

1.8.2 Fugacity

Much confusion surrounds fugacity. As with Z, some think it is simply a test for ideal-gas behavior, while others call it a "corrected pressure." Limiting the fugacity to this is like defining the number “45” to be “the age of the author when he first set out to write this book”: it is true, but coincidently so. Instead, we must recognize fugacity[14] as a convenient change of variables introduced by G.N. Lewis. Let’s examine this by starting with our current understanding of equilibrium in a two-phase mixture, given by Eqs. 1.60 through 1.62:

$$\mu_i^I(T, P, \overline{n}^I) = \mu_i^{II}(T, P, \overline{n}^{II}) \tag{1.65}$$

If we return to Sect. 1.4, we recall that we do not have a practical means of computing chemical potentials directly from an EOS, but can compute them relative to an ideal gas. We proceed via Eq. 1.34:

$$\mu_i(T, V, \overline{n}) = \mu_i^{IG}(T, V, \overline{n}) + \left(\frac{\partial A^{Res}}{\partial n_i}\right)_{T,V,n_{k\neq i}} \tag{1.66}$$

Note that the fluid and the ideal-gas states are evaluated at the same T,V and $\overline{n}$. This is made clear on the right side of Eq. 1.66. Particularly in the case of multiphase equilibria, we are interested in computations at a specified pressure (P). The values of V and P correspond via the fluid’s EOS. That is to say $P = EOS(T, V, \overline{n})$, and as a consequence $\mu_i(T, V, \overline{n}) = \mu_i(T, P, \overline{n})$. This is not true for the ideal-gas state; the ideal gas has a pressure $P^{IG} = \frac{nR_gT}{V}$. In terms of pressure, Eq. 1.66 must be written as:

$$\mu_i(T, P, \overline{n}) = \mu_i^{IG}(T, P^{IG}, \overline{n}) + \left(\frac{\partial A^{Res}}{\partial n_i}\right)_{T,V,n_{k\neq i}} \tag{1.67}$$

Formally recognizing the different pressures between the EOS and ideal-gas states (at the same T,V and $\overline{n}$), we have via Eq. 1.20

$$\mu_i^{IG}(T, P^{IG}, \overline{n}) = \mu_i^{IG}(T, P, \overline{n}) + \int_P^{P^{IG}} \frac{R_gT}{P}dP$$

$$= \mu_i^{IG}(T, P, \overline{n}) + R_gTln(\frac{P^{IG}}{P}) = \mu_i^{IG}(T, P, \overline{n}) + R_gTln(\frac{nR_gT}{PV})$$

$$= \mu_i^{IG}(T, P, \overline{n}) - R_gTlnZ \tag{1.68}$$

[14] From Latin: *fugere* to escape, giving the sense of volatility.

The functional form of $\mu_i^{IG}(T, P, \overline{n})$ was developed in Sect. 1.6 (Eq. 1.43), and substitution of Eq. 1.67 into Eq. 1.68 yields:

$$\mu_i(T, P, \overline{n}) = \left[g_i^{IG}(T, Po) - R_g Tln(Po)\right]$$

$$+R_g Tlnc_i + R_g TlnP + \left(\frac{\partial A^{Res}}{\partial n_i}\right)_{T,V,n_{k\neq i}} - R_g TlnZ \quad (1.69)$$

The terms within the braces on the right side of the above equation would be common[15] to all phases (i.e., they would cancel when equating chemical potentials across coexisting phases). Therefore, we might be forgiven for omitting these terms from our modeling of multiphase equilibria. Indeed, if we properly reference, it is unnecessary to calculate these terms. Proceeding accordingly, we could phrase our expression of multiphase equilibria as:

$$\left[R_g Tlnc_i + R_g TlnP + \left(\frac{\partial A^{Res}}{\partial n_i}\right)_{T,V,n_{k\neq i}} - R_g TlnZ\right]^I$$

$$\left[R_g Tlnc_i + R_g TlnP + \left(\frac{\partial A^{Res}}{\partial n_i}\right)_{T,V,n_{k\neq i}} - R_g TlnZ\right]^{II}$$

Note that the superscripts I and II represent two coexisting phases, and each of these labels applies to all of the quantities in respective the braces. The above equation is not the equality of μ_i; it equates a new quantity. To establish this, we recognize that the final two terms in this quantity would be zero for an ideal gas.

Thus, if we define;

$$R_g TlnK_i = \left(\frac{\partial A^{Res}}{\partial n_i}\right)_{T,V,n_{k\neq i}} - R_g TlnZ$$

maintaining the $R_g Tln()$ form for all terms, we can then define our equilibrium relationship as:

$$(c_i P K_i)^I = (c_i P K_i)^{II} \quad (1.70)$$

[15] In reacting systems, a different equilibrium criterion from Eq. 1.65 applies, and these terms cannot be neglected.

This quantity, which is equal across the coexisting phases, is called the *fugacity*, and K_i is called the fugacity *coefficient* with the symbol:

$$ln\hat{\varphi}_i \equiv \frac{1}{R_g T}\left(\frac{\partial A^{Res}}{\partial n_i}\right)_{T,V,n_{k\neq i}} - lnZ \tag{1.71}$$

That is, the fugacity of species i in phase j is given by:

$$\hat{f_i}^j \equiv c_i^j P^j \hat{\varphi_i}^j \tag{1.72}$$

and the equilibrium relationship for species i coexisting between two phases I and II is:

$$\hat{f_i}^I = \hat{f_i}^{II} \tag{1.73}$$

As above, the fugacity coefficient of an ideal gas is by definition one, therefore,

$$\hat{f_i}^{IG} = c_i P \tag{1.74}$$

Equation 1.74 only applies when approximating fluid behavior with the Ideal-Gas EOS and gives rise to the generally incorrect use of partial pressure as a proxy for rigorous treatment. Note that the circumflex (^) is used to indicate properties of a mixture. The fugacity of a pure substance is given by:

$$f_i = P\varphi_i \tag{1.75}$$

The fugacity coefficient of a pure substance *could* be calculated from:

$$ln\varphi_i = A^{Res}(T, v) - R_g T lnZ \tag{1.76}$$

In computational practice, however, the same function is used to calculate φ_i or $\hat{\varphi}_i$ (Eq. 1.71). Also note that the equilibrium relationship written in terms of fugacity (Eq. 1.73) is often written with the logarithms:

$$lnc_i^I + lnP^I + ln\hat{\varphi_i^I} = lnc_i^{II} + lnP^{II} + ln\hat{\varphi_i^{II}} \tag{1.77}$$

Notice that, with each phase having the same pressure (Eq. 1.61), the lnP terms will cancel. Also, as will be seen in subsequent sections, the retention of the $ln()$ form of the equilibrium relationship provides a measure of scaling between variables such as T, P and $\hat{\varphi}_i$. The following statements summarize the concept of fugacity applied to non-reacting systems:

- The equilibrium condition is equality of chemical potential at the same temperature and pressure:

$$\mu_i^I(T, P, \overline{n}^I) = \mu_i^{II}(T, P, \overline{n}^{II})$$

- This condition can be equivalently phrased in terms of fugacity:

$$\hat{f_i}^I = \hat{f_i}^{II}$$

- Fugacity derives from referencing chemical potential to ideal-gas behavior, and with the fugacity of an ideal gas being its "partial pressure," the fugacity of a substance can be expressed in terms of the fugacity coefficient:

$$\hat{f_i} \equiv c_i P \hat{\varphi}_i$$

- The fugacity coefficient can be calculated from an EOS according to:

$$ln\hat{\varphi}_i \equiv \frac{1}{R_g T}\left(\frac{\partial A^{Res}}{\partial n_i}\right)_{T,V,n_{k\neq i}} - lnZ$$

- Some useful relationships between chemical potential and fugacity are:

$$\mu_i^B - \mu_i^B = R_g T ln \frac{\hat{f_i}^B}{\hat{f_i}^A} \tag{1.78}$$

$$\mu_i = \mu_i^{IG} + R_g T ln \frac{\hat{f_i}}{\hat{f_i}^{IG}} = g_i^o + R_g T ln \frac{P}{P_o} + R_g T ln c_i + R_g T ln \hat{\varphi}_i \tag{1.79}$$

$$\mu_i = \mu_i^{pure} + R_g T ln \frac{\hat{f_i}}{f_i} = g_i + R_g T ln c_i + R_g T ln \hat{\varphi}_i - R_g T ln \varphi_i \tag{1.80}$$

1.8.3 Density

While this text uses the volume (V) as a central variable in thermodynamic modeling, the density (ρ) is nonetheless a familiar and useful quantity in chemical processes;

$$\rho = \frac{mass}{volume} = \frac{\sum_r m_r}{V} = \frac{\sum_r M_r n_r}{V} = \frac{\sum_r M_r c_r}{v} \tag{1.81}$$

where M_i, n_i and c_i are the molar mass, mole number, and mole fraction of species i in the phase, while v is the molar volume of the phase.

1.8.4 Concluding Remarks

The remaining chapters of this text sequentially develop the main ideas of phase equilibrium. Chapter 2 outlines how to implement an EOS in terms of the variables $T, V, \overline{n}$, including two important examples: SRK (thereby illustrating an approach for other cubics) and PC-SAFT. Important techniques for developing and validating the derivatives required from an EOS are also presented. Chapter 3 builds on this, outlining how thermodynamic quantities such as fugacity coefficients are computed in the variables $T, V, \overline{n}$ from an implemented EOS. A key message of this chapter is that expressions for computing these thermodynamic quantities are independent of the EOS choice. The validation techniques introduced in Chap. 2 are also applied to the calculation of thermodynamic quantities. Chapter 4 presents a key concept in thermodynamic modeling; the translation of a $T, P, \overline{n}$ specification into $T, V, \overline{n}$ using a root solver. Chapter 5 then applies this concept to the material in Chap. 3, developing the required derivatives at fixed pressure, adapting the previously developed validation techniques. Chapter 6 then applies the content of the preceding chapters, beginning with the simple example of pure component saturation. Next, an algorithm is presented for computing the multicomponent TP phase boundary, which is an essential thermodynamic tool. The lesser known but equally important concept of stability is then presented in both $T, V, \overline{n}$ and $T, P, \overline{n}$ specifications. This is fortified through an algorithm for computing the spinodal boundary, which leads into a discussion of multicomponent critical points. Finally, flash calculations are examined in both $T, V, \overline{n}$ and $T, P, \overline{n}$ specifications. Chapter 7 is a short glossary for the text, while Chap. 8 is an extensive appendix to the text. This covers a range of mathematical techniques as well as the EOS parameters used in this text, the association contribution to EOS modeling, the Rachford–Rice equation and the Wilson correlation. It also provides further details on solving multicomponent, multiphase equilibrium using successive substitution, acceleration, and Newton's method.

Chapter 2
Implementation of an EOS

Various EOS find use in computational thermodynamics. There are the cubics such as Soave–Redlich–Kwong (SRK) and Peng–Robinson (PR), the more powerful Statistical Associating Fluid Theory (SAFT)-type EOS, and various empirical EOS. These all have unique structures and parameters and may even require internal computations as is the case with association models. The approach for implementing such equations varies, but here we seek to generalize as much as possible. Our frame of reference is that the residual Helmholtz free energy (A^{Res}) is a known function of T, V, and $\overline{n}$.

2.1 Implementation of the SRK EOS

We have already seen a simple example of an EOS with the SRK EOS (Eq. 1.30). We now look at the SRK EOS in greater detail, so that we can see it in terms of T, V, and $\overline{n}$. Note that these equations are written with T in K, V in cm^3, n in moles, and A^{Res} in J.

We begin with an examination of the A^{Res} expression given by Eq. 1.30:

$$A^{Res}(T, V, \overline{n}) = -nR_g T ln\left[\frac{V - nb}{V}\right] - \frac{n^2 a}{nb} ln\left[\frac{V + nb}{V}\right] \tag{2.1}$$

Equation 2.1 contains two parameters n^2a and nb:

$$n^2 a = \sum_r \sum_s n_r n_s a_{rs} \tag{2.2}$$

$$nb = \sum_r n_r b_r \tag{2.3}$$

B. R. Greenhalgh, *An Introduction to Computational Thermodynamics*,
https://doi.org/10.1007/978-3-032-30317-2_2

Based on the logarithm in Eq. 2.1 (or in the pressure expression given in Eq. 1.25), we require that $V > nb$ for the SRK EOS . One can therefore restrict:

$$V_{min}^{SRK} = 1.01nb \tag{2.4}$$

The a_{ij} on the right side of Eq. 2.2 are defined by:

$$a_{ij} = \sqrt{a_i a_j}(1 - K_{ij}) \tag{2.5}$$

In Eq. 2.5, we see that binary interaction parameters (K_{ij}) can be used in the calculation of the mixture of parameters (a_{ij}). The SRK EOS does not (nor do most other EOS) specify the functional form of K_{ij}, which can be a function of T, V, and $\overline{n}$, but is typically limited to a constant or a function of temperature.[1] In any case, pure component considerations require $K_{ii} = 0$, while symmetry gives $K_{ij} = K_{ji}$. In this text, we consider $K_{ij} = KA_{ij} + KB_{ij}T$.

The SRK model further defines the parameters via:

$$a_i = a_{ci}\alpha_i \tag{2.6}$$

$$a_{ci} = \frac{4.2747E5 R_g^2 {T_{ci}}^2}{P_{ci}} \tag{2.7}$$

$$\alpha_i = \left(1 + m_i\left[1 - \sqrt{\frac{T}{T_{ci}}}\right]\right)^2 \tag{2.8}$$

$$m_i = 0.480 + 1.574w_i - 0.175w_i^2 \tag{2.9}$$

$$b_i = \frac{8.664E4 R_g T_{ci}}{P_{ci}} \tag{2.10}$$

We see that, although the SRK EOS is a *two-parameter* EOS (a and b), it is parameterized in terms of critical temperature, critical pressure, and acentric factor (T_c, P_c, and ω). Values of these parameters used in this text can be found in Appendix 8.5. Glancing ahead at Chap. 3, we see that specific derivatives of A^{Res} are required for thermodynamic calculations. These require the corresponding

[1] Homogeneity requirements, as outlined in Sect. 8.6, govern the functional form of K_{ij}. Also, care should be taken in overfitting K_{ij} in terms of temperature, particularly if data are only available over a narrow range. Also, if large magnitude values are needed to match data (i.e., $|K_{ij}| > 0.5$), consideration should be given to whether the EOS in use is appropriate for the system being modeled.

derivatives of the a and b terms given above. Using the same ordering convention as outlined in Chap. 3, we have:

$$\frac{d\alpha_i}{dT} = -\sqrt{\frac{\alpha_i}{TT_{ci}}} m_i \tag{2.11}$$

$$\frac{da_i}{dT} = a_{ci} \frac{d\alpha_i}{dT} \tag{2.12}$$

As discussed above, the functional form of K_{ij} that we have selected yields:

$$\left(\frac{dK_{ij}}{dT} \right) = KB_{ij} \tag{2.13}$$

$$\left(\frac{da_{ij}}{dT} \right) = 0.5 \left[\frac{\frac{da_i}{dT} a_j + a_i \frac{da_j}{dT}}{\sqrt{a_i a_j}} \right] (1 - K_{ij}) - \sqrt{a_i a_j} \left(\frac{dK_{ij}}{dT} \right) \tag{2.14}$$

Notice that, even though a simple expression for K_{ij} has been assumed, we nonetheless leave the representation of $\left(\frac{\partial K_{ij}}{\partial T} \right)_{V,\overline{n}}$ in general form. This results in a modular structure to our code, allowing us to easily modify our code (i.e., to consider a more complex form for K_{ij}) without having to rewrite large sections, following the suggestion of Michelsen and Mollerup [Michelsen book]. The final T derivative required is:

$$\left(\frac{\partial [n^2 a]}{\partial T} \right)_{\overline{n}} = \sum_r \sum_s n_r n_s \left(\frac{\partial a_{rs}}{\partial T} \right) \tag{2.15}$$

Using these structures, we can compute the temperature derivative of the SRK A^{Res} expression:

$$\left(\frac{\partial A^{Res}}{\partial T} \right)_{V,\overline{n}} = -nR_g ln \left[\frac{V - nb}{V} \right] - \frac{\left(\frac{\partial [n^2 a]}{\partial T} \right)_{\overline{n}}}{nb} ln \left[\frac{V + nb}{V} \right] \tag{2.16}$$

In keeping with the modular approach advocated so far, we do not substitute the expression for $\left(\frac{\partial [n^2 a]}{\partial T} \right)_{\overline{n}}$ into the expression for $\left(\frac{\partial A^{Res}}{\partial T} \right)_{V,\overline{n}}$. Doing so would make the resulting expressions and code harder to interpret and maintain.

The volume derivative of A^{Res} is more straightforward for SRK, as the a and b parameters do not depend on volume:

$$\left(\frac{\partial A^{Res}}{\partial V}\right)_{T,\overline{n}} = -nR_gT\left(\frac{1}{V-nb}-\frac{1}{V}\right)-\frac{[n^2a]}{nb}\left(\frac{1}{V+nb}-\frac{1}{V}\right) \quad (2.17)$$

Also, we see by inspection that Eq. 2.17 is consistent with Eq. 1.27. The mole number derivative of A^{Res} is more complicated; thus, we begin[2] with the a and b terms.

$$\left(\frac{\partial[n^2a]}{\partial n_i}\right)_{T,n_{k\neq i}} = \left(\frac{\partial\sum_r\sum_s n_r n_s a_{rs}}{\partial n_i}\right)_{T,n_{k\neq i}} \quad (2.18)$$

We will use the development above to demonstrate another important technique, which relates to differentiation over nested sums. To illustrate this, let us clarify that the differentiation in Eq. 2.18 is on n_i, and that all the moles numbers are independent of each other. This means that $\frac{\partial n_j}{\partial n_i} = \delta_{ij}$ (See Appendix 8.4). Equation 2.18 becomes:

$$\begin{aligned}\left(\frac{\partial[n^2a]}{\partial n_i}\right)_{T,n_{k\neq i}} &= \sum_r\sum_s(\delta_{ri}n_s+n_r\delta_{si})a_{rs}\\ &= \sum_r\sum_s\delta_{ri}n_s a_{rs}+\sum_r\sum_s n_r\delta_{si}a_{rs}\end{aligned} \quad (2.19)$$

A helpful way to interpret the summations in Eq. 2.19 is to remind yourself that the δ_{ij} symbol (Appendix 8.4) produces 0 until $i = j$. Therefore, Eq. 2.19 becomes:

$$\left(\frac{\partial[n^2a]}{\partial n_i}\right)_{T,n_{k\neq i}} = \sum_s n_s a_{is}+\sum_r n_r a_{ri} \quad (2.20)$$

We now recognize that the index s in the first term on the right side of Eq. 2.20 is arbitrary; we can use any index we want (except i) in place of s, even r. The corresponding statement can be made for index r in the second term on the right side. Also, since a_{ij} is a symmetric matrix ($a_{ij} = a_{ji}$, see Eq. 2.5), we conclude that:

$$\left(\frac{\partial[n^2a]}{\partial n_i}\right)_{T,n_{k\neq i}} = 2\sum_r n_r a_{ir} \quad (2.21)$$

[2] This is an important technique. It is unwise to attempt derivation and implementation of the expression for $\left(\frac{\partial A^{Res}}{\partial n_i}\right)_{T,V,n_{k\neq i}}$ without first accomplishing such tasks for the subordinate expressions (e.g., $\left(\frac{\partial n^2a}{\partial n_i}\right)_{T,n_{k\neq i}}$).

Applying these techniques to Eq. 2.3, we can immediately write:

$$\left(\frac{\partial[nb]}{\partial n_i}\right)_{n_{k\neq i}} = b_i \tag{2.22}$$

Now, with the requisite pieces derived, we can write:

$$\left(\frac{\partial A^{Res}}{\partial n_i}\right)_{T,V,n_{k\neq i}} = -R_g T ln\left[\frac{V-nb}{V}\right] + nR_g T\left[\frac{b_i}{V-nb}\right]$$
$$-\left[\frac{\left(\frac{\partial[n^2a]}{\partial n_i}\right)_{T,n_{k\neq i}}}{nb} - \frac{(n^2a)b_i}{(nb)^2}\right] ln\left[\frac{V+nb}{V}\right] - \frac{n^2a}{nb}\left[\frac{b_i}{V+nb}\right] \tag{2.23}$$

We can now move on to higher order derivatives such as $\left(\frac{\partial^2 A^{Res}}{\partial T^2}\right)_{V,\overline{n}}$. Looking at Eq. 2.16, we see that this is a straightforward derivative, requiring only the derivation of $\left(\frac{\partial^2[n^2a]}{\partial T^2}\right)_{\overline{n}}$. After tracing back through the various pieces, we begin:

$$\frac{d^2\alpha_i}{dT^2} = -(1/2)m_i\left(\frac{\frac{d\alpha_i}{dT}}{\sqrt{TT_{ci}\alpha_i}} - \frac{1}{T}\sqrt{\frac{\alpha_i}{TT_{ci}}}\right) \tag{2.24}$$

$$\frac{d^2a_i}{dT^2} = a_{ci}\left(\frac{d^2\alpha_i}{dT^2}\right) \tag{2.25}$$

With $K_{ij} = KA_{ij} + KB_{ij}T$, $\left(\frac{d^2K_{ij}}{dT^2}\right) = 0$. Of course, if a more complex form of K_{ij} were selected, this would have to be consistently incorporated into our implementation of the SRK EOS. Moving on, we have:

$$\left(\frac{d^2a_{ij}}{dT^2}\right) = \frac{1}{2}\left(\frac{\frac{d^2a_i}{dT^2}a_j + 2\frac{da_i}{dT}\frac{da_j}{dT} + a_i\frac{d^2a_j}{dT^2}}{\sqrt{a_ia_j}}\right)(1-K_{ij})$$
$$-\frac{1}{4}\left(\frac{(\frac{da_i}{dT}a_j + a_i\frac{da_j}{dT})^2}{(a_ia_j)^{\frac{3}{2}}}\right)(1-K_{ij})$$
$$-\frac{\frac{da_i}{dT}a_j + a_i\frac{da_j}{dT}}{\sqrt{a_ia_j}}\left(\frac{dK_{ij}}{dT}\right) \tag{2.26}$$

A particularly relevant but challenging derivative is the second-order mole number set. From Eq. 2.21:

$$\left(\frac{\partial^2[n^2a]}{\partial n_i \partial n_j}\right) = 2a_{ij} \tag{2.27}$$

$$\begin{aligned}\left(\frac{\partial^2 A^{Res}}{\partial n_i \partial n_j}\right) &= \frac{R_g T}{V-nb}\left[b_i + b_j + \frac{nb_i b_j}{V-nb}\right]\\ &-\left[\frac{\left(\frac{\partial^2[n^2a]}{\partial n_i \partial n_j}\right)}{nb} - \frac{\left(\frac{\partial[n^2a]}{\partial n_i}\right)_{T,n_{k\neq i}} b_j}{(nb)^2} - \frac{\left(\frac{\partial[n^2a]}{\partial n_j}\right)_{T,n_{k\neq j}} b_i}{(nb)^2} + \frac{2[n^2a]b_i b_j}{(nb)^3}\right] ln\left[\frac{V+nb}{nb}\right]\\ &-\left[\frac{\left(\frac{\partial[n^2a]}{\partial n_i}\right)_{T,n_{k\neq i}} b_j}{nb} + \frac{\left(\frac{\partial[n^2a]}{\partial n_j}\right)_{T,n_{k\neq j}} b_i}{nb} - \frac{2[n^2a]b_i b_j}{(nb)^2}\right]\frac{1}{V+nb}\\ &+\frac{[n^2a]}{nb}\frac{b_i b_j}{(V+nb)^2}\end{aligned} \tag{2.28}$$

Other derivatives can be developed as required.

2.2 Implementation of the PC-SAFT EOS

Section 2.1 demonstrates how to compute the key derivatives for cubic EOS, which are typically stated in the form $P = f(T, V, \overline{n})$. As shown in Sect. 1.5, when presented with an EOS in the form $P = f(T, V, \overline{n})$, the first step is to integrate the equation in order to derive the form of A^{Res}. For other classes of EOS, A^{Res} is given directly. Such is the case with the Perturbed Chain–Statistical Associating Fluid Theory Equation of State (PC-SAFT EOS) [PC SAFT paper]. This EOS has been very successful in describing a range of chemical systems, including hydrocarbons, polar species, and polymers. Despite its utility in describing complex phase behavior, the difficulty in implementing the PC-SAFT EOS (relative to a cubic EOS) and its computational cost are seen as barriers to its use. This section is nonetheless a primer for the implementation of this EOS and others of its kind.

Although there are other forms,[3] the PC-SAFT-EOS can be stated in terms of a *hard chain*, *dispersion*, and an *associating* contribution:

$$A^{Res} = A^{hc} + A^{disp} + A^{assoc} \tag{2.29}$$

[3] There are other *SAFT* type EOS, and many modifications of these, including terms to describe ionic species, etc. Though their treatment is beyond the scope of this text, it is hoped that the methods outlined here will be useful in implementing these models.

The hard chain term is further composed of *hard sphere* contribution:

$$A^{hc} = R_g T\left(n\overline{m}a^{hs} - \sum_r n_r(m_r - 1)lng_{rr}^{hs}\right) \tag{2.30}$$

Here we reiterate one of the main concepts of this text, the striking of a balance between conceptual understanding and computational abstraction. The purpose of this section is not to provide an explanation of the physical details the PC-SAFT EOS, which can be found elsewhere [PC SAFT paper], but rather to guide the reader in methodically structuring the equations for implementation in a computational model. To that end, we will elaborate on mathematical details as they are required. In Eq. 2.30, we see the temperature (T) and the total mole numbers (n), as well as the universal gas constant (R_g). For now we state that one of the parameters of the PC-SAFT EOS is the number of segments on a given molecule (m), which has a mole average value of:

$$\overline{m} = \frac{\sum_r n_r m_r}{n}$$

Also, as the PC-SAFT EOS is quite useful for the description of high-molecular-weight polymers, one often finds PC-SAFT m parameters in ratio to molecular weight. Adopting these two formalisms, Eq. 2.30 is rewritten as:

$$A^{hc} = R_g T\left[(n\overline{m})a^{hs} - \sum_r n_r([\frac{m}{M}]_r M_r - 1)lng_{rr}^{hs}\right] \tag{2.31}$$

Note that in our approach we have introduced the product of n and $\overline{m}$;

$$n\overline{m} = \sum_r n_r[\frac{m}{M}]_r M_r \tag{2.32}$$

which is written in terms of mole numbers rather than mole fractions. We now consider the definition of a^{hs}:

$$a^{hs} = \frac{3\xi_1\xi_2}{\xi_0(1-\xi_3)} + \frac{\xi_2^3}{\xi_0\xi_3(1-\xi_3)^2} + \frac{\xi_2^3 ln(1-\xi_3)}{\xi_0\xi_3^2} - ln(1-\xi_3) \tag{2.33}$$

Next we consider the definition of the ξ terms, which is stated in the original PC-SAFT paper as [PC SAFT paper]:

$$\xi_s = \frac{\pi}{6}\rho\sum_r c_r m_r d_r^s,\ s = 0, 1, 2, 3 \tag{2.34}$$

In Eq. 2.34 ρ is the *molecular* density (not to be confused with Eq. 1.81), c_i is the mole fraction and as will be shown the d terms are in Angstroms (Å), with

10^{10}Å $= 1m$. Recalling that in this text volumes are given in cm^3, we rewrite this equation as:

$$\xi_s = \frac{\pi N_{av} \sum_r n_r [\frac{m}{M}]_r M_r d_r^s}{6E24V}, s = 0, 1, 2, 3 \tag{2.35}$$

The conversion between molecular and molar densities is the reason for Avogadro's number (N_{av}) in Eq. 2.35. Note that according to Eq. 2.35 only ξ_3 is dimensionless; the dimension of ξ_s is Å^{s-3}. Also note that ξ_3 represents the molecular packing and can be used to set V_{min} for the PC-SAFT EOS since $\xi_3 < 0.7405$ [PC SAFT paper]:

$$V_{min}^{\text{PC-SAFT}} = \frac{\pi N_{av} \sum_r n_r [\frac{m}{M}]_r M_r d_r^3}{6E24(0.7405)} \tag{2.36}$$

The d terms are defined by:

$$d_i = \sigma_i \left[1 - 0.12exp(-\frac{3}{T}[\frac{\epsilon_i}{k}]) \right] \tag{2.37}$$

Equation 2.37 has introduced two additional PC-SAFT parameters; the segment size (σ), which as above is typically given in Å, and the segment energy ($\frac{\epsilon}{k}$), having units of K. With these parameters given, it is also worth noting that, despite the apparent complexity of the expressions given, a^{hs} is nonetheless an explicit function of T, V and $\overline{n}$, just as we require for EOS calculations. We keep this in mind as we consider the definition of g^{hs}, which is also required for the calculation of A^{hc}:

$$g_{ij}^{hs} = \frac{1}{1 - \xi_3} + \frac{3d_{ij}\xi_2}{(1 - \xi_3)^2} + \frac{2d_{ij}^2 \xi_2^2}{(1 - \xi_3)^3} \tag{2.38}$$

For compactness of notation, we have written d_{ij}, which is defined as:

$$d_{ij} = \frac{d_i d_j}{d_i + d_j} \tag{2.39}$$

The above expressions allow for the computation of the hard chain contribution to A^{Res} using the PC-SAFT EOS. The next term to be considered is the dispersion term, which we write as follows:

$$A^{disp} = -\frac{R_g T \pi N_{av}}{1E24V} \left(2I_1[nm\epsilon\sigma_1] + [n\overline{m}]C_{1n} I_2[nm\epsilon\sigma_2] \right) \tag{2.40}$$

Equation 2.40 introduces a number terms, and we begin with I_1 and I_2:

$$I_1 = \sum_{s=0}^{6} a_s \xi_3^s$$
$$I_2 = \sum_{s=0}^{6} b_s \xi_3^s \tag{2.41}$$

The a and b terms (not to be confused with those in the SRK EOS) are in turn defined as:

$$a_t = a_{0t} + \left(\frac{[n\overline{m}] - n}{[n\overline{m}]}\right) a_{1t} + \left(\frac{[n\overline{m}] - n}{[n\overline{m}]}\right)\left(\frac{[n\overline{m}] - 2n}{[n\overline{m}]}\right) a_{2t}$$
$$b_t = b_{0t} + \left(\frac{[n\overline{m}] - n}{[n\overline{m}]}\right) b_{1t} + \left(\frac{[n\overline{m}] - n}{[n\overline{m}]}\right)\left(\frac{[n\overline{m}] - 2n}{[n\overline{m}]}\right) b_{2t} \tag{2.42}$$

where a_{kt} and b_{kt} are constants (rather than parameters) of the PC-SAFT model, given in Table 2.1.

The term C_{1n} is defined[4] as:

$$C_{1n} = \frac{1}{n + (n\overline{m})F_1 + [n - (n\overline{m})]F_2}$$
$$F_1 = \frac{8\xi_3 - 2\xi_3^2}{(1 - \xi_3)^4}$$
$$F_2 = \frac{20\xi_3 - 27\xi_3^2 + 12\xi_3^3 - 2\xi_3^4}{[(1 - \xi_3)(2 - \xi_3)]^2} \tag{2.43}$$

The last terms in Eq. 2.40 are the $(nm\epsilon\sigma)_i$ terms:

$$[nm\epsilon\sigma_1] = \sum_r \sum_s n_r n_s \left[\frac{m}{M}\right]_r M_r \left[\frac{m}{M}\right]_s M_s \left[\frac{\epsilon_{rs}}{kT}\right] \sigma_{rs}^3$$
$$[nm\epsilon\sigma_2] = \sum_r \sum_s n_r n_s \left[\frac{m}{M}\right]_r M_r \left[\frac{m}{M}\right]_s M_s \left[\frac{\epsilon_{rs}}{kT}\right]^2 \sigma_{rs}^3 \tag{2.44}$$

[4] This form differs from C_1 as defined in Equation A.11 in the original PC-SAFT paper [PC SAFT paper], which has a typo omitting an exponent. Also, we have structured C_{1n} to retain the TVn formalism used throughout this text.

Table 2.1 Model constants of the PC-SAFT EOS [PC SAFT paper] for use in Eq. 2.42

t	a_{0t}	a_{1t}	a_{2t}	b_{0t}	b_{1t}	b_{2t}
0	0.9105631445	−0.3084016918	−0.0906148351	0.7240946941	−0.5755498075	0.0976883116
1	0.6361281449	0.1860531159	0.4527842806	2.2382791861	0.6995095521	−0.2557574982
2	2.6861347891	−2.5030047259	0.5962700728	−4.0025849485	3.892567339	−9.155856153
3	−26.547362491	21.419793629	−1.7241829131	−21.003576815	−17.215471648	20.642075974
4	97.759208784	−65.25588533	−4.1302112531	26.855641363	192.67226447	−38.804430052
5	−159.59154087	83.318680481	13.77663187	206.55133841	−161.82646165	93.626774077
6	91.297774084	−33.74692293	−8.6728470368	−355.60235612	−165.20769346	−29.666905585

Equation 2.44 has introduced mixture parameters, which can be calculated as follows:

$$\sigma_{ij} = \frac{\sigma_i + \sigma_2}{2} \tag{2.45}$$

$$\frac{\epsilon_{ij}}{k} = \sqrt{\left(\frac{\epsilon_i}{k}\right)\left(\frac{\epsilon_j}{k}\right)}(1 - K_{ij}) \tag{2.46}$$

Note that it is the dispersion term (a structure that can be seen in SRK in Eq. 2.5) that contains the binary interaction parameters (K_{ij}). For non-associating systems, these equations define A^{Res} for the PC-SAFT EOS. For such systems, A^{assoc} and all of its derivatives are zero, and for the remainder of this section, these terms will be omitted. The details of the association contribution to A^{Res} for the PC-SAFT EOS are given in Appendix 8.1.

As stated above, the underlying equations in the PC-SAFT model do seem complicated, especially compared to simpler models such as SRK (Sect. 2.1). The key to handling such a computational model is abstraction; we recognize that, given T, V and $\overline{n}$, we can determine A^{Res} by methodically calculating each of required terms as explicit functions of these variables. The following example demonstrates the calculation of A^{Res} using the PC-SAFT EOS.

Example 2 (Computation of A^{Res} for the PC-SAFT EOS for Non-associating Systems) Consider a mixture containing 1 mole of propane + 1 mole of n-hexane + 1 mole of water at 350K and $V_{min}^{\text{PC-SAFT}}$, using the parameters in Table 2.2 (excerpted from Table 8.3):

Starting with the hard chain terms, we calculate d_i via Eq. 2.37:

$$\overline{d} = 3.545, 3.738, 2.400 (\text{Å})$$

This allows us to calculate V_{min} via Eq. 2.36:

$$V = V_{min}^{\text{PC-SAFT}} = 116.381 (\text{cm}^3)$$

We can now calculate ξ_s (Eq. 2.35);

$$\xi_0, \xi_1, \xi_2, \xi_3 = 1.848E-2, 6.165E-2, 0.211, 0.7405$$

Table 2.2 Parameters used for propane, n-hexane, and water for PC-SAFT EOS calculations, ignoring association

Species	M $[\frac{g}{\text{mol}}]$	$\frac{m}{M}$ $[\frac{\text{mol}}{g}]$	σ [Å]	$\frac{\epsilon}{k}$ [K]
Propane	44.10	0.045	3.618	208.11
n-Hexane	86.18	0.035	3.798	236.77
Water	18.02	0.098	2.460	186.75

where ξ_3 is consistent with our calculation at $V = V_{min}$ (see the comment about units under Eq. 2.35). The hard sphere terms are (Eq. 2.33);

$$a^{hs} = 18.505(-)$$

and (Eq. 2.39);

$$\overline{\overline{d}} = \begin{bmatrix} 1.773 & 1.820 & 1.431 \\ 1.820 & 1.869 & 1.462 \\ 1.431 & 1.462 & 1.200 \end{bmatrix} (\text{Å})$$

and (Eq. 2.38):

$$\overline{\overline{g}}^{hs} = \begin{bmatrix} 36.630 & 37.935 & 27.818 \\ 37.935 & 39.336 & 28.556 \\ 27.818 & 28.556 & 22.529 \end{bmatrix} (-)$$

With (Eq. 2.32);

$$n\overline{m} = 6.820(\text{moles})$$

we then have via Eq. 2.31:

$$A^{hc} = 327855.833(\text{J})$$

Next we calculate the dispersion terms, starting with a and b using Eq. 2.42:

$$\overline{a} = 0.732, 0.771, 1.324, -14.666, 60.932, -111.998, 71.813$$

$$\overline{b} = 0.408, 2.613, -2.439, -29.256, 132.158, 122.216, -450.131$$

These allow computation of the I terms (Eq. 2.41):

$$I_1 = 1.298$$

$$I_2 = -18.140$$

The C_{1n} term is computed via Eq. 2.43:

$$F_1 = 1064.527(-)$$

$$F_2 = 40.028(-)$$

$$C_{1n} = 1.407E - 4(\frac{1}{\text{mole}})$$

The $nm\epsilon\sigma$ terms are determined by Eq. 2.45;

$$\overline{\overline{\sigma}} = \begin{bmatrix} 3.618 & 3.708 & 3.039 \\ 3.708 & 3.798 & 3.129 \\ 3.039 & 3.129 & 2.460 \end{bmatrix} (\text{Å})$$

Equation 2.46 (with the binary interaction parameter in this example assumed to be zero);

$$\frac{\overline{\overline{\epsilon}}}{k} = \begin{bmatrix} 208.110 & 221.978 & 197.141 \\ 221.978 & 236.770 & 210.278 \\ 197.141 & 210.278 & 186.750 \end{bmatrix} (K)$$

and Eq. 2.44:

$$(nm\epsilon\sigma_1) = 1189.589$$

$$(nm\epsilon\sigma_2) = 747.631$$

The preceding terms combine to give A^{disp} via Eq. 2.40:

$$A^{disp} = -145448.140(\text{J})$$

Having neglected the association term in this example, one would conclude that $A^{Res} = 182407.693$J. The reader is referred to Example 28 in Appendix 8.1, where the association term has been included in A^{Res}.

Example 2 demonstrates that the underlying PC-SAFT equations can be evaluated explicitly as functions of T, V, and $\overline{n}$. A more pressing challenge is to compute the underlying derivatives that are required in thermodynamic routines (see Chap. 3). Though it is not practical to list all possible derivatives related to A^{Res}, some strategies for the determination of these are outlined in the remainder of this section.

The temperature derivative of A^{Res} (Eq. 2.29) is (neglecting association) simply:

$$\left(\frac{\partial A^{Res}}{\partial T}\right)_{V,\overline{n}} = \left(\frac{\partial A^{hc}}{\partial T}\right)_{V,\overline{n}} + \left(\frac{\partial A^{disp}}{\partial T}\right)_{V,\overline{n}} \tag{2.47}$$

The temperature derivative of A^{hc} is also straightforward, by Eq. 2.31:

$$\left(\frac{\partial A^{hc}}{\partial T}\right)_{V,\overline{n}} = \frac{A^{hc}}{T} + R_g T\left[(n\overline{m})(\frac{\partial a^{hs}}{\partial T})_{V,\overline{n}} - \sum_r n_r([\frac{m}{M}]_r M_r - 1)\frac{(\frac{\partial g^{hs}_{rr}}{\partial T})_{V,\overline{n}}}{g^{hs}_{rr}}\right] \tag{2.48}$$

Now we face a more serious challenge, the determination of $(\frac{\partial a^{hs}}{\partial T})_{V,\bar{n}}$. We have to differentiate Eq. 2.33 with respect to temperature. We will of course require $(\frac{\partial \xi_s}{\partial T})$, which are non-zero except for $s = 0$;

$$(\frac{\partial \xi_s}{\partial T})_{V,\bar{n}} = \frac{\pi N_{av} \sum_r n_r [\frac{m}{M}]_r M_r s d_r^{s-1} \frac{d(d_i)}{dT}}{6E24V}, s = 0, 1, 2, 3 \tag{2.49}$$

in which:

$$\frac{d(d_i)}{dT} = \frac{3(d_i - \sigma_i)[\frac{\epsilon_i}{k}]}{T^2} \tag{2.50}$$

Given the number of terms in Eq. 2.33, the risk of making an error in writing out the temperature derivative (in terms of $(\frac{\partial \xi_s}{\partial T})_{V,\bar{n}}$) is large. Here we apply the build-up technique and numerical testing (introduced in Sect. 2.3). The result is:

$$\begin{aligned}(\frac{\partial a^{hs}}{\partial T})_{V,\bar{n}} = {} & \frac{3}{\xi_0}\Big[\frac{\xi_2(\frac{\partial \xi_1}{\partial T})_{V,\bar{n}}}{(1-\xi_3)} + \frac{\xi_1(\frac{\partial \xi_2}{\partial T})_{V,\bar{n}}}{(1-\xi_3)} + \frac{\xi_1\xi_2(\frac{\partial \xi_3}{\partial T})_{V,\bar{n}}}{(1-\xi_3)^2}\Big] \\ & + \frac{1}{\xi_0}\Big[\frac{3\xi_2^2(\frac{\partial \xi_2}{\partial T})_{V,\bar{n}}}{\xi_3(1-\xi_3)^2} - \frac{\xi_2^3(\frac{\partial \xi_3}{\partial T})_{V,\bar{n}}}{\xi_3^2(1-\xi_3)^2} + \frac{2\xi_2^3(\frac{\partial \xi_3}{\partial T})_{V,\bar{n}}}{\xi_3(1-\xi_3)^3}\Big] \\ & + \frac{1}{\xi_0}\Big[\frac{3\xi_2^2 ln(1-\xi_3)(\frac{\partial \xi_2}{\partial T})_{V,\bar{n}}}{\xi_3^2} - \frac{\xi_2^3(\frac{\partial \xi_3}{\partial T})_{V,\bar{n}}}{\xi_3^2(1-\xi_3)} - \frac{2(\xi_2^3) ln(1-\xi_3)(\frac{\partial \xi_3}{\partial T})_{V,\bar{n}}}{\xi_3^3}\Big] \\ & + \frac{(\frac{\partial \xi_3}{\partial T})_{V,\bar{n}}}{1-\xi_3}\end{aligned} \tag{2.51}$$

Equation 2.48 refers to $(\frac{\partial g_{ij}^{hs}}{\partial T})_{V,\bar{n}}$, which follows from Eq. 2.38:

$$\begin{aligned}(\frac{\partial g_{ij}^{hs}}{\partial T})_{V,\bar{n}} = {} & \frac{(\frac{\partial \xi_3}{\partial T})_{V,\bar{n}}}{(1-\xi_3)^2} + \frac{3\xi_2}{(1-\xi_3)^2}(\frac{d(d_{ij})}{dT}) + \frac{3d_{ij}}{(1-\xi_3)^2}(\frac{\partial \xi_2}{\partial T})_{V,\bar{n}} \\ & + \frac{6d_{ij}\xi_2}{(1-\xi_3)^3}(\frac{\partial \xi_3}{\partial T})_{V,\bar{n}} + \frac{4d_{ij}\xi_2^2}{(1-\xi_3)^3}(\frac{d(d_{ij})}{dT}) + \frac{4d_{ij}^2\xi_2}{(1-\xi_3)^3}(\frac{\partial \xi_2}{\partial T})_{V,\bar{n}} \\ & + \frac{6d_{ij}^2\xi_2^2}{(1-\xi_3)^4}(\frac{\partial \xi_3}{\partial T})_{V,\bar{n}}\end{aligned} \tag{2.52}$$

$(\frac{\partial g_{ij}^{hs}}{\partial T})_{V,\overline{n}}$ requires $(\frac{d[d_{ij}]}{dT})$. Referring to Eq. 2.39, this is:

$$(\frac{d(d_{ij})}{dT}) = \left(\frac{\frac{d(d_i)}{dT}}{d_i^2} + \frac{\frac{d(d_j)}{dT}}{d_j^2}\right) d_{ij}^2 \tag{2.53}$$

The temperature derivative of the A^{disp} follows from Eq. 2.40:

$$(\frac{\partial A^{disp}}{\partial T})_{V,\overline{n}} = \frac{A^{disp}}{T} - \frac{R_g T \pi Nav}{10^{24} V}\Bigg(2(\frac{\partial I_1}{\partial T})_{V,\overline{n}}[nm\epsilon\sigma_1] + 2I_1(\frac{\partial [nm\epsilon\sigma_1]}{\partial T})_{\overline{n}} + (n\overline{m})\left[(\frac{\partial C_{1n}}{\partial T})_{V,\overline{n}} I_2[nm\epsilon\sigma_2] + C_{1n}(\frac{\partial I_2}{\partial T})_{V,\overline{n}}[nm\epsilon\sigma_2] + C_{1n} I_2(\frac{\partial [nm\epsilon\sigma_2]}{\partial T})_{\overline{n}}\right] \Bigg) \tag{2.54}$$

The temperature derivatives of I_1 and I_2 are, via Eq. 2.41:

$$\begin{aligned} (\frac{\partial I_1}{\partial T})_{V,\overline{n}} &= \sum_{s=0}^{6} s a_s (\frac{\partial \xi_3}{\partial T})_{V,\overline{n}}^{s-1} \\ (\frac{\partial I_2}{\partial T})_{V,\overline{n}} &= \sum_{s=0}^{6} s b_s (\frac{\partial \xi_3}{\partial T})_{V,\overline{n}}^{s-1} \end{aligned} \tag{2.55}$$

The temperature derivatives of the C_{1n} terms are, via Eq. 2.43:

$$\begin{aligned} (\frac{\partial C_{1n}}{\partial T})_{V,\overline{n}} &= -(C_{1n})^2\left[n\overline{m}(\frac{\partial F_1}{\partial T})_{V,\overline{n}} + [n - (n\overline{m})](\frac{\partial F_2}{\partial T})_{V,\overline{n}}\right] \\ (\frac{\partial F_1}{\partial T})_{V,\overline{n}} &= (\frac{\partial \xi_3}{\partial T})_{V,\overline{n}}\left[\frac{8 + 20\xi_3 - 4\xi_3^2}{(1-\xi_3)^5}\right] \\ (\frac{\partial F_2}{\partial T})_{V,\overline{n}} &= (\frac{\partial \xi_3}{\partial T})_{V,\overline{n}}\left[\frac{40 - 48\xi_3 + 12\xi_3^2 + 2\xi_3^3}{(2 - 3\xi_3 + \xi_3^2)^3}\right] \end{aligned} \tag{2.56}$$

The temperature derivatives of the $(nm\epsilon\sigma)_i$ terms follow from Eq. 2.44 :

$$(\frac{\partial nm\epsilon\sigma_1}{\partial T})_{\overline{n}} = -\frac{nm\epsilon\sigma_1}{T} + \sum_r \sum_s n_r n_s [\frac{m}{M}]_r M_r [\frac{m}{M}]_s M_s \left(\frac{d[\frac{\epsilon_{rs}}{k}]}{dT}\right) \frac{\sigma_{rs}^3}{T}$$

$$(\frac{\partial nm\epsilon\sigma_2}{\partial T})_{\overline{n}} = -\frac{2nm\epsilon\sigma_2}{T} + 2\sum_r \sum_s n_r n_s [\frac{m}{M}]_r M_r [\frac{m}{M}]_s M_s [\frac{\epsilon_{rs}}{kT}] \left(\frac{d[\frac{\epsilon_{rs}}{k}]}{dT}\right) \frac{\sigma_{rs}^3}{T} \tag{2.57}$$

One can of course simplify or restructure such expressions as they wish. Example 3 provides numerical output for the temperature derivatives of the PC-SAFT EOS, neglecting association.

Example 3 (Computation of $(\frac{\partial A^{Res}}{\partial T})_{V,\overline{n}}$ Using the PC-SAFT EOS for Non-Associating Systems) Making use of the results given in Example 2, we can calculate $(\frac{\partial A^{hc}}{\partial T})_{V,\overline{n}}$ and $(\frac{\partial A^{disp}}{\partial T})_{V,\overline{n}}$ for a mixture containing one mole of propane, one mole of n-hexane, and one mole of water at 350K and $V = 116.38\text{cm}^3$.

Starting with the hard chain terms, we calculate via Eq. 2.50:

$$\frac{d(\overline{d})}{dT} = -3.72E-4, -3.47E-4, -2.72E-4(\frac{\text{Å}}{K})$$

The temperature derivatives of ξ_s are (Eq. 2.49);

$$(\frac{\partial \xi_0}{\partial T})_{V,\overline{n}}, (\frac{\partial \xi_1}{\partial T})_{V,\overline{n}}, (\frac{\partial \xi_2}{\partial T})_{V,\overline{n}}, (\frac{\partial \xi_3}{\partial T})_{V,\overline{n}} = 0, -6.19E-6, -4.20E-5, -2.19E-4$$

The derivative of a^{hs} is via Eq. 2.51:

$$(\frac{\partial a^{hs}}{\partial T})_{V,\overline{n}} = -2.98E-2(\frac{1}{K})$$

$$(\frac{d[\overline{\overline{d}}]}{dT}) = \begin{bmatrix} -1.86E-4 & -1.80E-4 & -1.57E-4 \\ -1.80E-4 & -1.74E-4 & -1.54E-4 \\ -1.57E-4 & -1.54E-4 & -1.36E-4 \end{bmatrix} (\frac{\text{Å}}{K})$$

which gives:

$$(\frac{\partial \overline{\overline{g}}^{hs}}{\partial T})_{V,\partial n} = \begin{bmatrix} -0.0870 & -0.0903 & -0.0632 \\ -0.0903 & -0.0938 & -0.0650 \\ -0.0632 & -0.0650 & -0.0492 \end{bmatrix} (\frac{1}{K})$$

We then have:

$$(\frac{\partial A^{hc}}{\partial T})_{V,\overline{n}} = 371.623\frac{\text{J}}{\text{K}}$$

Next the derivatives of the dispersion terms are calculated.
The I derivatives are:

$$(\frac{\partial I_1}{\partial T})_{V,\overline{n}} = -1.124E - 3$$

$$(\frac{\partial I_2}{\partial T})_{V,\overline{n}} = 5.523E - 2$$

Assuming $K_{ij} = 0$, $\frac{\partial(\frac{\epsilon_{ij}}{k})}{\partial T} = 0$, and:

$$(\frac{\partial[nm\epsilon\sigma_1]}{\partial T})_{V,\overline{n}} = -3.399$$

$$(\frac{\partial[nm\epsilon\sigma_2]}{\partial T})_{V,\overline{n}} = -4.272$$

Finally:

$$(\frac{\partial F_1}{\partial T})_{V,\overline{n}} = -3.839(\frac{1}{K})$$

$$(\frac{\partial F_2}{\partial T})_{V,\overline{n}} = -7.436E - 2(\frac{1}{K})$$

$$(\frac{\partial C_{1n}}{\partial T})_{V,\overline{n}} = 5.123E - 7(\frac{1}{mole * K})$$

Via Eq. 2.54:

$$(\frac{\partial A^{disp}}{\partial T})_{V,\overline{n}} = 125.102(\frac{J}{K})$$

With the association contribution neglected, Eq. 2.47 gives:

$$(\frac{\partial A^{Res}}{\partial T})_{V,\overline{n}} = 496.725(\frac{J}{K})$$

The mole number derivatives of A^{Res} for the PC-SAFT EOS (neglecting association) are given by:

$$\left(\frac{\partial A^{Res}}{\partial n_i}\right)_{T,V,n_{k\neq i}} = \left(\frac{\partial A^{hc}}{\partial n_i}\right)_{T,V,n_{k\neq i}} + \left(\frac{\partial A^{disp}}{\partial n_i}\right)_{T,V,n_{k\neq i}} \tag{2.58}$$

The hard chain contribution follows from Eq. 2.31:

$$\left(\frac{\partial A^{hc}}{\partial n_i}\right)_{T,V,n_{k\neq i}} = R_g T\Bigg[$$

$$[\frac{m}{M}]_i M_i a^{hs} + (n\overline{m})(\frac{\partial a^{hs}}{\partial n_i})_{T,V,n_{k\neq i}}$$

$$-([\frac{m}{M}]_i M_i - 1) ln g_{ii}^{hs} - \sum_r \frac{n_r([\frac{m}{M}]_r M_r - 1)}{g_{rr}^{hs}}\left(\frac{\partial g_{rr}^{hs}}{\partial n_i}\right)_{T,V,n_{k\neq i}}\Bigg] \quad (2.59)$$

We then require $(\frac{\partial a^{hs}}{\partial n_i})_{T,V,n_{k\neq i}}$, which follows from Eq. 2.33:

$$(\frac{\partial a^{hs}}{\partial n_i})_{T,V,n_{k\neq i}} = \frac{3(\frac{\partial \xi_1}{\partial n_i})_{T,V,n_{k\neq i}}\xi_2}{\xi_0(1-\xi_3)} + \frac{3\xi_1(\frac{\partial \xi_2}{\partial n_i})_{T,V,n_{k\neq i}}}{\xi_0(1-\xi_3)} - \frac{3(\frac{\partial \xi_0}{\partial n_i})_{T,V,n_{k\neq i}}\xi_1\xi_2}{\xi_0^2(1-\xi_3)}$$

$$+\frac{3\xi_1\xi_2(\frac{\partial \xi_3}{\partial n_i})_{T,V,n_{k\neq i}}}{\xi_0(1-\xi_3)^2} + \frac{3\xi_2^2(\frac{\partial \xi_2}{\partial n_i})_{T,V,n_{k\neq i}}}{\xi_0\xi_3(1-\xi_3)^2} - \frac{(\frac{\partial \xi_0}{\partial n_i})_{T,V,n_{k\neq i}}\xi_2^3}{\xi_0^2\xi_3(1-\xi_3)^2}$$

$$-\frac{\xi_2^3(\xi_3^2 - 5\xi_3 + 2)(\frac{\partial \xi_3}{\partial n_i})_{T,V,n_{k\neq i}}}{\xi_0\xi_3^2(1-\xi_3)^3} + \frac{3\xi_2^2(\frac{\partial \xi_2}{\partial n_i})_{T,V,n_{k\neq i}} ln(1-\xi_3)}{\xi_0\xi_3^2}$$

$$-\frac{(\frac{\partial \xi_0}{\partial n_i})_{T,V,n_{k\neq i}}\xi_2^3 ln(1-\xi_3)}{\xi_0^2\xi_3^2} - \frac{2\xi_2^3 ln(1-\xi_3)(\frac{\partial \xi_3}{\partial n_i})_{T,V,n_{k\neq i}}}{\xi_0\xi_3^3} + \frac{(\frac{\partial \xi_3}{\partial n_i})_{T,V,n_{k\neq i}}}{1-\xi_3} \quad (2.60)$$

The mole number derivatives of g^{hs} follow from Eq. 2.38:

$$(\frac{\partial g_{ij}^{hs}}{\partial n_k})_{T,V,n_{q\neq k}} = \frac{g_{ij}^{hs}}{1-\xi_3}(\frac{\partial \xi_3}{\partial n_k})_{T,V,n_{q\neq k}}$$

$$+\frac{3d_{ij}}{(1-\xi_3)^2}(\frac{\partial \xi_2}{\partial n_k})_{T,V,n_{q\neq k}} + \frac{3d_{ij}\xi_2}{(1-\xi_3)^3}(\frac{\partial \xi_3}{\partial n_k})_{T,V,n_{q\neq k}}$$

$$+\frac{4d_{ij}^2\xi_2}{(1-\xi_3)^3}(\frac{\partial \xi_2}{\partial n_k})_{T,V,n_{q\neq k}} + \frac{4d_{ij}^2\xi_2^2}{(1-\xi_3)^4}(\frac{\partial \xi_3}{\partial n_k})_{T,V,n_{q\neq k}} \quad (2.61)$$

The form of Eq. 2.61 results from first rewriting Eq. 2.38 as $g_{ij}^{hs}(1-\xi_3) = 1 + \frac{3d_{ij}\xi_2}{1-\xi_3} + ...$ Also, do not be confused by the choice of the index k for the mole number of differentiation in Eq. 2.61 vs. i in Eq. 2.59. The use of k in the former results from having already used i in Eq. 2.38.

The mole number derivatives of the ξ terms, required in both Eq. 2.60 and Eq. 2.61, follow from Eq. 2.35:

$$\left(\frac{\partial \xi_s}{\partial n_i}\right)_{T,V,n_{k\neq i}} = \frac{\pi N_{av}\left[\frac{m}{M}\right]_i M_i d_i^s}{6E24V} \tag{2.62}$$

The mole number derivative of the dispersion term follows from Eq. 2.40:

$$\begin{aligned}\left(\frac{\partial A^{disp}}{\partial n_i}\right)_{T,V,n_{k\neq i}} &= -\frac{R_g T\pi N_{av}}{1E24V}\Big[\\ 2\left(\frac{\partial I_1}{\partial n_i}\right)_{T,V,n_{k\neq i}}&[nm\epsilon\sigma 1] + 2I_1\left(\frac{\partial [nm\epsilon\sigma 1]}{\partial n_i}\right)_{T,V,n_{k\neq i}}\\ +\left[\frac{m}{M}\right]_i M_i C_{1n} I_2[nm\epsilon\sigma 2] &+ (n\overline{m})\left(\frac{\partial C_{1n}}{\partial n_i}\right)_{T,V,n_{k\neq i}} I_2[nm\epsilon\sigma 2]\\ +(n\overline{m})C_{1n}\left(\frac{\partial I_2}{\partial n_i}\right)_{T,V,n_{k\neq i}}[nm\epsilon\sigma 2] &+ (n\overline{m})C_{1n}I_2\left(\frac{\partial [nm\epsilon\sigma 2]}{\partial n_i}\right)_{T,V,n_{k\neq i}}\Big]\end{aligned} \tag{2.63}$$

The mole number derivatives of the I terms are from Eq. 2.41

$$\begin{aligned}\left(\frac{\partial I_1}{\partial n_i}\right)_{T,V,n_{k\neq i}} &= \sum_{s=0}^{6}\left[\left(\frac{\partial a_s}{\partial n_i}\right)_{T,V,n_{k\neq i}}\xi_3^s + sa_s\left(\frac{\partial \xi_3}{\partial n_i}\right)_{T,V,n_{k\neq i}}\xi_3^{s-1}\right]\\ \left(\frac{\partial I_2}{\partial n_i}\right)_{T,V,n_{k\neq i}} &= \sum_{s=0}^{6}\left[\left(\frac{\partial b_s}{\partial n_i}\right)_{T,V,n_{k\neq i}}\xi_3^s + sb_s\left(\frac{\partial \xi_3}{\partial n_i}\right)_{T,V,n_{k\neq i}}\xi_3^{s-1}\right]\end{aligned} \tag{2.64}$$

The required mole number derivatives of the a and b terms follow from Eq. 2.42:

$$\begin{aligned}\left(\frac{\partial a}{\partial n_i}\right)_{T,V,n_{k\neq i}} &= \left(\frac{n\left[\frac{m}{M}\right]_i M_i - [n\overline{m}]}{[n\overline{m}]^2}\right)a_{1t} + \left(\frac{[4n-3n\overline{m}][n\overline{m}-n[\frac{m}{M}]_i M_i]}{[n\overline{m}]^3}\right)a_{2t}\\ \left(\frac{\partial b}{\partial n_i}\right)_{T,V,n_{k\neq i}} &= \left(\frac{n\left[\frac{m}{M}\right]_i M_i - [n\overline{m}]}{[n\overline{m}]^2}\right)b_{1t} + \left(\frac{[4n-3n\overline{m}][n\overline{m}-n[\frac{m}{M}]_i M_i]}{[n\overline{m}]^3}\right)b_{2t}\end{aligned} \tag{2.65}$$

The mole number derivatives of the $(nm\epsilon\sigma)_i$ terms are, from Eq. 2.44:

$$\begin{aligned}\left(\frac{\partial [nm\epsilon\sigma_1]}{\partial n_i}\right)_{T,V,n_{k\neq i}} &= 2\left[\frac{m}{M}\right]_i M_i \sum_r n_r\left[\frac{m}{M}\right]_r M_r\left[\frac{\epsilon_{ir}}{kT}\right]\sigma_{ir}^3\\ \left(\frac{\partial [nm\epsilon\sigma_2]}{\partial n_i}\right)_{T,V,n_{k\neq i}} &= 2\left[\frac{m}{M}\right]_i M_i \sum_r n_r\left[\frac{m}{M}\right]_r M_r\left[\frac{\epsilon_{ir}}{kT}\right]^2\sigma_{ir}^3\end{aligned} \tag{2.66}$$

The mole number derivatives of C_{1n} follow from Eq. 2.43:

$$\left(\frac{\partial F_1}{\partial n_i}\right)_{T,V,n_{k\neq i}} = \frac{8+20\xi_3-4\xi_3^2}{(1-\xi_3)^5}\left(\frac{\partial \xi_3}{\partial n_i}\right)_{T,V,n_{k\neq i}}$$

$$\left(\frac{\partial F_2}{\partial n_i}\right)_{T,V,n_{k\neq i}} = \frac{40-48\xi_3+12\xi_3^2+2\xi_3^3}{(2-3\xi_3+\xi_3^2)^3}\left(\frac{\partial \xi_3}{\partial n_i}\right)_{T,V,n_{k\neq i}}$$

$$\left(\frac{\partial C_{1n}}{\partial n_i}\right)_{T,V,n_{k\neq i}} = -C_{1n}^2\Big[1+[\frac{m}{M}]_i M_i F_1 + n\overline{m}\left(\frac{\partial F_1}{\partial n_i}\right)_{T,V,n_{k\neq i}}$$
$$+(1-[\frac{m}{M}]_i M_i)F_2 + (n-n\overline{m})\left(\frac{\partial F_2}{\partial n_i}\right)_{T,V,n_{k\neq i}}\Big] \tag{2.67}$$

Example 4 demonstrates the computation of $(\frac{\partial A^{Res}}{\partial n_i})_{T,V,n_{k\neq i}}$ neglecting association.

Example 4 (Calculation of $(\frac{\partial A^{Res}}{\partial \overline{n}})_{T,V,n}$) We continue from Example 2, which considers a mixture containing 1 mole of propane + 1 mole of n-hexane + 1 mole of water at 350K and $V = 116.38\text{cm}^3$.

The required hard chain terms are:

$$(\frac{\partial \overline{\xi}}{\partial \overline{n}})_{T,V} = \begin{bmatrix} 5.424E-3 & 8.284E-3 & 4.768E-3 \\ 1.923E-2 & 3.097E-2 & 1.145E-2 \\ 6.818E-2 & 0.116 & 2.748E-2 \\ 0.242 & 0.433 & 6.595E-2 \end{bmatrix}$$

$$(\frac{\partial a^{hs}}{\partial \overline{n}})_{T,V} = 33.090, 59.065, 8.270(\frac{1}{\text{mole}})$$

$$\begin{bmatrix} 95.386 & 169.328 & 28.076 \\ 99.322 & 176.309 & 29.240 \\ 69.110 & 122.711 & 20.302 \end{bmatrix}$$

$$\left(\frac{\partial \overline{g}^{hs}}{\partial \overline{n}}\right)_{T,V} = \begin{bmatrix} 99.322 & 176.309 & 29.240 \\ 103.555 & 183.820 & 30.493 \\ 71.28 & 126.575 & 20.946 \end{bmatrix} \left(\frac{1}{\text{mol}}\right)$$

$$\begin{bmatrix} 69.110 & 122.7108 & 20.302 \\ 71.288 & 126.575 & 20.946 \\ 53.648 & 95.278 & 15.730 \end{bmatrix}$$

Note the indexing carefully: g_{ij}^{hs} has two indices corresponding to species i and species j. By contrast $(\frac{\partial g_{ij}^{hs}}{\partial n_k})_{T,V,n_{q\neq k}}$ has three; the change in g_{ij}^{hs} with respect to n_k. This raises the question of how to list a high-dimensional (more than two) array on a page. For clarity, consider that:

$$(\frac{\partial \overline{\overline{g}}^{hs}}{\partial n_{propane}}) = \begin{bmatrix} 95.386 & 99.322 & 69.110 \\ 99.322 & 103.555 & 71.288 \\ 69.110 & 71.288 & 53.648 \end{bmatrix}$$

With this understood, the hard-chain derivatives are:

$$(\frac{\partial A^{hc}}{\partial \overline{n}})_{T,V} = 725367.803, 1264024.298, 243595.097(\frac{\text{J}}{\text{mole}})$$

The required dispersion terms are:

$$(\frac{\partial \overline{a}}{\partial \overline{n}}) = \begin{bmatrix} 7.362E-3 & -0.021 & 0.014 \\ -0.013 & 0.038 & -0.025 \\ 0.031 & -0.089 & 0.058 \\ -0.337 & 0.976 & -0.638 \\ 1.231 & -3.561 & 2.330 \\ -1.757 & 5.080 & -3.324 \\ 0.779 & -2.252 & 1.473 \end{bmatrix}$$

$$(\frac{\partial \overline{b}}{\partial \overline{n}}) = \begin{bmatrix} 7.949E-3 & -0.023 & 0.015 \\ -6.688E-3 & 0.019 & -0.013 \\ 0.131 & -0.378 & 0.247 \\ -0.147 & 0.424 & -0.278 \\ -2.529 & 7.314 & -4.785 \\ 0.799 & -2.312 & 1.513 \\ 3.534 & -10.221 & 6.687 \end{bmatrix}$$

$$(\frac{\partial I_1}{\partial \overline{n}})_{T,V} = 1.225, 2.263, 0.310$$

$$(\frac{\partial I_2}{\partial \overline{n}})_{T,V} = -60.919, -109.145, -16.596$$

$$\left(\frac{\partial F_1}{\partial n}\right)_{T,V} = 4235.154, 7582.798, 1155.517(\frac{1}{mol})$$

$$\left(\frac{\partial F_2}{\partial n}\right)_{T,V} = 82.032, 146.873, 22.382(\frac{1}{mol})$$

$$\left(\frac{\partial C_{1n}}{\partial n}\right)_{T,V} = -6.066E-4, -1.075E-3, -1.907E-4(\frac{1}{mol^2})$$

$$(\frac{\partial[nm\epsilon\sigma_1]}{\partial \overline{n}})_{T,V} = 733.200, 1287.221, 358.758$$

$$(\frac{\partial[nm\epsilon\sigma_2]}{\partial \overline{n}})_{T,V} = 448.157, 839.056, 208.049$$

$$(\frac{\partial A^{Res}}{\partial n})_{T,V} = -227933.128, -412785.117, -78891.554(\frac{J}{mol})$$

With the association term neglected, Eq. 2.58 gives:

$$(\frac{\partial A^{Res}}{\partial n})_{V,\overline{n}} = 497434.675, 851239.180, 164703.543(\frac{J}{mol})$$

It is not practical to list the all the required derivatives of the PC-SAFT EOS. By applying the techniques demonstrated here one can systematically develop them.

2.3 Numerical Testing of an EOS Implementation

Computational thermodynamics might be called *applied differential calculus*. As we saw in Sects. 2.1 and 2.2, the differentiation often involves expressions more complicated than simple polynomials. Making the occasional error in either a derivation or coding is inevitable, especially when we begin working with a complicated EOS and/or higher-order derivatives. Fortunately, a simple yet powerful technique is available to aid in the detection of such errors. This section will explain that concept in basic form.

Consider a simple function;

$$F = x^2 \tag{2.68}$$

the derivative of which is:

$$\frac{dF}{dx} = 2x \tag{2.69}$$

While this is a trivial example, how could we check whether our expressions for $F(x)$ and $\frac{dF}{dx}$ correspond? To address this, let's consider perturbing the function $F(x)$ from its value at x, by evaluating it at $x_{pert} = x + \Delta x$. We can represent perturbations in F via a Taylor series (Appendix 8.19):

$$F(x\_pert) = F(x + \Delta x) = F(x) + \frac{dF}{dx}\Delta x + \frac{1}{2}\frac{d^2F}{dx^2}(\Delta x)^2 + \frac{1}{6}\frac{d^3F}{dx^3}(\Delta x)^3 + ... \quad (2.70)$$

By truncating Eq. 2.70, only including the initial term on the right-hand side, we can compute an error (because if F depends on x, then $F(x_{pert}) \neq F(x)$), which we call $E1$:

$$E1 = F(x\_pert) - F(x) \quad (2.71)$$

We next consider the error that results from including $\frac{dF}{dx}$ (the derivative we are attempting to test), defining $E2$:

$$E2 = F(x\_pert) - F(x) - \frac{dF}{dx}\Delta x = E1 - \frac{dF}{dx}\Delta x \quad (2.72)$$

In general, neither $E1$ nor $E2$ will be zero. However, if we compare $E1$ and $E2$ values, by generating a series of Δx which *decrease* in magnitude, the dominant term in $E1$ will be of order of Δx, while that in $E2$ will be of order $(\Delta x)^2$; higher powers of Δx multiplying higher (and likely unknown in our development) derivatives of F in Eq. 2.70 will be of limited contribution. This can be seen by considering the right-hand side of Eq. 2.70, and recognizing that x is constant in this context, while Δx is changing. As a result, the slope of $ln|E2|$ vs. $ln|E1|$ will trend to a value of 2. To illustrate this concept, let's return to our simple function ($F = x^2$):

$$E1 = (x + \Delta x)^2 - x^2 = 2x\Delta x + (\Delta x)^2 \quad (2.73)$$

$$E2 = E1 - \frac{dF}{dX}\Delta x = (\Delta x)^2 \quad (2.74)$$

At very small[5] values of Δx, one has:

$$E1 = 2x\Delta x + (\Delta x)^2 \approx 2x\Delta x \quad (2.75)$$

$$E2 = (\Delta x)^2 \quad (2.76)$$

[5] This is also a useful technique in modeling, whereby one term within a set is deemed negligible compared with others and is discarded. Note that this can be applied to $E1$ but not to $E2$, as the latter contains only a single term.

Therefore:

$$ln|E1| = ln(2x) + ln(\Delta x) \tag{2.77}$$

$$ln|E2| = 2ln(\Delta x) \tag{2.78}$$

The above results indicate that, if we have coded $\frac{dF}{dx}$ correctly, we expect that a plot of $ln|E2|$ vs. $ln|E1|$ will have a slope[6] of 2.

In general form, with

$$\begin{aligned} F1 &= ln|E1| \\ F2 &= ln|E2| \end{aligned} \tag{2.79}$$

we can compute the slope via Linear Least Squares (see Appendix 8.11) via:

$$slope = \frac{\sum_r F1_r(F2_r - \overline{F2})}{\sum_r F1_r(F1_r - \overline{F1})} \tag{2.80}$$

The pseudocode in Algorithm 1 illustrates how to structure a numerical test using this technique.

Algorithm 1 Numerical testing of the derivative of a scalar function

---

Calculate $F = fun\_F(x)$
Create list of N perturbations: $perts = [1E-6, ..., 1E-2]$
Initialize $E1$ and $E2$ (length N)
for p (N) **do**
  $x\_pert = x(1 + perts[p])$
  $F\_pert = fun\_F(x\_pert)$
  $E1[p] = F\_pert - F$
  $E2[p] = E1[p] - \frac{dF}{dx}px$
end for
$F1 = ln|E1|$
$F2 = ln|E2|$
From the sets $F1$ and $F2$, calculate the least squares slope:
$\overline{F1} = \sum_r F1_r/N$
$\overline{F2} = \sum_r F2_r/N$
$slope = \sum_r F1_r(F2_r - \overline{F2})/\sum_r F1_r(F1_r - \overline{F1})$

In practice, the slope will not be exactly two, as the value determined depends on the functions being tested and the range of perturbations. As a guide, you should expect slope values between 1.99 and 2.01. Also note that linear functions will *always* fail this test, because $E2$ will be identically zero. Other than for trivial cases,

[6] Recall that the variable changing in this context is Δx, not x.

a good practice for linear functions is to confirm that E_2 values are sufficiently close to zero (i.e., $1E - 16$). One way of doing so, which makes the testing appear consistent with other functions, is to use $slope = 2 - \sum_r E2_r$ for linear functions. Another point to consider is that conditions where $\frac{dF}{dx} \approx 0$ should be avoided. Aside from these points, a failed test ($slope \neq 2$) for a nonlinear function indicates an error in the derivative.[7] Repeatedly attempting to rederive/recode a derivative can be a frustrating experience, as you may simply introduce new errors on each attempt. Alternatively, you might be making a conceptual or mathematical mistake that continuously leads to the wrong result. We can do better by systematically "building up" the derivative term by term to assure a passing test. To demonstrate, let's look at a simple example from the implementation of the SRK EOS.

Suppose that, in attempting to code Eq. 2.17, we make a mistake, and instead we write:

$$\left(\frac{\partial A^{Res}}{\partial V}\right)_{T,\overline{n}} = -nR_gT\left(\frac{1}{V-nb} - \frac{1}{V}\right) - \frac{n^2a}{nb}\left(\frac{1}{V+nb}" + "\frac{1}{V}\right) \tag{2.81}$$

With the sign error in Eq. 2.81 (compare with Eq. 2.17), numerical testing of this derivative will fail. Having detected this by $slope \neq 2$, we use the "build-up" technique by truncating our definition of A^{Res};

$$A^{Res}(T, V, \overline{n}) = -nR_gTln\left[\frac{V-nb}{V}\right] \tag{2.82}$$

The corresponding derivative of Eq. 2.82 is:

$$\left(\frac{\partial A^{Res}}{\partial V}\right)_{T,\overline{n}} = -nR_gT\left(\frac{1}{V-nb} - \frac{1}{V}\right) \tag{2.83}$$

Numerical testing at this point will pass; an encouraging result indicating that the mistake in our original expression for $\left(\frac{\partial A^{Res}}{\partial V}\right)_{T,\overline{n}}$ is associated with the latter portion A^{Res}. For the purposes of illustration, let's assume that by visual inspection we do not see the simple sign error in Eq. 2.81. Instead, let's use "build-up" again by writing;

$$A^{Res}(T, V, \overline{n}) = -nR_gTln\left[\frac{V-nb}{V}\right] - \frac{n^2a}{nb}ln\left[V+nb\right] \tag{2.84}$$

[7] This actually indicates an inconsistency between coded expressions for a function and its derivative. Here we assume that the function itself has been coded correctly.

the derivative of which is:

$$\left(\frac{\partial A^{Res}}{\partial V}\right)_{T,\overline{n}} = -nR_gT\left(\frac{1}{V-nb}-\frac{1}{V}\right)-\frac{n^2a}{nb}\left(\frac{1}{V+nb}\right) \tag{2.85}$$

Again, numerical testing will pass, leading us further to recognize[8] our sign error in Eq. 2.81. Though this example is somewhat trivial, the derivatives encountered in thermodynamic modeling are not, and this technique will prove valuable in locating and correcting typos and mathematical errors.

The pseudocode in Algorithm 2 outlines how to test scalar derivatives such as the volume derivative of A^{Res}.

Algorithm 2 Numerical testing of an implementation of $\left(\frac{\partial A^{Res}}{\partial V}\right)_{T,\overline{n}}$

Calculate $A^{Res} = fun\_A^{Res}(T, V, \overline{n})$ (EOS-specific)
Create list of N perturbations between $1E-6$ and $1E-2$: $perts = [1E-6, ..., 1E-2]$
for p (N) **do**
  $V\_pert = V(1+p)$
  $A^{Res}\_pert = fun\_A^{Res}(T, V_{pert}, \overline{n})$ (EOS-specific)
  $E1[p] = A^{Res}\_pert - A^{Res}$
  $E2[p] = E1[p] - \frac{\partial A^{Res}}{\partial V} Vp$
end for
$F1 = ln|E1|$
$F2 = ln|E2|$
From the sets $F1$ and $F2$, calculate the least squares slope:
$\overline{F1} = \sum_r F1_r/N$
$\overline{F2} = \sum_r F2_r/N$
$slope = \sum_r F1_r(F2_r - \overline{F2})/\sum_r F1_r(F1_r - \overline{F1})$

Numerical testing can also be done on non-scalar variables in scalar functions. For example, we can test our expression for $\frac{\partial A^{Res}}{\partial n_i}$ as indicated in the pseudocode given in Algorithm 3.

The same approach can be used to test non-scalar functions with non-scalar variables, as demonstrated in Algorithm 4.

With these examples, it should be clear that one can test derivatives of functions with respect to variables without restriction to dimension. The author advises that all but the most trivial derivatives be tested. A useful approach is to embed the numerical test within the code. For example, a flag can be used to signal whether the numerical test should be run in the function call. Pseudocode for embedding a numerical test within the code for $\left(\frac{\partial \alpha_i}{\partial T}\right)_{V,\overline{n}}$ within the SRK EOS (see Eq. 2.11) is shown in Algorithm 5.

[8] Numerical testing does not reveal the correct form of a derivative, but it does allow us (when combined with the "build-up" technique) to *progressively* locate an error.

Algorithm 3 Numerical testing of $\left(\frac{\partial A^{Res}}{\partial n_i}\right)_{T,V,n_{k\neq i}}$

Calculate $A^{Res} = fun\_A^{Res}(T, V, \overline{n})$ (EOS-specific)
Create list of N perturbations between $1E-6$ and $1E-2$: $perts = [1E-6, .., 1E-2]$
Initialize $slope$ (species)
for i (species) **do**
  for p (N) **do**
    $\overline{n}\_pert = copy(\overline{n})$
    $n\_pert[i] = n[i](1+p)$
    $A^{Res}\_pert = fun\_A^{Res}(T, V, \overline{n}\_pert)$ (EOS-specific)
    $E1[p] = A^{Res}\_pert - A^{Res}$
    $E2[p] = E1[p] - \frac{\partial A^{Res}}{\partial n}[i]n[i]p$
  end for
  $F1 = ln|E1|$
  $F2 = ln|E2|$
  From the sets $F1$ and $F2$, calculate the least squares slope:
  $\overline{F1} = \sum_r F1_r/N$
  $\overline{F2} = \sum_r F2_r/N$
  $slope[i] = \sum_r F1_r(F2_r - \overline{F2})/\sum_r F1_r(F1_r - \overline{F1})$
end for

Algorithm 4 Numerical testing of $\left(\frac{\partial^2 A^{Res}}{\partial n_i \partial n_j}\right)$

Calculate $\frac{A^{Res}}{\partial n} = fun\_\frac{A^{Res}}{\partial n}(T, V, \overline{n})$ (EOS-specific)
Create list of N perturbations: $perts = [1E-6, ..., 1E-2]$
Initialize $slope$ (species x species)
for i (species) **do**
  for j (species) **do**
    for p (N) **do**
      $\overline{n}\_pert = copy(\overline{n})$
      $n\_pert[j] = n[j](1+p)$
      $\frac{\partial A^{Res}}{\partial n}\_pert[i] = fun\_\frac{A^{Res}}{\partial n}(T, V, \overline{n}_{pert})$ (EOS-specific)
      $E1[p] = (\frac{\partial A^{Res}}{\partial n}\_pert - \frac{\partial A^{Res}}{\partial n})[i]$
      $E2[p] = E1[p] - \frac{\partial^2 A^{Res}}{\partial n^2}[i, j]n[j]p$
    end for
    $F1 = ln|E1|$
    $F2 = ln|E2|$
    From the sets $F1$ and $F2$, calculate the least squares slope:
    $\overline{F1} = \sum_r F1_r/N$
    $\overline{F2} = \sum_r F2_r/N$
    $slope[i, j] = \sum_r F1_r(F2_r - \overline{F2})/\sum_r F1_r(F1_r - \overline{F1})$
  end for
end for

Algorithm 5 Embedding numerical tests within functions ($\frac{\partial \alpha}{\partial T}$, Eq. 2.11)

Definition: $fun\_\frac{\partial \overline{\alpha}}{\partial T}(\overline{Tc}, T, \overline{m}, \overline{\alpha}, Test)$
$\frac{\partial \overline{\alpha}}{\partial T} = -sqrt(\overline{\alpha}/(T\overline{Tc}))\overline{m}$
if $Test == True$ **then**
  Create list of N perturbations: $perts = [1E-6, ..., 1E-2]$
  Initialize $slope$ (species)
  for i (species) **do**
    for p (N) **do**
      $T\_pert = T(1+p)$
      $\overline{\alpha}\_pert = fun\_\overline{\alpha}(\overline{Tc}, \overline{m}, T\_pert)$ (Eq. 2.8)
      $E1[p] = (\overline{\alpha}\_pert - \overline{\alpha})[i]$
      $E2[p] = E1[p] - \frac{\partial \overline{\alpha}}{\partial T}[i]Tp$
    end for
    $F1 = ln|E1|$
    $F2 = ln|E2|$
    From the sets $F1$ and $F2$, calculate the least squares slope:
    $\overline{F1} = \sum_r F1_r/N$
    $\overline{F2} = \sum_r F2_r/N$
    $slope[i] = \sum_r F1_r(F2_r - \overline{F2})/\sum_r F1_r(F1_r - \overline{F1})$
  end for
  Print($slope$)
end if
Return $\frac{\partial \overline{\alpha}}{\partial T}$

A number of points should be observed in Algorithm 5. First, note that the required "pieces" for the calculation of $\left(\frac{\partial \alpha_i}{\partial T}\right)_{V,\overline{n}}$ are passed into the function. This increases the efficiency of the code, avoiding unnecessary recomputation of terms (e.g., m and α in the function for $\left(\frac{\partial \alpha_i}{\partial T}\right)_{V,\overline{n}}$). Second, the numerical test can be called (returning a print statement of the slope) when tracing bugs or other issues in your code, but by default (if your programming language allows), this test is skipped. Next, to test $\left(\frac{\partial \alpha_i}{\partial T}\right)_{V,\overline{n}}$ the value of α is needed. In the above case, α was passed as a required piece into the function for the calculation of $\left(\frac{\partial \alpha_i}{\partial T}\right)_{V,\overline{n}}$ itself (i.e., not for the test). In other cases your test will need such pieces, even in the function you're testing does not. A wholly different approach is to only use testing when implementing your derivatives the first time, and once passed, the testing code can be commented out. It is your preference, but it is the author's experience that having the testing feature available, even in "mature" code, is very helpful when debugging and developing code. Also, one needs to consider whether they are writing high-performance commercial code or something practical and transparent for a particular task at hand.

Chapter 3
Thermodynamic Properties from an Equation of State $(T, V, \overline{n})$

In this chapter, we will develop in general form the equations by which thermodynamic quantities can be obtained from an equation of state (EOS). The development will proceed assuming that we have an EOS providing the residual Helmholtz free energy as a function of temperature, volume, and mole numbers ($A^{Res} = A^{Res}(T, V, \overline{n})$), as well as its derivatives. In other words, this chapter assumes that, for the particular EOS you want to use, the content of Chap. 2 has been applied. For some this might be an abstract starting point for calculations. Recall that some EOS (e.g., SRK and other cubics) are stated as expressions of the form $P = P(T, V, \overline{n})$, as in Eq. 1.27. These can be integrated to give A^{Res} (Eq. 1.28). More modern examples such as PC-SAFT provide A^{Res} directly. In any case, the methods developed in the chapter shall apply to both, and the reader is encouraged to think of an EOS as a model for A^{Res}, from which the necessary derivatives can be derived, rather than an expression for P. Also, this section may come across with less development than others in the text. This is somewhat intentional, as it is intended to serve as a reference for subsequent sections. For example, Chap. 4 will require formulas for Z and its volume derivatives ($(\left(\frac{\partial Z}{\partial V}\right)_{T,\overline{n}}, \left(\frac{\partial^2 Z}{\partial V^2}\right)_{T,\overline{n}} \ldots$). This section simply presents a method of calculating these and other thermodynamic quantities in terms of A^{Res}. An important concept to understand from this chapter is that (nearly) all relevant thermodynamic quantities can be calculated from A^{Res} and its derivatives with respect to T, V, and $\overline{n}$. Note that there is no limit to the number of derivatives we could calculate. This section only references those required for the applications in this text.

Finally, in coding such formulas to perform thermodynamic calculations, it is helpful to apply a formalism for the order of variables. In writing $(T, V, \overline{n})$, we have chosen such an order. We can then organize derivatives into groups assigned to each variable. For example, the derivatives of Z can be listed in the following order:

B. R. Greenhalgh, *An Introduction to Computational Thermodynamics*,
https://doi.org/10.1007/978-3-032-30317-2_3

Variable series

$$\left(\frac{\partial Z}{\partial T}\right)_{V,\overline{n}}, \left(\frac{\partial Z}{\partial V}\right)_{T,\overline{n}}, \left(\frac{\partial Z}{\partial n_i}\right)_{T,V,n_{k\neq i}}$$

T series

$$\left(\frac{\partial^2 Z}{\partial T^2}\right)_{V,\overline{n}}, \left(\frac{\partial^2 Z}{\partial T \partial V}\right), \left(\frac{\partial^2 Z}{\partial T \partial n_i}\right)$$

V series

$$\left(\frac{\partial^2 Z}{\partial V^2}\right)_{V,\overline{n}}, \left(\frac{\partial^2 Z}{\partial V \partial n_i}\right)$$

n_i series

$$\left(\frac{\partial^2 Z}{\partial n_i \partial n_j}\right)_{T,V}$$

Note that some members of one series are contained within another. For example, the T series contains $\left(\frac{\partial^2 Z}{\partial T \partial V}\right)$, and based on the symmetry of mixed partial derivatives, we do not include $\left(\frac{\partial^2 Z}{\partial V \partial T}\right)$ in the V series. This ordering concept is not strictly necessary, but will assist you in organizing your code, including the order of variables in the syntax of your functions.

3.1 *Z* and Its $(T, V, \overline{n})$ Derivatives

Given an expression for A^{Res}, we can derive, using Eq. 1.33, the following result:

$$\left(\frac{\partial A^{Res}}{\partial V}\right)_{T,\overline{n}} = -\left(P - \frac{nR_gT}{V}\right) = -\frac{nR_gT}{V}(Z-1) \tag{3.1}$$

The compressibility factor is then given by:

$$Z = 1 - \frac{V}{nR_gT}\left(\frac{\partial A^{Res}}{\partial V}\right)_{T,\overline{n}} \tag{3.2}$$

The $(T, V, \overline{n})$ derivatives of Z are:

$$\left(\frac{\partial Z}{\partial T}\right)_{V,\overline{n}} = -\frac{1}{T}(Z-1) - \frac{V}{nR_gT}\left(\frac{\partial^2 A^{Res}}{\partial T \partial V}\right) \tag{3.3}$$

$$\left(\frac{\partial Z}{\partial V}\right)_{T,\overline{n}} = \frac{1}{V}(Z-1) - \frac{V}{nR_gT}\left(\frac{\partial^2 A^{Res}}{\partial V^2}\right)_{T,\overline{n}} \tag{3.4}$$

$$\left(\frac{\partial Z}{\partial n_i}\right)_{T,V,n_{k\neq i}} = -\frac{1}{n}(Z-1) - \frac{V}{nR_gT}\left(\frac{\partial^2 A^{Res}}{\partial V \partial n_i}\right) \tag{3.5}$$

Higher-order derivatives can be structured as follows:

$$\left(\frac{\partial^2 Z}{\partial T \partial V}\right)_{\overline{n}} = \frac{1}{TV}(Z-1) + \frac{1}{V}\left(\frac{\partial Z}{\partial T}\right)_{V,\overline{n}} - \frac{1}{T}\left(\frac{\partial Z}{\partial V}\right)_{T,\overline{n}} - \frac{V}{nR_gT}\left(\frac{\partial^3 A^{Res}}{\partial T \partial V^2}\right) \tag{3.6}$$

$$\left(\frac{\partial^2 Z}{\partial V^2}\right)_{T,\overline{n}} = -\frac{2}{V^2}(Z-1) + \frac{2}{V}\left(\frac{\partial Z}{\partial V}\right)_{T,\overline{n}} - \frac{V}{nR_gT}\left(\frac{\partial^3 A^{Res}}{\partial V^3}\right)_{T,\overline{n}} \tag{3.7}$$

$$\left(\frac{\partial^3 Z}{\partial V^3}\right)_{T,\overline{n}} = \frac{6}{V^3}(Z-1) - \frac{6}{V^2}\left(\frac{\partial Z}{\partial V}\right)_{T,\overline{n}} + \frac{3}{V}\left(\frac{\partial^2 Z}{\partial V^2}\right)_{T,\overline{n}} - \frac{V}{nR_gT}\left(\frac{\partial^4 A^{Res}}{\partial V^4}\right)_{T,\overline{n}} \tag{3.8}$$

3.2 P and Its $(T,V,\overline{n})$ Derivatives

You may find it convenient to develop expressions for the pressure and its derivatives. From Eq. 1.64 we have:[1]

$$P = \frac{ZnR_gT}{V} \tag{3.9}$$

Selected $(T, V, \overline{n})$ derivatives of P are:

$$\left(\frac{\partial P}{\partial V}\right)_{T,\overline{n}} = \frac{nR_gT}{V}\left[\left(\frac{\partial Z}{\partial V}\right)_{T,\overline{n}} - \frac{Z}{V}\right] \tag{3.10}$$

[1] Note that these expressions are written without regard to unit choice. For P in Pa and V in cm^3, a factor or $1E6$ must be applied.

$$\left(\frac{\partial P}{\partial n_i}\right)_{T,V,n_{k\neq i}} = \frac{nR_gT}{V}\left[\left(\frac{\partial Z}{\partial n_i}\right)_{T,V,n_{k\neq i}} + \frac{Z}{n}\right] \tag{3.11}$$

$$\left(\frac{\partial^2 P}{\partial V^2}\right)_{T,\overline{n}} = \frac{nR_gT}{V}\left[\left(\frac{\partial^2 Z}{\partial V^2}\right)_{T,\overline{n}} - \frac{2}{V}\left(\frac{\partial Z}{\partial V}\right)_{T,\overline{n}} + \frac{2}{V^2}Z\right] \tag{3.12}$$

$$\left(\frac{\partial^3 P}{\partial V^3}\right)_{T,\overline{n}} = \frac{nR_gT}{V}\left[\left(\frac{\partial^3 Z}{\partial V^3}\right)_{T,\overline{n}} - \frac{3}{V}\left(\frac{\partial^2 Z}{\partial V^2}\right)_{T,\overline{n}} + \frac{6}{V^2}\left(\frac{\partial Z}{\partial V}\right)_{T,\overline{n}} - \frac{6}{V^3}Z\right] \tag{3.13}$$

3.3 $ln\hat{\varphi}_i$ and Its $(T,V,\overline{n})$ Derivatives

As we have already seen, the fugacity coefficient is computed from Eq. 1.71:

$$ln\hat{\varphi}_i = \frac{1}{R_gT}\left(\frac{\partial A^{Res}}{\partial n_i}\right)_{T,V,n_{k\neq i}} - lnZ$$

Selected $(T, V, \overline{n})$ derivatives of $ln\hat{\varphi}_i$ are:

$$\left(\frac{\partial ln\hat{\varphi}_i}{\partial T}\right)_{V,\overline{n}} = -\frac{1}{R_gT^2}\left(\frac{\partial A^{Res}}{\partial n_i}\right)_{T,V,n_{k\neq i}} + \frac{1}{R_gT}\left(\frac{\partial^2 A^{Res}}{\partial T\partial n_i}\right) - \frac{1}{Z}\left(\frac{\partial Z}{\partial T}\right)_{V,\overline{n}} \tag{3.14}$$

$$\left(\frac{\partial ln\hat{\varphi}_i}{\partial V}\right)_{V,\overline{n}} = \frac{1}{R_gT}\left(\frac{\partial^2 A^{Res}}{\partial V\partial n_i}\right) - \frac{1}{Z}\left(\frac{\partial Z}{\partial V}\right)_{T,\overline{n}} \tag{3.15}$$

$$\left(\frac{\partial ln\hat{\varphi}_i}{\partial n_j}\right)_{T,V,n_{k\neq j}} = \frac{1}{R_gT}\left(\frac{\partial^2 A^{Res}}{\partial n_i\partial n_j}\right) - \frac{1}{Z}\left(\frac{\partial Z}{\partial n_j}\right)_{T,V,n_{k\neq j}} \tag{3.16}$$

3.4 Other Potentials and Their $(T,V,\overline{n})$ Derivatives

The Helmholtz free energy is the primary basis of EOS calculations, and correspondingly, the other potentials (S, U, H, and G) can be coded in terms of A^{Res} and its derivatives. As S derives from the temperature derivative of A, we begin there (Eq. 1.32).

$$S^{Res} = -\left(\frac{\partial A^{Res}}{\partial T}\right)_{V,\overline{n}} \tag{3.17}$$

Selected $(T, V, \overline{n})$ derivatives of S^{Res} are:

$$\left(\frac{\partial S^{Res}}{\partial T}\right)_{V,\overline{n}} = -\left(\frac{\partial^2 A^{Res}}{\partial T^2}\right)_{V,n_{k\neq i}} \tag{3.18}$$

$$\left(\frac{\partial S^{Res}}{\partial V}\right)_{V,\overline{n}} = -\left(\frac{\partial^2 A^{Res}}{\partial T \partial V}\right) \tag{3.19}$$

$$\left(\frac{\partial S^{Res}}{\partial n_i}\right)_{T,V,n_{k\neq i}} = -\left(\frac{\partial^2 A^{Res}}{\partial T \partial n_i}\right) \tag{3.20}$$

With $U = A + TS$, U^{Res} is given by:

$$U^{Res} = A^{Res} + TS^{Res} = A^{Res} - T\left(\frac{\partial A^{Res}}{\partial T}\right)_{V,\overline{n}} \tag{3.21}$$

Selected $(T, V, \overline{n})$ derivatives of U^{Res} are:

$$\left(\frac{\partial U^{Res}}{\partial T}\right)_{V,\overline{n}} = -T\left(\frac{\partial^2 A^{Res}}{\partial T^2}\right)_{V,\overline{n}} \tag{3.22}$$

$$\left(\frac{\partial U^{Res}}{\partial V}\right)_{V,\overline{n}} = \left(\frac{\partial A^{Res}}{\partial V}\right)_{T,\overline{n}} - T\left(\frac{\partial^2 A^{Res}}{\partial T \partial V}\right) \tag{3.23}$$

$$\left(\frac{\partial U^{Res}}{\partial n_i}\right)_{T,V,n_{k\neq i}} = \left(\frac{\partial A^{Res}}{\partial n_i}\right)_{T,V,n_{k\neq i}} - T\left(\frac{\partial^2 A^{Res}}{\partial T \partial n_i}\right) \tag{3.24}$$

With $H = U + PV$, H^{Res} is given by:

$$H^{Res} = U^{Res} + P^{Res}V = A^{Res} - T\left(\frac{\partial A^{Res}}{\partial T}\right)_{V,\overline{n}} - V\left(\frac{\partial A^{Res}}{\partial V}\right)_{T,\overline{n}} \tag{3.25}$$

Selected $(T, V, \overline{n})$ derivatives of H^{Res} are:

$$\left(\frac{\partial H^{Res}}{\partial T}\right)_{V,\overline{n}} = -T\left(\frac{\partial^2 A^{Res}}{\partial T^2}\right) - V\left(\frac{\partial^2 A^{Res}}{\partial T \partial V}\right) \tag{3.26}$$

$$\left(\frac{\partial H^{Res}}{\partial V}\right)_{V,\overline{n}} = -T\left(\frac{\partial^2 A^{Res}}{\partial T \partial V}\right) - V\left(\frac{\partial^2 A^{Res}}{\partial V^2}\right)_{T,\overline{n}} \tag{3.27}$$

$$\left(\frac{\partial H^{Res}}{\partial n_i}\right)_{T,V,n_{k\neq i}} = \left(\frac{\partial A^{Res}}{\partial n_i}\right)_{T,V,n_{k\neq i}} - T\left(\frac{\partial^2 A^{Res}}{\partial T \partial n_i}\right) - V\left(\frac{\partial^2 A^{Res}}{\partial V \partial n_i}\right) \tag{3.28}$$

Finally, with $G = H - TS$ (or $G = A + PV$), we have:

$$G^{Res} = H^{Res} - TS^{Res} = A^{Res} - V\left(\frac{\partial A^{Res}}{\partial V}\right)_{T,\overline{n}} \tag{3.29}$$

Selected $(T, V, \overline{n})$ derivatives of G^{Res} are:

$$\left(\frac{\partial G^{Res}}{\partial T}\right)_{V,\overline{n}} = \left(\frac{\partial A^{Res}}{\partial T}\right) - V\left(\frac{\partial^2 A^{Res}}{\partial T \partial V}\right) \tag{3.30}$$

$$\left(\frac{\partial G^{Res}}{\partial V}\right)_{V,\overline{n}} = -V\left(\frac{\partial^2 A^{Res}}{\partial V^2}\right)_{T,\overline{n}} \tag{3.31}$$

$$\left(\frac{\partial G^{Res}}{\partial n_i}\right)_{T,V,n_{k\neq i}} = \left(\frac{\partial A^{Res}}{\partial n_i}\right)_{T,V,n_{k\neq i}} - V\left(\frac{\partial^2 A^{Res}}{\partial V \partial n_i}\right) \tag{3.32}$$

3.5 Numerical Testing for $(T, V, \overline{n})$ Equation of State Calculations

Just as we did with functions within a specific EOS, we can apply numerical testing to the functions that use the implemented EOS. For example, the pseudocode given in Algorithm 6 demonstrates how to test the implementation of $\left(\frac{\partial Z}{\partial V}\right)_{T,\overline{n}}$, including the underlying EOS.[2]

[2] Inside Algorithm 6, one could even pass a Test flag into the call for $\left(\frac{\partial A^{Res}}{\partial V}\right)_{T,\overline{n}}$, in order to ensure that this test is also correctly passed.

Algorithm 6 Embedded numerical test within $\left(\frac{\partial Z}{\partial V}\right)_{T,\overline{n}}$, (Eq. 3.4)

Definition: $fun\_\frac{\partial Z}{\partial V}(T, V, \overline{n}, \frac{\partial^2 A}{\partial V^2}, Z, Test)$
$\frac{\partial Z}{\partial V} = (Z-1)/V - \frac{\partial^2 A^{Res}}{\partial V^2}(V/(nR_gT))$
if $Test == True$ **then**
    Create list of N perturbations: $perts = [1E-6, \ldots, 1E-2]$
    for p (N) **do**
        $V\_pert = V(1+p)$
        $(\frac{\partial A^{Res}}{\partial V})\_pert = fun\_(\frac{\partial A^{Res}}{\partial V})(T, V\_pert, \overline{n})$ (EOS-specific)
        $Z\_pert = fun\_Z(T, V\_pert, \overline{n}, (\frac{\partial A^{Res}}{\partial V})\_pert)$ (Eq. 3.2)
        $E1[p] = Z\_pert - Z$
        $E2[p] = E1[p] - \frac{\partial Z}{\partial V} Vp$
    end for
    $F1 = ln|E1|$
    $F2 = ln|E2|$
    From the sets $F1$ and $F2$, calculate the least squares slope:
    $\overline{F1} = \sum_r F1_r/N$
    $\overline{F2} = \sum_r F2_r/N$
    $slope = \sum_r F1_r(F2_r - \overline{F2})/\sum_r F1_r(F1_r - \overline{F1})$
    Print($slope$)
end if
Return Z

Chapter 4
Volume Root Solver

This chapter bridges concepts with which you are likely familiar (the important role played by Gibbs free energy at a specified temperature, pressure, and mole numbers) and what this book aims to explain (the machinery of computational thermodynamics is based on Helmholtz free energy at a specified temperature, volume, and mole numbers). In simple form, we must do the following:

For a required property at $(T, P_*, \overline{n})$:

1. Solve all possible roots from the EOS:
   the set of volumes that produce the value P_*
2. Select the appropriate root:

   "Liquid" (smallest volume)
   "Vapor" (largest volume)
   "Stable" (volume with lowest Gibbs free energy)

3. Use that volume to perform the required calculation in terms of A^{Res} (Chap. 3)

In this chapter, we will focus on steps 1 and 2. If one is working with a cubic EOS, a reliable and efficient root solver can be based around solving a third-order (hence the term *cubic*) polynomial in Z. This is demonstrated below for the SRK EOS starting with Eq. 1.25:

$$P = \frac{nR_gT}{V - nb} - \frac{n^2a}{V(V + nb)}$$

Recalling from Eq. 1.64 that $Z = \frac{PV}{nR_gT}$, we seek to generate such groups:

$$\frac{P}{nR_gT}V(V + nb)(V - nb) = V(V + nb) - \frac{n^2a}{nR_gT}(V - nb)$$

B. R. Greenhalgh, *An Introduction to Computational Thermodynamics*,
https://doi.org/10.1007/978-3-032-30317-2_4

$$\frac{P^3}{(nR_gT)^3}\left[V^3 - (nb)^2V\right] = \frac{P^2}{(nR_gT)^2}\left[V^2 + (nb)V\right] - \frac{P^2}{(nR_gT)^2}\left[\frac{n^2a}{nR_gT}V - \frac{n^2a}{nR_gT}nb\right]$$

Gathering groups of $\frac{PV}{nR_gT}$ as Z, and further defining the coefficients:

$$A = \frac{P(n^2a)}{(nR_gT)^2} \tag{4.1}$$

and

$$B = \frac{Pnb}{nR_gT} \tag{4.2}$$

we have:

$$Z^3 - Z^2 + (A - B - B^2)Z - AB = 0 \tag{4.3}$$

With T, P, and $\overline{n}$ known (we will from this point only write P_* if such clarity is required), the coefficients are calculable. One can, by either Newton's method (Appendix 8.13) or more practically using a polynomial solver built into the coding language that you're using, solve the (at most three) roots of Eq. 4.3. We now look at an example.

Example 5 (Solving a Cubic EOS) Let us solve the roots for a mixture containing 1 mole of propane and 1 mole of n-hexane at 350K and 5 bar using the SRK EOS. Using the equations given in Sect. 2.1, we must calculate the coefficients in Eq. 4.3. We begin with the SRK parameters for the species of interest, excerpted from Table 8.2 and summarized in Table 4.1.

We will further assume in this example that $K_{ij} = 0$. The calculations proceed as follows:

$$\overline{m} = 0.717, 0.935(-) \text{ (Eq. 2.9)}$$

$$\overline{\alpha} = 1.039, 1.342(-) \text{ (Eq. 2.8)}$$

Table 4.1 Pure component SRK for propane and n-hexane

Species	T_c [K]	P_c [bar]	ω [–]
Propane	369.80	42.50	0.153
n-Hexane	507.50	30.10	0.299

$$\overline{a_c} = 9.509E5, 2.529E6\left(\frac{\text{Jcm}^3}{\text{mol}^2}\right) \text{ (Eq. 2.7)}$$

$$\overline{a} = 9.882E5, 3.394E6\left(\frac{\text{Jcm}^3}{\text{mol}^2}\right) \text{ (Eq. 2.6)}$$

$$K_{ij} = 0(-) \text{ (As above)}$$

$$\overline{\overline{a}} = \begin{bmatrix} 9.882E5 & 1.831E6 \\ 1.831E6 & 3.394E6 \end{bmatrix}\left(\frac{\text{Jcm}^3}{\text{mol}^2}\right) \text{ (Eq. 2.5)}$$

$$n^2a = 8.045E6\left(\text{Jcm}^3\right) \text{ (Eq. 2.2)}$$

$$\overline{b} = 62.681, 121.457\left(\frac{\text{cm}^3}{\text{mol}}\right) \text{ (Eq. 2.10)}$$

$$nb = 184.138\,(\text{cm}^3) \text{ (Eq. 2.3)}$$

The coefficients, which by Eq. 4.3 must be dimensionless, are:

$$A = 0.119$$
$$B = 0.016$$

Using a root solver,[1] we can solve Eq. 4.3, yielding:

$$\overline{Z} = 0.887, 8.984E - 2, 2.358E - 2$$

By virtue of Eq. 1.64:

$$\overline{V} = 10320.036, 1045.737, 274.526\,cm^3(cm^3)$$

With the set of roots determined, one must select the desired root. The smallest volume can be considered the liquid root, while the largest volume can be considered the vapor. Recalling that each root reproduces the target pressure (P), the volume with the lowest Gibbs free energy is the stable root.[2] We therefore need an expression for the absolute (not the residual) Gibbs free energy at a specified *volume*. By definition:

$$\begin{aligned} G(T, V, \overline{n}) &= G^{Res}(T, V, \overline{n}) + G^{IG}(T, V, \overline{n}) \\ &= G^{Res}(T, V, \overline{n}) + G^{IG}(T, P^{IG}, \overline{n}) \end{aligned}$$

[1] The author, using Python, used the "roots" function within the numpy package.

[2] Although the root solver uses A^{Res} and its volume derivatives, the search is nonetheless at fixed $T, P, \overline{n}$. According to in Table 1.2, G is the potential to be minimized.

$$= G^{Res}(T, V, \overline{n}) + nR_gTln(\frac{P^{IG}}{P}) + G^{IG}(T, P, \overline{n})$$

$$= A^{Res}(T, V, \overline{n}) + P^{Res}V + nR_gTln(\frac{P^{IG}}{P}) + G^{IG}(T, P, \overline{n})$$

$$= A^{Res}(T, V, \overline{n}) + nR_gT(Z - 1) - nR_gTln(Z) + G^{IG}(T, P, \overline{n})$$

The first term is the residual Helmholtz free energy of the root, while the second and third terms account for the fact that our fluid (with our EOS) does not have the same pressure as an ideal gas (P^{IG}) at the same $T, V, \overline{n}$. The final term in the above expression is the Gibbs free energy of an ideal gas mixture at $T, P, \overline{n}$. This value is common to every volume in the set of roots and cancels in the comparison of Gibbs free energy values. We therefore seek the minimum value of the following quantity among our candidate volumes:

$$A^{Res} + nR_gT[Z - 1 - ln(Z)] \quad (4.4)$$

Table 4.2 examines the three roots of Eq. 4.3, which are also plotted on the corresponding isotherm in Fig. 4.1.

Solving a cubic EOS, as demonstrated in Example 5, does not generalize to more advanced equations of state for the following reasons:

- Writing a cubic EOS as a 3rd order polynomial in Z:
  This only possible for a cubic EOS, and not for a more advanced EOS like PC-SAFT.

Table 4.2 Examination of roots for the SRK EOS of 1 mole of propane and 1 mole of n-Hexane at 350K and 5 bar. Z solved from Eq. 4.3, V calculated from Eq. 3.9, $\left(\frac{\partial P}{\partial V}\right)_{T,\overline{n}}$ calculated from Eq. 3.10, and Stability values calculated from Eq. 4.4

Z	V cm^3	$(\frac{\partial P}{\partial V})_{T,\overline{n}}$ Pa/cm^3	Stability J	Label
2.358E-2	274.526	−3.403E5	167.785	Liquid
8.984E-2	1045.737	3.227E3	2769.188	Unstable
0.887	10320.036	−42.395	−627.341	Vapor (Most Stable)

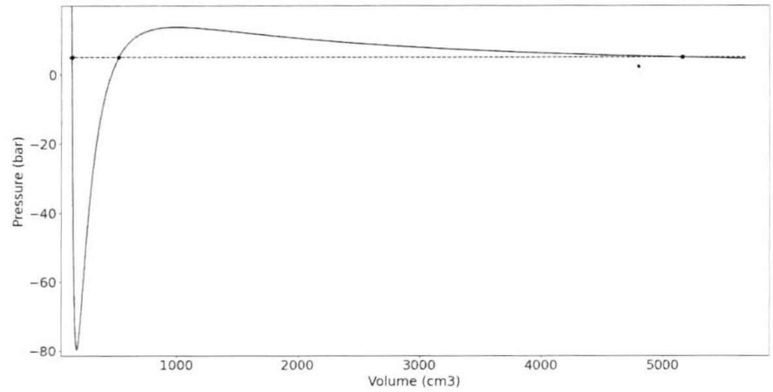

Fig. 4.1 Cubic EOS solver applied to 1 mole of propane + 1 mole of n-hexane at 350K and 5 bar, using the SRK EOS

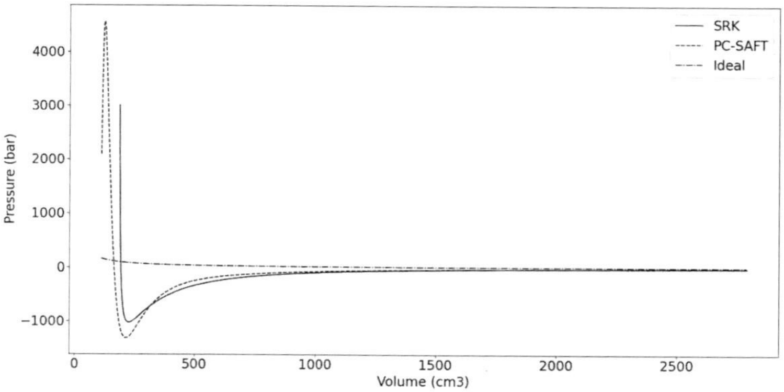

Fig. 4.2 Comparison of SRK, PC-SAFT, and Ideal Gas EOS isotherms for 1 mole of propane + 1 mole of n-Hexane at 110K

- Some texts suggest that vapor roots can solved by an initial guess of $Z = 1$:
  An EOS can yield $Z = 1$ at isolated points far from ideal gas behavior. (See Fig 4.2).
- Some texts suggest that liquid roots can be solved by an initial guess of $Z = Pb/R_gT$:
  In general this initialization only appies to cubics like SRK and PR, though other EOS do have 'co-volumes' (or minimum volumes). There may be roots at volumes less than the co-volume or other complications[3] with a non-cubic EOS.
- Discard the intermediate root (if it exists):
  Under severe (highly unlike species and/or very low temperatures), a non-cubic EOS can have more than 3 roots for a given pressure.

Some of these concepts are illustrated in Fig 4.2, which shows isotherms for a mixture of one mole of propane and one mole of n-hexane at 110K and 5 bar with three different EOS. Though an extreme[4] case, this illustrates the challenge of a non-cubic EOS. At low volumes, SRK shows typical liquid-like behavior, with a sharp increase in pressure trending to ∞ as the volume decreases toward V_{min} (in this case 185.98 cm^3). In contrast, the behavior of the PC-SAFT EOS is much more complex, with the most striking observation being that $(\frac{\partial P}{\partial V})_{T,\overline{n}} > 0$ at $V_{min} = 111.64\,cm^3$. Applying Newton's method blindly,[5] our algorithm may converge to volumes less than V_{min} (for $P < P(V_{min})$) or possibly get stuck, as $(\frac{\partial P}{\partial V})_{T,\overline{n}}$ changes sign before $(P - P_*)$ does. Also shown is the intersection of the

[3] The sign of $\left(\frac{\partial P}{\partial V}\right)_{T,\overline{n}}$

[4] At such a low temperature the formation of solids should be considered.

[5] When solving for the roots in Fig. 4.2, at first we only know T, P, and $\overline{n}$, and the only feature of the isotherm known to us is V_{min} (Eq. 2.36).

ideal gas equation with SRK (≈192 cm^3) and with PC-SAFT (≈164 cm^3), each of these occuring at conditions far from ideal behavior. Therefore, using the ideal gas EOS to initialize a Newton's method search for *vapor*-like roots at a pressure near these intersections may falsely converge to liquid-like or even unstable roots. Also indicated is that, while the differences between SRK and PC-SAFT are pronounced in liquid-like conditions, these two equations become indistinguishable as they both converge toward ideal-gas behavior as $V \to \infty$, with P trending to 0.

The main concepts for a general EOS root solver are outlined in Algorithm 7.

Algorithm 7 The main concepts in a general EOS solver

Select the minimum search volume (V_{min}) for the EOS (Sects. 2.1 and 2.2 provide examples).

Sequentially locate all of the points for which $\left(\frac{\partial P}{\partial V}\right)_{T,\overline{n}} = 0$. These roots are sometimes called "spinodal volumes" [Koningsveld book] (see Appendix 8.17). Their calculation will be covered in detail in the next sections. This process continues through increasing volumes and terminates when increasing volume produces *repeated* instances of Z sufficiently close to unity. At this point, ideal-gas behavior for the isotherm has been reached, and no further solutions of $\left[\frac{\partial P}{\partial V}\right]_{T,\overline{n}} = 0$ exist. The set of volumes (including V_{min}) define a set of "intervals": $V_{intervals} = [V_{min}, V_1, V_2, \ldots]$.

In each interval (e.g., $[V_{min} \leq V \leq V_1]$) ensure that $\left(\frac{\partial P}{\partial V}\right)_{T,\overline{n}} < 0$; otherwise, a root in this interval will be mechanically unstable (This is why we "ignore the middle root" when solving cubics) and is of limited interest to us. Next calculate the endpoint pressures, which in an interval where $\left[\frac{\partial P}{\partial V}\right]_{T,\overline{n}} < 0$ can be assigned as $P_{high} = P(V_{left})$ and $P_{low} = P(V_{right})$. A root exists in the interval if $[P_{high} \geq P_* \geq P_{low}]$, and Newton's method can be easily used to solve for it, as the interval is monotonic in P, the volume at the midpoint of the interval serves as the initial guess and endpoint volumes of the interval can be used for stepsize control. Thereafter move to the next interval.

When the last volume in $V_{intervals}$ (i.e., the last spinodal) is reached, check the pressure at this volume. If this pressure is lower than the target pressure, there are no further roots, and the search is terminated. However, if this pressure is higher than the target pressure, the volume is increased (e.g., by multiples of 10) until the calculated pressure is lower than the target pressure. This sets the final interval for solving the remaining root, which in all cases will be characterized by $\left(\frac{\partial P}{\partial V}\right)_{T,\overline{n}} < 0$, and the root is determined as above. Note that this approach also applies to the case where there are no spinodal volumes on the isotherm; $V_{intervals} = [V_{min}]$. A sufficiently large volume is found to yield $P < P_*$, and then the single root is found using Newton's method. This case is shown in Fig. 4.4. It is also possible for a given EOS and accompanying selection of V_{min} that no root can be found because $P*$ is too large (consider Fig. 4.2 with $P_* = 5000$ bar. One can adjust the minimum volume of the EOS to ensure at least one root is found over the widest range of conditions, but this is an area where some judgment on your part is required, as applying a fluid EOS at extremely high pressures may be of limited practical use.

With the set of roots determined, the desired root is selected. The smallest volume can be considered the "Liquid" root, while the largest volume can be considered the "Vapor." If the most stable root is desired, that with the lowest value of Eq. 4.4 is selected.

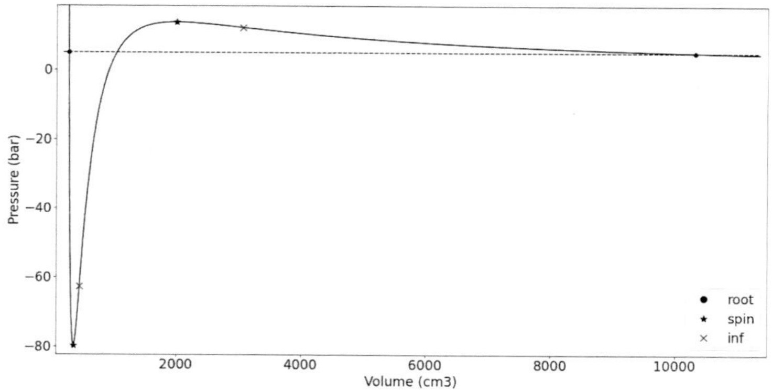

Fig. 4.3 Application of a general root solver to 1 mole of propane + 1 mole of n-hexane at 350K and 5 bar, using the SRK EOS

Table 4.3 Root solver output for 1 mole of propane and 1 mole n-hexane at 350K and 5 bar. All volumes are in cm^3

V_{min} (not shown in Fig. 4.3)	185.98
$V_{spin}1$	358.36
$V_{inf}1$	451.47
$V_{spin}2$	1998.17
$V_{inf}2$	3068.69
V_L	274.53
Equation 4.4 for V_L	167.78
V_V	10320.04
Equation 4.4 for V_V	−627.34

Example 6 provides an illustration of Algorithm 7 in detail.

Example 6 (Application of a General Root Solver) Let's revisit Example 5, applying a general root solver to a mixture containing 1 mole of propane and 1 mole of n-hexane at 350K and 5 bar. Using the SRK EOS, the isotherm, spinodal volumes, and roots are shown in Fig. 4.3, and the key output from the root solver is summarized in Table 4.3. Based on Eq. 4.4, the vapor is the most stable root. Note the agreement between these results and those of Example 5.

At higher temperatures, the isotherm yields no spinodal volumes, and only one root is detected. This is demonstrated in Fig. 4.4.

Note that finite volume solutions for $\left(\frac{\partial P}{\partial V}\right)_{T,\overline{n}} = 0$ cannot be found, but the isotherm does feature a point where $\left(\frac{\partial^2 P}{\partial V^2}\right)_{T,\overline{n}} = 0$. More on this will be discussed shortly. We now look in more detail at each of the key steps within the root solver.

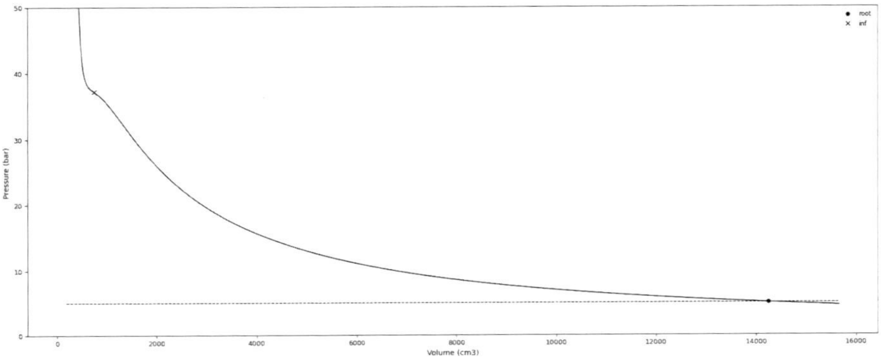

Fig. 4.4 Solving the single root for 1 mole of propane + 1 mole of n-hexane at 450K and 5 bar, using the SRK EOS

4.1 Solving the First Spinodal Volume

Referring to Eq. 3.10, we are seeking a root of the following equation:

$$F = (\frac{\partial P}{\partial V})_{T,\overline{n}} = 0 \tag{4.5}$$

for which we can use Newton's method (Appendix 8.13). It is important to understand that F is only a function of T, V and $\overline{n}$. That is, solutions to Eq. 4.5 are properties of the isotherm; they do not depend on the target pressure (this can be seen in Fig. 4.3). As will be clarified in the next section, the code we're developing here will actually be used to locate all spinodal volumes; for the first spinodal volume, the starting volume is V_{min}, while subsequent searches will use a different starting volume. With V the only unknown, the derivative of our target function has already been developed as Eq. 3.12:

$$H = \left(\frac{\partial^2 P}{\partial V^2}\right)_{T,\overline{n}} \tag{4.6}$$

We begin with a starting volume for the search, and we make updates via Newton's method (see Eq. 8.110):

$$dV = -\frac{\left(\frac{\partial P}{\partial V}\right)_{T,\overline{n}}}{\left(\frac{\partial^2 P}{\partial V^2}\right)_{T,\overline{n}}} \tag{4.7}$$

As is often required in Newton's method applied to functions with multiple roots, stepsize control is important. You might choose to restrict the size of dV to be a certain fraction of V, say 10%. There also arises the important issue of curvature,

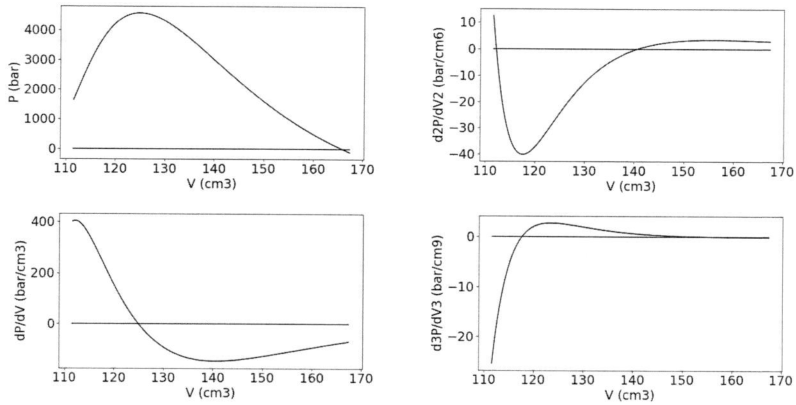

Fig. 4.5 Effect of curvature in solving $\left(\frac{\partial P}{\partial V}\right)_{T,\overline{n}} = 0$ with the PC-SAFT EOS, developed from Fig. 4.2

which from the perspective of the isotherm[6] is the sign of $\left(\frac{\partial^2 P}{\partial V^2}\right)_{T,\overline{n}}$. Figure 4.5 shows P and its dervatives with respect to V, where it can be seen that, in the vicinity of $V_{min} = 111.64\,\text{cm}^3$, $\left(\frac{\partial P}{\partial V}\right)_{T,\overline{n}} > 0$, and $\left(\frac{\partial^2 P}{\partial V^2}\right)_{T,\overline{n}} > 0$. The lower left and upper right of Fig. 4.5 make this much clearer. As indicated by Eq. 4.7, the initial Newton iterations will therefore attempt to reduce the volume ($dV < 0$). In this example, this will continue until $\left(\frac{\partial^2 P}{\partial V^2}\right)_{T,\overline{n}}$ changes sign, which occurs at around $V = 112\,\text{cm}^3$, as can be seen directly in the upper right portion of Fig. 4.5 and also intuitively in the lower left. The point of curvature change is called the *inflection* point, and we will return to this in the wider discussion of the root solver. For now, our strategy to avoid this issue is to capture the sign of $\left(\frac{\partial P}{\partial V}\right)_{T,\overline{n}}$ upon initialization (in this example it is positive, but this will not always be the case.). We then monitor the sign of $\left(\frac{\partial P}{\partial V}\right)_{T,\overline{n}}$ during the Newton iterations, and if we encounter $dV < 0$ by Eq. 4.7 without $\left(\frac{\partial P}{\partial V}\right)_{T,\overline{n}}$ having changed sign, we use the absolute value of dV. This will succeed because we began our search from V_{min} and work our way up in volume to ideal-gas behavior.

It is also important to understand that there are conditions for which no solutions to Eq. 4.5 can be found (see Fig. 4.4). In its present form, applying our algorithm for finding V_{spin} to such cases would simply move higher and higher in volume and ultimately crash. Therefore, our routine must include a check for ideal-gas behavior ($Z \approx 1$). While a single instance of this might simply indicate that your EOS and the ideal-gas EOS have intersected (see Fig. 4.2), repeated (e.g., more than 2) subsequent instances indicate that your routine has reached the low pressure

[6] You are likley more familiar with the (expected) shape of P vs. V than with derivatives of P.

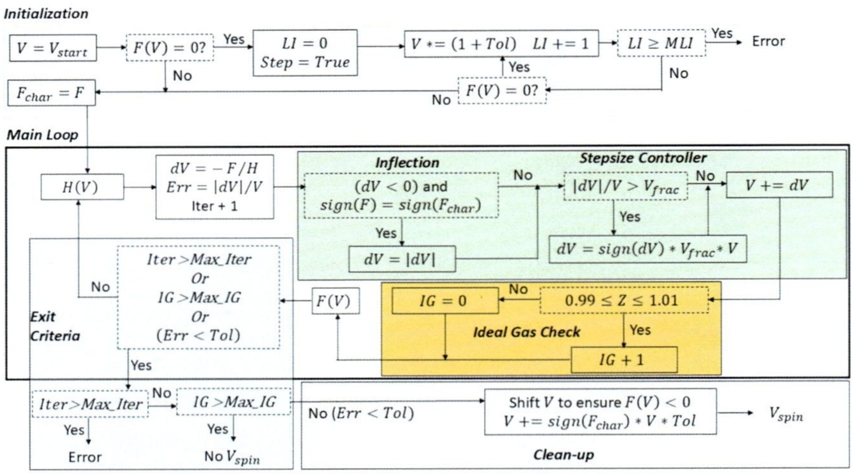

Fig. 4.6 Outline of algorithm for solving 1st spinodal volume

(high volume) region where the EOS is exhibiting ideal-gas behavior. Should this occur, your routine must terminate and appropriately indicate that no solution to Eq. 4.5. This is not an "error"; it simply indicates that a spinodal volume could not be found.[7] Also, Newton's method typically involves repeated iterations, and it is good practice to limit the number of iterations in any such routine. This avoids "infinite loops" that occur in the development of new code. As is typical in numerical (vs. analytical) methods, our solution to Eq. 4.5 will not produce F identically zero, but rather $|F| \leq Tol$, where Tol is a tolerance that we have selected for our routine. Accordingly, your routine should ensure that your roots to Eq. 4.5 are slightly to the correct side of the true V_{spin}. This will avoid intervals (as defined above) that contain mechanically unstable states: $\left(\frac{\partial P}{\partial V}\right)_{T,\overline{n}} > 0$. These ideas are summarized in Fig. 4.6. Notice that the final cleanup shifts our solution either left or right, depending on the characteristic sign of $\left(\frac{\partial P}{\partial V}\right)_{T,\overline{n}}$ (sign of F_{char}). Referring to Fig. 4.5, we would want to shift our volume to the *right* for the V_{spin} at 124.96 cm^3. A pseudocode representation of Fig. 4.6 is outlined in Algorithm 8, and Example 7 demonstrates the calculation of the first V_{spin} value.

[7] If this occurs in the 1st search, this means the isotherm has no spinodal volumes. As we shall soon see, this result inevitably occurs, even if spinodal volumes are found at lower volumes, as the search moves into the ideal-gas regime. Therefore, your code must be able to handle and report this condition without crashing.

Algorithm 8 Solving the spinodal volume

```
Set Tol = 10^-8 (Tolerance for Eq. 4.5)
Initialize Iter = 0 (Iteration counter) and set MaxIter = 100 (Max iterations)
Vfrac = 0.1 (Maximum allowed fractional change in volume)
MLI = 10 (Maximum iterations used in local controllers)
Initialize IG = 0 (Ideal-gas counter) and Set MaxIG = 3 (Max ideal gas detections)
Set Solver_Flag = True (Activate main loop)
Initialize volume: V = V_start

Calculate F (Eq. 4.5)

if (F == 0) then (In case F numerically zero, must shift to slightly larger volume).
    Initialize LI = 0 (This is a local iteration counter)
    Set Step_Flag = True
    while Step_Flag = True do
        V* = (1 + Tol) (This algorithm seeks V > V_start)
        Calculate F (Eq. 4.5)
        LI+ = 1
        if (LI ≥ MLI) or (F ≠ 0) then
            Step_Flag = False
        end if
    end while
    if
        thenRaise Error (Failure to initialize V_spin)
    end if
end if

Capture characteristic value of F:
F_char = F

while (Solver_Flag == True) do
    Newton:
    Calculate H (Eq. 4.6)
    dV = −F/H

    Iter and Err:
    Iter+ = 1
    Err = |dV/V|

    Inflection:
    if ( (dV ≤ 0) and (sign(F) == sign(F_char)) ) then (Reduction in V with same sign of F)
        dV = |dV|
    end if

    Stepsize Controller:
    if |dV/V| ≥ V_frac then
        dV = V_frac V sign(dV)
    end if

    Step and recalculate Objective:
    V+ = dV
    Calculate F (Eq. 4.5)
```

Algorithm 8 Continued.

```
    Ideal Gas Check:
    if 0.99 ≤ Z ≤ 1.01 then
        IG+ = 1
    else
        IG = 0 (We are looking for consecutive ideal gas states)
    end if
    Exit criteria
    if ((Err < Tol) or (IG ≥ MaxIG) or (Iter ≥ MaxIter)) then
        Solver_Flag = False
    end if

end while

Examine Solution
if IG ≥ MaxIG then
    No spinodal detected
else if ((Err ≥ Tol) or (Iter ≥ MaxIter)) then
    Raise Error (Failure in main loop)
else(Ensure solution is on correct side of V_spin (F < 0))
    if F ≥ 0 then
        Initialize LI = 0 and Shift_Flag = True
        while
            do V+ = sign(F_char) * V * Tol
            Calculate F (Eq. 4.5)
            LI+ = 1
            if ((LI ≥ MLI) or (F < 0)) then
                Shift_Flag = False
            end if
        end while
        if ((LI ≥ MLI) or (F >= 0)) then
            Raise Error (Failure in shift)
        end if
    end if
end if
```

Example 7 (Calculation of the 1st V_{spin} Value) Revisiting Example 6, we have 1 mole of propane and 1 mole of n-hexane at 350K (the pressure is not relevant to our algorithm for computing V_{spin} values.). We use the SRK EOS (Eq. 1.26) to calculate a value of $V_{min} = 185.98\,\text{cm}^3$. Starting with this volume, Table 4.4 summarizes the iterations to solve Eq. 4.5. As shown, this example does not encounter issues with curvature, but does require stepsize control. With the error taken to be $\frac{|dV|}{V}$, convergence is achieved. Also the value of V_{spin} does not require correction to produce $\left(\frac{\partial P}{\partial V}\right)_{V,\bar{n}} < 0$.

At this point, we have detected a spinodal value, which means there must be an inflection point on the isotherm. We now look in detail at how to determine such points.

Table 4.4 Iterations to solve Eq. 4.5 for 1 mole of propane + 1 mole of n-hexane at 350K using the SRK EOS. *a: Stepsize control* $(|dV| \leq 0.1V)$ *applied*

Iter	V cm^3	$\left(\frac{\partial P}{\partial V}\right)_{V,\overline{n}}$ bar/cm^3	$\left(\frac{\partial^2 P}{\partial V^2}\right)_{V,\overline{n}}$ bar/cm^6	dV cm^3
1	185.98	−17155.71	18643.67	0.92
2	186.90	−7622.39	5526.97	1.38
3	188.28	−3385.42	1639.51	2.06
4	190.34	−1502.42	486.99	3.09
5	193.43	−665.66	145.05	4.59
6	198.02	−293.95	43.43	6.77
7	204.79	−128.99	13.14	9.82
8	214.60	−55.96	4.04	13.85
9	228.45	−23.83	1.28	18.67
10	247.12	−9.87	0.42	23.55
11	270.67	−3.92	0.14	27.07[a]
12	297.74	−1.46	0.054	27.04
13	324.78	−0.49	0.023	21.22
14	346.01	−0.13	0.012	10.36
15	356.36	−0.017	9.01E-3	1.93
16	358.30	−4.87E-4	8.51E-3	0.057
17	358.36	−4.10e-7	8.50E-3	4.83E-5
18	358.36	−2.92E-13	8.40E-3	3.44E-11

4.2 Solving the Inflection Volume

If we find a spinodal volume, we continue searching for subsequent[8] roots to Eq. 4.5. As discussed in Sect. 4.1 and Appendix 8.13, care must be taken with Newton's method. If we make an initial guess for the next V_{spin} too close to our previous root, we will simply converge back to the 1st value. Also, if we make an arbitrary guess, we may miss a V_{spin} value and thereby totally misinterpret the form of the isotherm.[9] Finally, we really have no physical basis for making such a guess; the only precise information immediately available to us when considering an isotherm (i.e., given T, V, and $\overline{n}$) is V_{min}. We thus seek the inflection point of the isotherm, where $\frac{\partial^2 P}{\partial V^2} = 0$. Having identified this point, we can "step over" to a region of the isotherm where Newton's method will begin converging forward to the next V_{spin}. Referring back to Eq. 4.6, we seek the roots of the following equation (calculated from Eq. 3.12);

$$F = \left(\frac{\partial^2 P}{\partial V^2}\right)_{T,\overline{n}} = 0 \tag{4.8}$$

[8] For any useful EOS, the existence of the first V_{spin} guarantees the existence of an inflection point, and at least one more spinodal volume and one more inflection point. Otherwise the isotherm cannot trend to ideal gas behavior.

[9] Keep in mind we are designing an algorithm that *automatically* detects roots. We cannot plot the isotherm and look at its shape.

the derivative of which, calculable from Eq. 3.13, is:

$$H = \frac{dF}{dV} = \left(\frac{\partial^3 P}{\partial V^3}\right)_{T,\overline{n}} \tag{4.9}$$

and Newton updates are generated via Eq. 8.110:

$$dV = -\frac{F}{H}$$

In addition to solving Eq. 4.8 to a suitable tolerance, the routine must ensure that $\left(\frac{\partial^2 P}{\partial V^2}\right)_{T,\overline{n}}$ has changed sign. This is handled somewhat analogously as in Algorithm 8, with a few differences. These are described in Algorithm 9 and demonstrated in Example 8.

Algorithm 9 Solving the inflection volume

Set $Tol = 10^{-8}$ (Tolerance for Eq. 4.8)
Initialize $Iter = 0$ (Iteration counter) and set $MaxIter = 100$ (Max iterations)
$Vfrac = 0.1$ (Maximum allowed fractional change in volume)
$MLI = 10$ (Maximum iterations used in local controllers)
Initialize $IG = 0$ (Ideal gas counter) and Set $MaxIG = 3$ (Max ideal gas detections)
Set $Solver\_Flag = True$ (Activate main loop)
Initialize volume: $V = V_{start}$

Calculate F (Eq. 4.8)

if $(F == 0)$ **then** (In case F numerically zero, must shift to slightly larger volume).
  Initialize $LI = 0$ (This is a local iteration counter)
  Set $Step\_Flag = True$
  while $Step\_Flag = True$ **do**
    $V* = (1 + Tol)$ (This algorithm seeks $V > V_{start}$)
    Calculate F (Eq. 4.8)
    $LI+ = 1$
    if $(LI \geq MLI)$ or $(F \neq 0)$ **then**
      $Step\_Flag = False$
    end if
  end while
  if
    thenRaise Error (Failure to initialize V_{inf})
  end if
end if

Calculate H (Eq. 4.9)
Capture characteristic values of F and H:
$F_{char} = F$
$H_{char} = H$

Algorithm 9 Continued.

while $(Solver\_Flag == True)$ **do**
Newton:
$dV = -F/H$

Iter and Err:
$Iter + = 1$
$Err = |dV/V|$

Inflection:
if (($dV \leq 0$) and ($sign(F)$==$sign(F_{char})$)) **then** (Reduction in V with same sign of F)
$dV = |dV|$
end if

Stepsize Controller 1:
if $|dV/V| \geq V_{frac}$ **then**
$dV = V_{frac} * V * sign(dV)$
end if

$V_{old} = 1. * V$ Capture V_{old}
Step and recalculate:
$V = V_{old} + dV$
Calculate F (Eq. 4.8)
Calculate H (Eq. 4.9)
Stepsize Controller 2:
if ($sign$(F)$\neq$$sign(F_{char})$) and ($sign(H)\neq$$sign(H_{char})$) **then** (Do not allow both F and H to change sign)
Initialize $LI = 0$ (This is a local iteration counter)
Set $Step\_Flag = True$
while $Step\_Flag = True$ **do**
$dV / = 2$ (Cut the stepsize)
$V = V_{old} + dV$ (Restep)
Calculate F (Eq. 4.8)
Calculate H (Eq. 4.9)
$LI + = 1$
if ($LI \geq MLI$) or not(($sign$(F)$\neq$$sign(F_{char})$) and ($sign(H)\neq$$sign(H_{char})$)) **then**
$Step\_Flag = False$
end if
end while
if ($LI \geq MLI$) or(($sign$(F)$\neq$$sign(F_{char})$) and ($sign(H)\neq$$sign(H_{char})$)) **then**
Raise Error (Failure in stepsize controller 2)
end if
end if

Ideal Gas Check:
if $0.9 \leq Z \leq 1.0$ **then**
$IG + = 1$
else
$IG = 0$ (We are looking for *consecutive* ideal gas states)
end if
Exit criteria
if (($Err < Tol$) or ($IG \geq MaxIG$) or ($Iter \geq MaxIter$)) **then**
$Solver_Flag = False$H
end if

end while

Algorithm 9 Continued.

```
Examine Solution
if IG ≥ MaxIG then
    No inflection detected
else if ((Err ≥ Tol) or (Iter ≥ MaxIter)) then
    Raise Error (Failure in main loop)
else(Ensure F has changed sign)
    if sign(F)≠sign(F_char) and F ≠ 0 then
        Pass
    else:
        Initialize LI = 0 and Shift_Flag = True
        while Shift_Flag = True do
            V* = (1 + Tol)
            Calculate F (Eq. 4.8)
            LI+ = 1
            if ( (LI ≥ MLI) or (sign(F)≠sign(F_char) and F ≠ 0) ) then
                Shift_Flag = False
            end if
        end while
        if ((LI ≥ MLI) or sign(F)==sign(F_char) or F == 0 ) then
            Raise Error (Failure in shift)
        end if
    end if
end if
```

Example 8 (Calculation of the 1st V_{inf} Value) Continuing from Example 7, we use $V_{start} = 358.356\,\text{cm}^3$. Table 4.5 summarizes the iterations to solve Eq. 4.8.

Note that the sign of $\left(\frac{\partial^2 P}{\partial V^2}\right)_{V,\bar{n}}$ has not changed. The final stage of 9 handles this.

We have now detected V_{inf}, we can step over this value to search for the next V_{spin}.

Table 4.5 Iterations to solve Eq. 4.8 for 1 mole of propane and 1 mole of n-hexane at 350K using the SRK EOS

Iter	V cm^3	$\left(\frac{\partial^2 P}{\partial V^2}\right)_{V,\bar{n}}$ $\frac{\text{bar}}{\text{cm}^6}$	$\left(\frac{\partial^3 P}{\partial V^3}\right)_{V,\bar{n}}$ $\frac{\text{bar}}{\text{cm}^9}$	dV (cm^3)
1	358.36	849.91	−25.03	33.95
2	392.30	299.42	−9.90	30.23
3	422.53	92.20	−4.52	20.38
4	442.91	20.48	−2.67	7.67
5	450.58	1.92	−2.18	0.88
6	451.46	0.02	−2.13	0.01
7	451.47	3.07E-6	−2.13	1.44E-6
8	451.47	4.37E-14	−2.13	2.05E-14

4.3 Subsequent Spinodal Volumes

Having passed the inflection point, we use this volume as the initial guess for subsequent V_{spin} values in the algorithm described in Sect. 4.1.

4.4 Determining the Intervals

The volume intervals, described in Chap. 4, are *monotonic* sections of the isotherm. In other words, within each interval the sign of $\left(\frac{\partial P}{\partial V}\right)_{T,\bar{n}}$ does not change. The lower boundary is V_{min} (dictated by the EOS), after which each successfully solved V_{spin} is added to the set. Once the V_{spin} routine returns no value, the search terminates, yielding for the example considered in this chapter the results given in Table 4.3.

4.5 Seeking the Roots

With the volume intervals determined, the next stage in the root solver checks each interval to ensure that:

$$\left(\frac{\partial P}{\partial V}\right)_{T,\bar{n}} < 0$$

otherwise a root in this interval would be unstable. With the correct slope detected, the minimum and maximum pressures of the intervals (i.e., the endpoints of the interval) are then checked against the target pressure (P_*) to confirm that:

$$P_{min} \leq P_* \leq P_{max}$$

otherwise no root exists in the interval. Meeting these two conditions, Newton's method is used to solve for the root. The objective function is:

$$F = P - P_* \tag{4.10}$$

The derivative is of course:

$$H = \left(\frac{\partial P}{\partial V}\right)_{T,\bar{n}} \tag{4.11}$$

This approach can be used to find all the roots on the isotherm, except for the final root. There are no other V_{spin} values beyond this, but we can use the target

Table 4.6 Assessment of the initial interval for 1 mole of propane and 1 mole of n-hexane at 350K and 5 bar using the SRK EOS

V (cm^3)	P (bar)	$\Delta P/\Delta V$ (bar/cm^3)	Stable root exists?
185.98	30438.83		
358.36	−79.74	−177.046	Yes

Table 4.7 Solving the initial root for 1 mole of propane and 1 mole of n-hexane at 350K and 5 bar using the SRK EOS

Iter	V cm^3	F bar	H bar/cm^3	dV cm^3
1	272.17	8.39	−3.71	2.26
2	274.43	0.338	−3.42	0.099
3	274.53	6.12E-4	−3.40	1.80E-4
4	274.53	2.01E-9	−3.40	5.91E-10

pressure P_*. If the pressure of the final V_{spin} exceeds P_* (i.e., a root exists[10] in the final interval), we seek higher and higher volumes until the calculated pressure is less than P_*. Having found the upper limit of the volume, Newton's method is used a final time to seek the root in this interval. Let's look at an example.

Example 9 (Solving the Roots within the Intervals) Continuing from Example 8, we have $V_{ints} = [185.98, 358.36, 1998.17]\,\text{cm}^3$. Starting with the first interval, we can calculate the pressures at the low and high volumes, asses the slope, and confirm the presence of a root:

The results in Table 4.6 confirm that the slope of the isotherm is negative and that the target pressure lies within the interval. We therefore proceed to solve the root.

We begin by initializing the volume as the midpoing of the interval:

$$V = (V_{low} + V_{high})/2 = (185.98 + 358.36)/2 = 272.168\,\text{cm}^3$$

Applying Newton's method with Eqs. 4.10 and 4.11, the iterations are summarized in Table 4.7.

As shown in Table 4.7, the convergence to the solution in a monotonic interval is straightforward. In this particular example, no stepsize control is required. Other strategies that may be considered is the use of bisection (exploiting the fact that the intervals are monotonic), thereby avoiding the need to compute a Jacobian.

Though we recognize the alternating nature of the sign of $\left(\frac{\partial P}{\partial V}\right)_{T,\overline{n}}$, a formal check of the next interval demonstrates its instability. Our algorithm then skips this interval and moves to the next. In this example, this was the final interval. As stated in Algorithm 7, we now check for vapor-like behavior. As shown in Table 4.8, the last spinodal pressure (13.64 bar at $V = 1998.17$) exceeds the target pressure (5 bar), guaranteeing a root (refer to Fig. 4.3). We now increase volume from the final

[10] The slope is guaranteed to have the correct sign, as we have detected the final V_{spin} value, and the EOS must trend to ideal-gas behavior.

Table 4.8 Assessment of the second interval for 1 mole of propane and 1 mole of n-hexane at 350K and 5 bar using the SRK EOS

V cm^3	P bar	$\Delta P/\Delta V$ bar/cm^3	Stable root exists?
358.36	−79.74		
1998.17	13.64	0.057	No

Table 4.9 Solving the final root for 1 mole of propane and 1 mole of n-hexane at 350K and 5 bar using the SRK EOS

Iter	V cm^3	F bar	H bar/cm^3	dV cm^3
1	10989.91	−0.27	−3.80E-4	−707.39
2	10282.53	0.016	−4.27E-4	37.39
3	10319.92	4.97E-5	−4.24E-4	0.12
4	10320.04	4.86E-10	−4.24E-4	1.15E-6

V_{spin} until the pressure falls below the target. Taking multiples of 10, we find $V = 19981.66\,\text{cm}^3$, where $P = 2.74\,\text{bar}$. This provides a monotonic interval in which we seek the final root, with the initial guess being the midpoint (Table 4.9). These results can be compared with those of Examples 5 and 6.

Having solved all the roots for the target pressure, the final task of the root solver is to select the desired root.

4.6 Selecting the Root

The desired root from the values detected by the root solver might be the liquid (the minimum value of V), the vapor (the maximum value of V), or the most stable (the root with the minimum value of Eq. 4.4). As stated in Table 4.3, the vapor is the more stable root, in agreement Table 4.2. As will be seen, this selection depends on the algorithm in use. In most cases, the most stable root is desired, but there are scenarios where a "liquid-like" or a "vapor-like" root is desired.

Chapter 5
Thermodynamic Properties from an Equation of State (T, P, $\overline{n}$)

This chapter formally assumes that a robust root solver is available and recasts Chap. 3 from the $T, V, \overline{n}$ to a $T, P, \overline{n}$ framework. There are two main ideas in this chapter:

- How are derivatives at fixed *pressure* calculated for functions that are canonical in *volume*?
- Can we avoid unnecessarily calling our root solver (which costs us time and iterations) when we perform calculations at fixed *pressure*?

We begin with the first question. We can establish the basic concept of derivatives at specified pressure by considering how we might compute $\left[\frac{\partial Z}{\partial T}\right]_{P,\overline{n}}$. Note carefully this information is not provided by Eq. 3.3. It may seem at this time that we cannot compute this quantity, because the pressure is not an independent variable within A^{Res}. We also cannot expect to identify some clever rearrangement of variables to achieve this, as volume is in general not a continuous function[1] of pressure. Instead, let's consider a change in T with a corresponding change in V that *keeps the pressure constant*. How would we achieve this? The answer can be seen from a 1st order Taylor series (Appendix 8.19) of Z:

$$dZ = \left(\frac{\partial Z}{\partial T}\right)_{V,\overline{n}} dT + \left(\frac{\partial Z}{\partial V}\right)_{T,\overline{n}} dV + \sum_r \left(\frac{\partial Z}{\partial n_r}\right)_{T,V,n_{k\neq r}} dn_r \tag{5.1}$$

[1] The mathematical definition of *function* requires a unique output value for a given input. Of course, a simple EOS, like $PV = nR_gT$, allows V as a *function* of P, but it does not describe real fluid behavior. Regardless of which equation of state we use, a substance at its boiling point (e.g., saturated steam at 100 °C) exhibits properties of both liquid and vapor at the same pressure, so which state (volume) are we specifying?

B. R. Greenhalgh, *An Introduction to Computational Thermodynamics*,
https://doi.org/10.1007/978-3-032-30317-2_5

As above, we are considering a change in T that is compensated by a change in V to keep P constant. What would the change in V need to be to produce this? This is given by:

$$dV = \left[\frac{\partial V}{\partial T}\right]_{P,\overline{n}} dT \tag{5.2}$$

Also, in Eq. 5.1 the mole numbers are all constant ($dn_i = 0$). Therefore we can write Eq. 5.1 as:

$$dZ = \left(\frac{\partial Z}{\partial T}\right)_{V,\overline{n}} dT + \left(\frac{\partial Z}{\partial V}\right)_{T,\overline{n}} \left[\frac{\partial V}{\partial T}\right]_{P,\overline{n}} dT \tag{5.3}$$

With the imposed conditions, we can interpret dZ as taking place at fixed P and $\overline{n}$. Therefore:

$$\left[\frac{\partial Z}{\partial T}\right]_{P,\overline{n}} = \left(\frac{\partial Z}{\partial T}\right)_{V,\overline{n}} + \left(\frac{\partial Z}{\partial V}\right)_{T,\overline{n}} \left[\frac{\partial V}{\partial T}\right]_{P,\overline{n}} \tag{5.4}$$

Note that we have used [] to designate $T, P, \overline{n}$ derivatives, while () designates $T, V, \overline{n}$ derivatives. We now turn to Eq. 1.64 ($V = \frac{ZnR_gT}{P}$) to generate the required V derivative:

$$\begin{aligned}\left[\frac{\partial V}{\partial T}\right]_{P,\overline{n}} &= \frac{nR_gT}{P}\left[\frac{\partial Z}{\partial T}\right]_{P,\overline{n}} + \frac{nR_gZ}{P}\\ &= V\left(\frac{1}{Z}\left[\frac{\partial Z}{\partial T}\right]_{P,\overline{n}} + \frac{1}{T}\right)\end{aligned} \tag{5.5}$$

If we substitute Eq. 5.5 into Eq. 5.4, we have:

$$\begin{aligned}\left[\frac{\partial Z}{\partial T}\right]_{P,\overline{n}} &= \left(\frac{\partial Z}{\partial T}\right)_{V,\overline{n}} + \left(\frac{\partial Z}{\partial V}\right)_{T,\overline{n}} \frac{V}{T}\left(\frac{T}{Z}\left[\frac{\partial Z}{\partial T}\right]_{P,\overline{n}} + 1\right)\\ &= \frac{\left(\frac{\partial Z}{\partial T}\right)_{V,\overline{n}} + \left(\frac{\partial Z}{\partial V}\right)_{T,\overline{n}} \frac{V}{T}}{1 - \left(\frac{\partial Z}{\partial V}\right)_{T,\overline{n}} \frac{V}{Z}}\end{aligned} \tag{5.6}$$

Note that all terms on the right side of Eq. 5.6 are known functions of $(T, V, \overline{n})$ given in Sect. 3.1. We shall develop the other $T, P, \overline{n}$ derivatives of Z in the next section. Before doing so, it will be convenient to have the remaining $T, P, \overline{n}$ derivatives of V:

$$\left[\frac{\partial V}{\partial P}\right]_{T,\overline{n}} = \frac{V}{Z}\left[\frac{\partial Z}{\partial P}\right]_{T,\overline{n}} - \frac{V}{P}$$

$$= V\left(\frac{1}{Z}\left[\frac{\partial Z}{\partial P}\right]_{T,\overline{n}} - \frac{V}{ZnR_gT}\right) \tag{5.7}$$

$$\left[\frac{\partial V}{\partial n_i}\right]_{T,P,n_{k\neq i}} = \frac{V}{Z}\left[\frac{\partial Z}{\partial n_i}\right]_{T,P,n_{k\neq i}} + \frac{V}{n}\left[\frac{\partial n}{\partial n_i}\right]_{T,P,n_{k\neq i}}$$

$$= V\left(\frac{1}{Z}\left[\frac{\partial Z}{\partial n_i}\right]_{T,P,n_{k\neq i}} + \frac{1}{n}\right) \tag{5.8}$$

5.1 Z and its $(T, P, \overline{n})$ Derivatives

We now derive the functional forms of the compressibility factor and its $T, P, \overline{n}$ derivatives. Beginning with Z, we recognize that we already have developed Eq. 3.2, which provides Z as a function of *volume*, and we have already derived $\left[\frac{\partial Z}{\partial T}\right]_{P,\overline{n}}$. The pressure derivative of Z is developed as follows:

$$\left[\frac{\partial Z}{\partial P}\right]_{T,\overline{n}} = \left(\frac{\partial Z}{\partial V}\right)_{T,\overline{n}}\left[\frac{\partial V}{\partial P}\right]_{T,\overline{n}}$$

$$= \left(\frac{\partial Z}{\partial V}\right)_{T,\overline{n}} V\left(\frac{1}{Z}\left[\frac{\partial Z}{\partial P}\right]_{T,\overline{n}} - \frac{V}{ZnR_gT}\right)$$

$$= -\frac{\left(\frac{\partial Z}{\partial V}\right)_{T,\overline{n}}\frac{V^2}{ZnR_gT}}{1-\left(\frac{\partial Z}{\partial V}\right)_{T,\overline{n}}\frac{V}{Z}} \tag{5.9}$$

For the mole number derivative, we return Eq. 5.1:

$$dZ = \left(\frac{\partial Z}{\partial T}\right)_{V,\overline{n}} dT + \left(\frac{\partial Z}{\partial V}\right)_{T,\overline{n}} dV + \sum_r \left(\frac{\partial Z}{\partial n_r}\right)_{T,V,n_{k\neq r}} dn_r \tag{5.10}$$

For the required derivative, we are changing one of the mole numbers (n_i), with T, P, and all the other mole numbers fixed. Analogous to the reasoning that led to the expressions for $\left[\frac{\partial Z}{\partial T}\right]_{P,\overline{n}}$ and $\left[\frac{\partial Z}{\partial P}\right]_{T,\overline{n}}$, we consider the volume to be changing so that the pressure is held constant. This change in V is described by Eq. 5.8. The response of n_r to changes in n_i is based on the δ_{ij} operator (Appendix 8.4)

$$dZ = \left(\frac{\partial Z}{\partial V}\right)_{T,\overline{n}} V\left(\frac{1}{Z}\left[\frac{\partial Z}{\partial n_i}\right]_{T,P,n_{k\neq i}} + \frac{1}{n}\right) dn_i + \left(\frac{\partial Z}{\partial n_i}\right)_{T,V,n_{k\neq r}} dn_i \quad (5.11)$$

We can then rearrange Eq. 5.11 to yield:

$$\begin{aligned}\left[\frac{\partial Z}{\partial n_i}\right]_{T,P,n_{k\neq i}} &= \left(\frac{\partial Z}{\partial V}\right)_{T,\overline{n}} V\left(\frac{1}{Z}\left[\frac{\partial Z}{\partial n_i}\right]_{T,P,n_{k\neq i}} + \frac{1}{n}\right) + \left(\frac{\partial Z}{\partial n_i}\right)_{T,V,n_{k\neq r}} \\ &= \frac{\left(\frac{\partial Z}{\partial V}\right)_{T,\overline{n}} \frac{V}{n} + \left(\frac{\partial Z}{\partial n_i}\right)_{T,V,n_{k\neq r}}}{1 - \left(\frac{\partial Z}{\partial V}\right)_{T,\overline{n}} \frac{V}{Z}}\end{aligned} \quad (5.12)$$

5.2 $ln\hat{\varphi}_i$ and its $(T, P, \overline{n})$ Derivatives

The $T, P, \overline{n}$ derivatives of $ln\hat{\varphi}_i$ are derived using a similar approach as we used with Z:

$$dln\hat{\varphi}_i = \left(\frac{\partial ln\hat{\varphi}_i}{\partial T}\right)_{V,\overline{n}} dT + \left(\frac{\partial ln\hat{\varphi}_i}{\partial V}\right)_{T,\overline{n}} dV + \sum_r \left(\frac{\partial ln\hat{\varphi}_i}{\partial n_r}\right)_{T,V,n_{k\neq r}} dn_r \quad (5.13)$$

The T derivative of $ln\hat{\varphi}_i$ at specified $T, P, \overline{n}$ is:

$$\left[\frac{\partial ln\hat{\varphi}_i}{\partial T}\right]_{P,\overline{n}} = \left(\frac{\partial ln\hat{\varphi}_i}{\partial T}\right)_{V,\overline{n}} + \left(\frac{\partial ln\hat{\varphi}_i}{\partial V}\right)_{T,\overline{n}} \left[\frac{\partial V}{\partial T}\right]_{P,\overline{n}} \quad (5.14)$$

The expression for $\left[\frac{\partial V}{\partial T}\right]_{P,\overline{n}}$, along with the P and n_i derivatives of V, can be found in Sect. 5.1. There is no need to expand these expressions to extract the Z derivatives, in contrast to the development of the Z derivatives themselves. The P derivative of $ln\hat{\varphi}_i$ at specified $T, P, \overline{n}$ is:

$$\left[\frac{\partial ln\hat{\varphi}_i}{\partial P}\right]_{T,\overline{n}} = \left(\frac{\partial ln\hat{\varphi}_i}{\partial V}\right)_{T,\overline{n}} \left[\frac{\partial V}{\partial P}\right]_{T,\overline{n}} \quad (5.15)$$

For the sake of generality,[2] we will derive the expression for the derivative of $ln\hat{\varphi}_i$ with respect to moles of species j.

$$\begin{aligned}\left[\frac{\partial ln\hat{\varphi}_i}{\partial n_j}\right]_{T,P,n_{k\neq j}} &= \left(\frac{\partial ln\hat{\varphi}_i}{\partial V}\right)_{T,\overline{n}}\left[\frac{\partial V}{\partial n_j}\right]_{T,P,n_{k\neq j}} \\ &\quad + \sum_r \left(\frac{\partial ln\hat{\varphi}_i}{\partial n_r}\right)_{T,V,n_{k\neq r}}\left[\frac{\partial n_r}{\partial n_j}\right]_{T,P,n_{k\neq j}} \\ &= \left(\frac{\partial ln\hat{\varphi}_i}{\partial V}\right)_{T,\overline{n}}\left[\frac{\partial V}{\partial n_j}\right]_{T,P,n_{k\neq j}} + \left(\frac{\partial ln\hat{\varphi}_i}{\partial n_j}\right)_{T,V,n_{k\neq j}} \end{aligned} \quad (5.16)$$

Equation 5.16 gives further clarification of how the δ_{ij} operator is applied within summations.

5.3 Other Potentials and their $(T, P, \overline{n})$ Derivatives

Just as for other thermodynamic quantities at specified $T, P, \overline{n}$, thermodynamic potentials are calculated by first determining the volume, and then using the expressions in Sect. 3.4. For example, calculating G^{Res} at fixed T,P,$\overline{n}$ finds use in Thermodynamic modeling (e.g., within flash calculations at T,P,$\overline{n}$). The expressions for the $T, P, \overline{n}$ derivatives of the other potentials can be developed similarly to how those of $ln\hat{\varphi}_i$ in Sect. 5.2. For example:

$$\left[\frac{\partial G^{Res}}{\partial T}\right]_{P,\overline{n}} = \left(\frac{\partial G^{Res}}{\partial T}\right)_{V,\overline{n}} + \left(\frac{\partial G^{Res}}{\partial V}\right)_{T,\overline{n}}\left[\frac{\partial V}{\partial T}\right]_{P,\overline{n}} \quad (5.17)$$

$$\left[\frac{\partial G^{Res}}{\partial P}\right]_{T,\overline{n}} = \left(\frac{\partial G^{Res}}{\partial V}\right)_{T,\overline{n}}\left[\frac{\partial V}{\partial P}\right]_{P,\overline{n}} \quad (5.18)$$

$$\left[\frac{\partial G^{Res}}{\partial n_i}\right]_{T,P,n_{k\neq i}} = \left(\frac{\partial G^{Res}}{\partial V}\right)_{T,n_{k\neq i}}\left[\frac{\partial V}{\partial n_i}\right]_{T,P,n_{k\neq i}} + \left(\frac{\partial G^{Res}}{\partial n_i}\right)_{T,V,n_{k\neq i}} \quad (5.19)$$

[2] $\left[\frac{\partial ln\hat{\varphi}_i}{\partial n_j}\right]_{T,P,n_{k\neq j}}$ is a two dimensional matrix, while $\left[\frac{\partial ln\hat{\varphi}_i}{\partial n_i}\right]_{T,P,n_{k\neq i}}$ is its one dimensional diagonal.

5.4 A Note about (T,P,$\overline{n}$) Calculations and the Root Solver

Consider the calculation of Z in Sect. 5.1. At fixed $T, P, \overline{n}$ we can run our root solver, thereby determining the V corresponding to P. However, suppose we have already called the root solver for another purpose within the context of our calculation. It is then unnecessary (and inefficient) to call it again. We can then simply input this volume[3] into Eq. 3.2, or we can (for convenience) structure a function $fun\_Z\_TPn(T, P, n, root, V, \left(\frac{\partial A^{Res}}{\partial V}\right)_{T,\overline{n}})$, which calls the root solver if V is unknown,[4] or otherwise calls Eq. 3.2. Peudocode for this might look like:

Algorithm 10 Optional calling of the root solver for a function $fun\_Z\_TPn(T, P, n, root, V, \left(\frac{\partial A^{Res}}{\partial V}\right)_{T,\overline{n}})$

```
if V is unknown then
    Get V from your root solver, using T, P, n, root (e.g., L,V,Stable)
    Compute (∂A^Res/∂V)_{T,n̄} (EOS-specific)
end if
Calculate Z = Z(T, V, n, (∂A^Res/∂V)_{T,n̄} from Eq. 3.2
```

Note that the use of Algorithm 10 requires the V and the $\left(\frac{\partial A^{Res}}{\partial V}\right)_{T,\overline{n}}$ values to be consistent; otherwise, an incorrect value of Z will be returned. Also, the function called to compute $\left(\frac{\partial A^{Res}}{\partial V}\right)_{T,\overline{n}}$) depends on the EOS you're using.[5] The idea in Algorithm 10 may seem overly complicated. However, we can apply the above concept of bypassing the root solver on functions that are much more complicated than that for Z. Here we are illustrating the idea in a simple way. The same principle of avoiding unnecessary calls to the root solver is applied to T,P,$\overline{n}$ derivatives of thermodynamic quantities, except for them we must calculate the constant *pressure* derivatives. That is, there is only one expression for Z, Eq. (3.2), but $\left(\frac{\partial Z}{\partial T}\right)_{V,\overline{n}}$ (Eq. 3.3) and $\left[\frac{\partial Z}{\partial T}\right]_{P,\overline{n}}$ (Eq. 5.6) are not equivalent.

[3] Do not place expressions like $Z = \frac{PV}{ZnR_gT}$ in different places in your code, as they are uncontrolled duplicates of formal expressions like Eq. 3.2. Doing so makes your code less modular, and debugging becomes very difficult.

[4] Your way of signaling this will depend on the programming language you are using.

[5] How you do this will depend on the computing language and approach you're using. You may have only one equation of state that you use, in which case you only have one expression for $\left(\frac{\partial A^{Res}}{\partial V}\right)_{T,\overline{n}}$) or you may have many EOS options, and you select the one of interest using a class structure, for example.

5.5 Numerical Testing for $(T, P, \overline{n})$ Equation of State Calculations

Numerical testing of $T, P, \overline{n}$ calculations is conducted exactly as in Sects. 2.3 and 3.5, with one important difference. Since the functions being tested are not *continuous functions* of pressure (they can return multiple volumes for the given $T, P, \overline{n}$), one must avoid conditions where changes in stability occur. In practice, this means conducting numerical tests on derivatives for the liquid root at high pressure, for example, avoiding the transition in the most stable root. A plot of the isotherm for the system of interest is helpful in selecting such conditions.

This aside, it is helpful to conduct numerical tests at specified $T, P, \overline{n}$, and Algorithm 11 provides an outline of the procedure.

Algorithm 11 Embedded numerical test within $\left[\frac{\partial Z}{\partial P}\right]_{T,\overline{n}}$ (Eq. 5.9)

Definition: $fun\_\frac{\partial Z\_TPn}{\partial P}(T, P, n, root, V, Z, \frac{\partial Z}{\partial V}, Test)$
if V is unknown **then**
  Get V from your root solver, using $T, P, n, root$ (e.g. L,V,Stable)
  Compute $\left(\frac{\partial A^{Res}}{\partial V}\right)_{T,\overline{n}}$ (EOS-specific)
  Compute $\left(\frac{\partial^2 A^{Res}}{\partial V^2}\right)_{T,\overline{n}}$ (EOS-specific)
  Compute Z from Eq. 3.2
  Compute $\left(\frac{\partial Z}{\partial V}\right)_{T,\overline{n}}$ from Eq. 3.7
end if
$\left[\frac{\partial Z}{\partial P}\right]_{T,\overline{n}} = -\frac{\partial Z}{\partial V} V^2/[(ZnRgT)(1 - \frac{\partial Z}{\partial V}(V/Z))]$
if $Test == True$ **then**
  Create list of N perturbations: $perts = [1E-6, \ldots, 1E-2]$
  for p (N) **do**
    $P_{pert} = P(1+p)$
    $Z_{pert} = fun\_Z\_TPn(T, P_{pert}, \overline{n}, root, dummy, dummy)$
    $E1[p] = Z_{pert} - Z$
    $E2[p] = E1[p] - \left[\frac{\partial Z}{\partial P}\right]_{T,\overline{n}} Pp$
  end for
  $F1 = ln|E1|$
  $F2 = ln|E2|$
  From the sets $F1$ and $F2$, calculate the least squares slope:
  $\overline{F1} = \sum_r F1_r/N$
  $\overline{F2} = \sum_r F2_r/N$
  $slope = \sum_r F1_r(F2_r - \overline{F2})/\sum_r F1_r(F1_r - \overline{F1})$
  Print($slope$)
end if
Return $\left[\frac{\partial Z}{\partial P}\right]_{T,\overline{n}}$

A possible structure for the function fun_Z_TPn in Algorithm 11 was outlined in Algorithm 10. In Algorithm 11, we are formally assuming that this function has been coded in such a way that when "dummy" values are entered for V and/or $\left(\frac{\partial A^{Res}}{\partial V}\right)_{T,\overline{n}}$ the root solver is called automatically. This type of structure makes numerical testing at fixed pressure particularly straightforward.

Chapter 6
Phase Equilibrium Calculations

In this chapter, we cover a number of standard routines useful in computational thermodynamics. These draw on Chaps. 3 through 5; they do not depend on the choice of EOS (Sect. 2).

6.1 Pure Component Saturation Pressure

A pure substance boils at saturation conditions between the triple and critical points. That is, there is a temperature (T_{sat}) and pressure (P_{sat}) at which the liquid and vapor states coexist in equilibrium. For pure water, this occurs at 100°C (373.15K) and 1.01325 bar. For a given temperature (below the critical temperature), the pressure at which a pure substance boils is called the *vapor pressure*. Looking at an EOS (e.g., SRK), how could we determine this condition? The answer lies in Eqs. 1.60 through 1.62. Suppose we are seeking the vapor pressure of a substance at a particular T. We then require that both the liquid and vapor phases have this temperature (Eq. 1.60), and whatever the vapor pressure is, both phases have this pressure (Eq. 1.61). Any application of our root solver will correspond to these constraints, whether one or more distinct volumes are found. It is the inclusion Eq. 1.62 that *defines* the saturation pressure. Via Sect. 1.8.2, we can say:

$$ln y_i + ln\hat{\varphi}_i{}^{v} + ln P^{v} = ln x_i + ln\hat{\varphi}_i{}^{\ell} + ln P^{\ell} \tag{6.1}$$

B. R. Greenhalgh, *An Introduction to Computational Thermodynamics*,
https://doi.org/10.1007/978-3-032-30317-2_6

With the temperature and pressure common to both phases, and the fact that we are dealing with a pure substance, we can simplify[1] our target function:

$$F(P) = ln\varphi_i{}^{\nu} - ln\varphi_i{}^{\ell} \tag{6.2}$$

Our derivative is then:

$$H(P) = \frac{dF}{dP} = \left[\frac{\partial ln\varphi_i{}^{\nu}}{\partial P}\right]_T - \left[\frac{\partial ln\varphi_i{}^{\ell}}{\partial P}\right]_T \tag{6.3}$$

This problem can be solved by Newton's method (Appendix 8.13), for which we require an initial guess. Here we cannot use our "chemical familiarity," as we want our Psat routine to work even for substances with which we are unfamiliar. Fortunately, we can use the Wilson correlation (Eq. 8.200) to estimate P_{sat}:

$$P^W_{sat\,i} = P_{ci} exp(5.373(1+\omega_i)(1-\frac{T_{ci}}{T}))$$

As the Wilson equation is only an approximation, and does not match exactly the particular EOS we might be using, it is therefore a good idea to check that the Wilson estimate of P_{sat} yields distinct liquid and vapor volumes. One can reduce the pressure from the Wilson approximation, or one could consider using spinodal volumes (and their pressures) to bracket P_{sat}. Once the Newton iterations begin, stepsize control is also advised so that when seeking very low vapor pressures (at low T) the pressure remains positive. An outline of these concepts is given in Algorithm 12, which offers several opportunities for passing in volumes (bypassing the root solver) to improve the efficiency of the calculation of P_{sat}. For example, the calculation of F (Eq. 6.2) requires V_L and V_V, as does H (Eq. 6.3). There are other quantities that can be passed in as well (both F and H require Z values). The degree to which this is built into your code is up to you, and you may choose to forgo this entirely.

Note that reporting $P_{sat} = P_c$ at $T = T_c$ can be a misleading result. While for a cubic EOS parameterized to the critical point, this is an exact result; for non-cubics like PC-SAFT, this is not so. This raises the question of whether to compute "EOS critical points" or simply accept this as an inconsistency.[2] Also, there is some flexibility on how to handle errors or failures to converge. Putting these matters aside, Example 10 looks at the calculation of P_{sat} for a pure substance.

Example 10 (Calculation of Psat for a Pure Substance) We will calculate P_{sat} for water at 600K using the SRK EOS. With $T_c = 647.30$K, $P_c = 221.2$bar and $\omega = 0.344$, we expect to find a valid $P_{sat} < P_c$ because $T < T_c$. As shown in

[1] Though here we write the pure component fugacity coefficients, as discussed below Eq. 1.76, you will likely not have distinct expressions for pure vs. mixture fugacity coefficients.

[2] Note that there are no such inconsistencies with computing the mixture critical points.

Algorithm 12 Calculation of pure component vapor pressure

```
if T > T_c then
    Return statement that there is no P_sat at this temperature.
else
    if T = T_C then
        Return P_sat = P_c
    else
        Set Tol = 10^-6, Iter = 0, MaxIter = 100, Solve = True
        Initialize P via Wilson (Eq. 8.200)
        Solve roots (Chap. 4): V_ℓ and V_v
        Ensure two volume roots: V_ℓ, V_v (reduce P or bracket with spinodal pressures)
        Compute F (Eq. 6.2)

        while ( doSolve==True)

            Compute H (Eq. 6.3)
            Newton dP = −F/H
            Iter += 1 and Err = |dP|/P
            P_new = P + dP
            Stepsize Controller: limit dP so that P_new > 0 and |dP|/P < 0.1
            Step P = P_new

            Exit
            if ((Iter ≥ MaxIter)or(Err < Tol)) then
                Solve = False
            else
                Set Tol = 10^-6, Iter = 0, MaxIter = 100, Solve = True
                Solve roots (Chap. 4): V_ℓ and V_v
                Compute F (Eq. 6.2)

            end if
        end while
        if Iter >= MaxIter then
            Raise Error
        else
            Psat = P
        end if
    end if
end if
```

Fig. 6.1, the Wilson approximation (Eq. 8.200) yields $P_{sat} \approx 125.18$bar at which our root solver finds $V_L = 41.46\text{cm}^3$ and $V_V = 264.61\text{cm}^3$. Numerical outputs from the Newton iterations solving Eq. 6.2 are summarized in Table 6.1.

As shown in Fig. 6.1, the Wilson approximation is quite close to the converged result ($P_{sat} = 126.43$ bar) and the function F (Eq. 6.2) is nearly linear[3] in the vicinity of the initial estimate and solution.

[3] This statement is somewhat misleading, as the root solver, which is nonlinear, also must be run among the iterations.

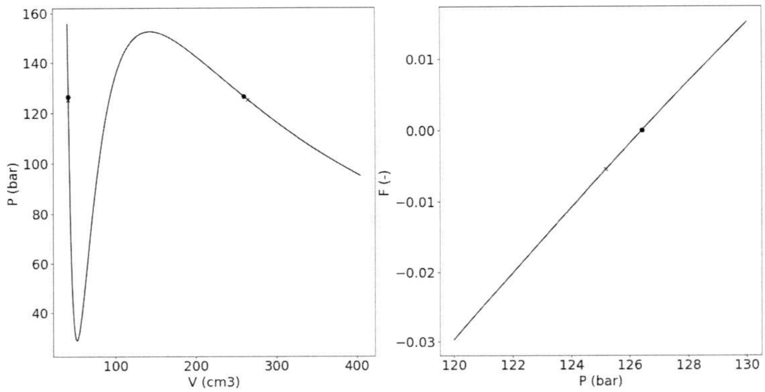

Fig. 6.1 Left: Isotherm for water at 600K Right: Plot of Eq. 6.2, calculated with the SRK EOS. The x symbols mark initialization by Wilson approximation, while the dots show the iterations

Table 6.1 Newton iterations for P_{sat} for water at 600K using the SRK EOS

Iter	P bar	V_ℓ $\frac{cm^3}{mol}$	V_ν $\frac{cm^3}{mol}$	F -	H $\frac{1}{bar}$	dP bar
1	125.18	41.46	264.61	−5.51E−3	4.47E−8	1.23
2	126.42	41.41	259.91	−5.75E−5	4.38E−8	1.31E−2
3	126.43	41.41	259.86	−6.47E−9	4.38E−8	1.45E−6

6.2 Pure Component Saturation Temperature

We can apply a very similar approach as outlined in Algorithm 12 for the determination of T_{sat} for a given P_{sat}. Our target function is unchanged;

$$F(T) = ln\varphi_i{}^{\nu} - ln\varphi_i{}^{\ell}$$

our derivative becomes:

$$H(T) = \frac{dF}{dT} = \left[\frac{\partial ln\varphi_i{}^{\nu}}{\partial T}\right]_P - \left[\frac{\partial ln\varphi_i{}^{\ell}}{\partial T}\right]_P \tag{6.4}$$

initialization is based on Wilson equation (Eq. 8.199), and the stepsize controller prevents negative temperature. Applying such a concept to the conditions examined in Example 10 yields $T_{sat} = 600$K for a given $P_{sat} = 126.43$.

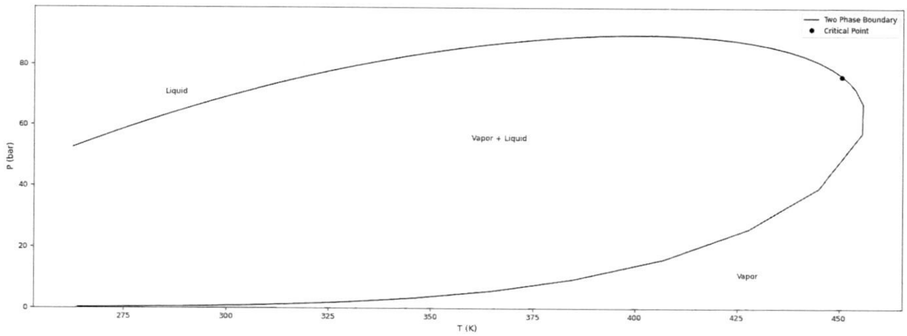

Fig. 6.2 Two-phase TP boundary for an equimolar mixture of methane, n-butane, and n-hexane calculated with the SRK EOS

6.3 Mixture TP Envelopes

In this section we look at a very useful algorithm in thermodynamic modeling: the PT phase envelope for a mixture. We will follow the work of Michelsen and Mollerup [Michelsen book]. Formally,[4] this calculation considers a mixture of fixed (known) composition, and determines the boundary (set of T and P values) dividing two-phase (vapor and liquid coexistence) from single phase (vapor *or* liquid) behavior. This is demonstrated in Fig. 6.2.

At temperatures and pressures that lie above the boundary (e.g., $T = 250$K and $P = 80$bar) in Fig. 6.2, the mixture exists as a liquid. We therefore refer to this part of the boundary as the bubble point curve. Below boundary (e.g., $T = 400$K and $P = 10$bar), the mixture exists as a vapor, and correspondingly, this part of the boundary is called the dewpoint curve. The bubble- and dew-point curves[5] meet at the critical point, which is discussed in Sect. 6.6. Inside the boundary, the mixture exists in a state of vapor liquid equilibrium, the relative amounts of these phases depending on the particular temperature and pressure, and in the general case composition.[6]

For a C-component mixture, the equilibrium relationship is given by Eqs. 1.60 through 1.62, or via Sect. 1.8.2, Eq. 1.77. With x and y used to represent the mole fractions in the liquid and vapor phases, respectively, we must therefore satisfy the following set of equations:

$$lny_i + ln\hat{\varphi}_i{}^{v}(T, P, \overline{y}) - lnx_i - ln\hat{\varphi}_i{}^{\ell}(T, P, \overline{x}) = 0 \tag{6.5}$$

[4] We will develop this algorithm to model vapor–liquid equilibrium, though sometimes it is capable of returning solutions representing liquid–liquid equilibrium.

[5] The terms bubble- and dew-point are synonymous for a pure substance.

[6] See Sect. 1.7.2.

Note carefully that the lnP terms in Eq. 1.77 cancel when we write Eq. 6.5 for equilibrium at stated pressure, but the $ln\hat{\varphi}$ terms are nonetheless functions of P. We therefore have C (one for each component) equations in 2C ($\overline{x}$ and $\overline{y}$) + 2 (T and P) unknowns. We therefore require additional equations, one of which derives from the context of a "boundary." On the bubble-point curve, the amount of vapor is exactly zero, as is the amount of liquid on the dew-point curve. As shown in Appendix 8.16 (Eqs. 8.139 and 8.140), we can relate[7] the liquid and vapor compositions in terms of f, $\overline{k}$, and $\overline{z}$ values:

$$x_i = \frac{z_i}{1 + f(k_i - 1)}$$

$$y_i = \frac{k_i z_i}{1 + f(k_i - 1)}$$

This change of variables allows us to rewrite our equilibrium expression as:

$$EQ_i = lnk_i + ln\hat{\varphi}_i{}^{v}(T, P, \overline{y}) - ln\hat{\varphi}_i{}^{\ell}(T, P, \overline{x}) = 0 \tag{6.6}$$

While this is certainly helpful in reducing the number of independent unknowns, we must restrict the $\overline{k}$ values such that $\overline{x}$ and $\overline{y}$ are normalized. This is achieved by adding the Rachford–Rice equation (Appendix 8.16, specifically Eq. 8.142) to our set:

$$\overline{F} = \begin{Bmatrix} \overline{EQ} \text{ C equations} \\ RR \text{ 1 equation} \end{Bmatrix} = \overline{0} \tag{6.7}$$

Our unknowns[8] are:

$$\overline{\alpha} = \begin{cases} T \text{ 1 unknown} \\ P \text{ 1 unknown} \\ \overline{k} \text{ C unknowns} \end{cases} \tag{6.8}$$

The feed composition and phase fraction are not *unknowns*; they are parameters. As in the work of Michelsen [Michelsen book], we represent these compactly as:

$$\overline{\beta} = \begin{cases} \overline{z} \text{ C values} \\ f \text{ 1 value} \end{cases} \tag{6.9}$$

[7] As we will see, x_i and y_i are calculated within the algorithm, but formally the k_i values are the unknowns.

[8] Here we have attempted to preserve the *order* of the variables. Since in other parts of this text, we have written TVn and TPn; here we write TPk to follow that order.

As the number of unknowns exceeds the number of equations, we require an additional specification. We can further understand this by looking back at Fig. 6.2; which point (T,P) on the boundary are we attempting to calculate? We might therefore specify a particular temperature or pressure, but look again at Fig. 6.2. We must recognize that, for an arbitrary mixture, we would not know where the maximum temperature (called the *cricondentherm*, which in this example is about 450K) lies. What if we had naively tried to compute the "first" point at $T = 500\text{K}$? A safer approach might be to say all (most) mixtures that we might consider will at sufficiently low temperature exhibit dew-point behavior at ambient pressure ($P \approx 1\text{bar}$). If we specify this pressure, we *could* rewrite our system (Eqs. 6.7 and 6.8) into a C + 1 by C + 1 system. However, as we will see shortly, calculation of not one point but the full envelope (or at least a set of points) is enabled by retaining T, P, and $\overline{k}$ as our set of unknowns. We therefore need an additional equation in $\overline{F}$, an example of which is to state that the pressure matches a particular specification:

$$Spec = P - P_{spec} = 0$$

Our full set becomes:

$$\overline{F} = \begin{cases} \overline{EQ} & \text{C equations} \\ RR & \text{1 equation} \\ Spec & \text{1 equation} \end{cases} \tag{6.10}$$

In principle we can solve $\overline{F}(\overline{\alpha}, \overline{\beta}) = \overline{0}$ via Newton's method (Appendix 8.13), but before doing so, we will incorporate another technique from the work of Michelsen and Mollerup [Michelsen book]. Our independent variables are T, P, and $\overline{k}$. Depending on your choice of units, these will have wildly different magnitudes.[9] We can address this potential scaling issue by placing the disparate quantities on a log scale. After adjusting our current specification to be:

$$Spec = ln P - ln P_{spec} \tag{6.11}$$

our system of equations becomes:

$$\overline{F}(\overline{\alpha}, \overline{\beta}) = \left\{ \begin{matrix} \overline{EQ} & \text{C equations} \\ RR & \text{1 equation} \\ Spec & \text{1 equation} \end{matrix} \right\} = \overline{0} \tag{6.12}$$

[9] At $T = 300\text{K}$ and $P = 10^5\text{Pa}$, the Wilson k values for methane, n-butane, and n-hexane values are approximately 300, 3, and 0.2.

$$\overline{\alpha} = \begin{cases} lnT & 1 \text{ unknown} \\ lnP & 1 \text{ unknown} \\ ln\overline{k} & \text{C unknowns} \end{cases} \tag{6.13}$$

$$\overline{\beta} = \begin{cases} \overline{z} & \text{C values} \\ f & 1 \text{ value} \end{cases} \tag{6.14}$$

To proceed, we need an initial guess. If we select P_{spec} = 1 bar (10^5 Pa), we can use the Wilson approximation to first estimate the temperature and lnk_i values. As this initial step is somewhat of a departure from the main algorithm, it is considered in detail in Example 11.

Example 11 (Initializing the PT Phase Envelope with the Wilson Equation) For the case shown in Fig. 6.2, the first point we seek to find on the dew point is characterized by $P = 10^5$Pa and $\overline{z} = \frac{1}{3}, \frac{1}{3}, \frac{1}{3}$ and $f = 1$. As shown in Appendix 8.16.1, our unknown is the temperature, which we initialize via the Wilson approximation (Eq. 8.199) and follow Algorithm 33. Using the parameters given in Appendix 8.5, we have an initial estimate of $T = 241.52K$.

The target function is Eq. 8.143;

$$F = \sum_r \frac{z_r({k_r}^W(T, P) - 1)}{1 + f({k_r}^W(T, P) - 1)} = 0$$

while the derivative is given by Eq. 8.144

$$H = \left[\frac{\partial F}{\partial T}\right]_P = \sum_r \frac{z_r \left[\frac{\partial {k_r}^W}{\partial T}\right]_P}{(1 + f({k_r}^W - 1))^2}$$

Numerical outputs for the initial estimate of T_{sat} are shown below in Table 6.2.

Table 6.2 Newton iterations for T_{sat} with $P = 10^5$Pa and $\overline{z} = \frac{1}{3}, \frac{1}{3}, \frac{1}{3}$ and $f = 1$ via the Wilson approximation

Iter	T [K]	F [-]	H $[\frac{1}{K}]$	dT [K]
1	241.516	−24.310	1.521	15.987
2	257.503	−9.291	0.542	17.123
3	274.625	−3.419	0.204	16.735
4	291.361	−1.133	0.087	12.970
5	304.330	−0.284	0.048	5.897
6	310.227	−0.034	0.037	0.902
7	311.129	−6.4E−4	0.036	0.018
8	311.147	−2.4E−7	0.036	6.7E−6

From the result in Table 6.2 and the Wilson approximation, we can estimate $ln\overline{k} = 5.937, 1.276, -0.999$.

With the results of Example 11, we now have our initialization of $\overline{\alpha}$, and with $\overline{\beta}$ defined, we can attempt to solve $\overline{F}(\overline{\alpha}, \overline{\beta}) = \overline{0}$. This also proceeds via Newton's method, for which we require the derivative (Jacobian) of $\overline{F}$ with respect to $\overline{\alpha}$. We now develop that structure.

$$\overline{\overline{H}} = \frac{\partial \overline{F}}{\partial \overline{\alpha}} = \begin{bmatrix} \frac{\partial \overline{EQ}}{\partial lnT} & \frac{\partial \overline{EQ}}{\partial lnP} & \frac{\partial \overline{EQ}}{\partial ln\overline{k}} \\ \frac{\partial RR}{\partial lnT} & \frac{\partial RR}{\partial lnP} & \frac{\partial RR}{\partial ln\overline{k}} \\ \frac{\partial Spec}{\partial lnT} & \frac{\partial Spec}{\partial lnP} & \frac{\partial Spec}{\partial ln\overline{k}} \end{bmatrix} \tag{6.15}$$

On the one hand, the required Jacobian seems very complicated; it is certainly difficult to visualize in physical terms the meaning of this structure as a whole. However, using abstraction, whereby we reduce the Jacobian into its component parts, we can build the required structure from simpler elements. Figure 6.3 clarifies the layout of the Jacobian.

Also note that for clarity and ease of implementation, we will depart from our practice of keeping the code entirely modular compared with the rest of the text; substitutions of one equation into another will be made where this results in simpler forms.

We begin with the derivatives of $\overline{EQ}$, which we derive from Eqs. 6.6 and 6.13. As we referred to in Appendix 8.16 (i.e., other than in our *approximate*, initialization of

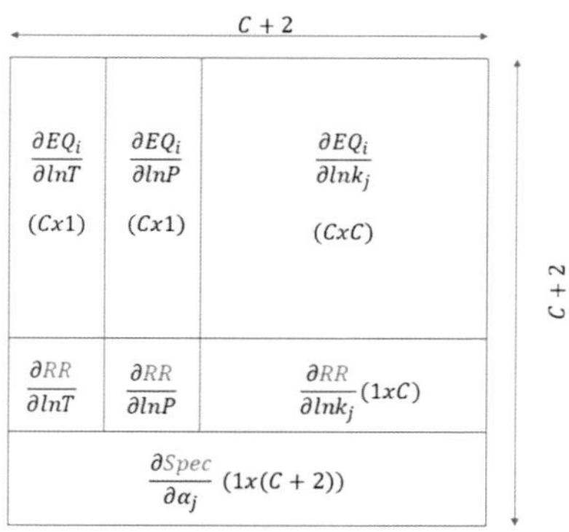

Fig. 6.3 Layout of Jacobian used in VLE Boundary Calculations

T_{sat}), lnT, lnP, and lnk_i are *independent* of each other. This means that $\frac{\partial lnk_i}{\partial lnT} = 0$ and:

$$\frac{\partial EQ_i}{\partial lnT} = \left[\frac{\partial ln\hat{\varphi}_i{}^{\nu}(T, P, \overline{y})}{\partial lnT}\right]_{P,\overline{y}} - \left[\frac{\partial ln\hat{\varphi}_i{}^{\ell}(T, P, \overline{x})}{\partial lnT}\right]_{P,\overline{x}}$$
$$= T\left(\left[\frac{\partial ln\hat{\varphi}_i{}^{\nu}(T, P, \overline{y})}{\partial T}\right]_{P,\overline{y}} - \left[\frac{\partial ln\hat{\varphi}_i{}^{\ell}(T, P, \overline{x})}{\partial T}\right]_{P,\overline{x}}\right) \tag{6.16}$$

In Eq. 6.16, we have made use of the identity $dlnT = \frac{1}{T}dT$. Also, the required derivatives are available[10] via Eq. 5.14. By the same logic, and using Eq. 5.15, we can write:

$$\frac{\partial EQ_i}{\partial lnP} = P\left(\left[\frac{\partial ln\hat{\varphi}_i{}^{\nu}(T, P, \overline{y})}{\partial P}\right]_{T,\overline{y}} - \left[\frac{\partial ln\hat{\varphi}_i{}^{\ell}(T, P, \overline{x})}{\partial P}\right]_{T,\overline{x}}\right) \tag{6.17}$$

The $\frac{\partial \overline{EQ}}{\partial ln\overline{k}}$ term has not been developed so far in this text. Based on Eq. 6.6:

$$\frac{\partial EQ_i}{\partial lnk_j} = \frac{\partial lnk_i}{\partial lnk_j} + \left[\frac{\partial ln\hat{\varphi}_i{}^{\nu}(T, P, \overline{y})}{\partial lnk_j}\right] - \left[\frac{\partial ln\hat{\varphi}_i{}^{\ell}(T, P, \overline{x})}{\partial lnk_j}\right] \tag{6.18}$$

The first derivative is trivial (Appendix 8.4), but the other two are more difficult. As our implementation of $ln\hat{\varphi}_i$ does not lend itself to lnk_j values, we proceed rather pedantically, noting that if lnk_j affects $ln\hat{\varphi}_i$, it would be through the composition, writing:

$$\left[\frac{\partial ln\hat{\varphi}_i{}^{\nu}(T, P, \overline{y})}{\partial lnk_j}\right] = \sum_r \left[\frac{\partial ln\hat{\varphi}_i{}^{\nu}(T, P, \overline{y})}{\partial y_r}\right]_{T,P,y_{k\neq r}} \left[\frac{\partial y_r}{\partial k_j}\right] k_j \tag{6.19}$$

According to Eq. 8.147, $\left[\frac{\partial y_r}{\partial k_j}\right] = \delta_{rj}\left(x_j - \frac{x_j y_j}{z_j} f\right)$. This eliminates the summation, yielding:

$$\left[\frac{\partial ln\hat{\varphi}_i{}^{\nu}(T, P, \overline{y})}{\partial lnk_j}\right] = \left[\frac{\partial ln\hat{\varphi}_i{}^{\nu}(T, P, \overline{y})}{\partial y_j}\right]_{T,P,y_{k\neq j}} \left(y_j - \frac{y_j^2}{z_j} f\right) \tag{6.20}$$

[10] Because $\left[\frac{\partial ln\hat{\varphi}_i}{\partial T}\right]_{P,\overline{n}}$ is a degree 0 homogenous function of $\overline{n}$ (refer to Appendix 8.6), we can simply use $\overline{y}$ (or $\overline{x}$) in place of $\overline{n}^{\nu}$ (or $\overline{n}^{\ell}$) in the calculation.

By the same logic:

$$\left[\frac{\partial ln\hat{\varphi}_i{}^{\ell}(T,P,\overline{x})}{\partial lnk_j}\right]=\sum_r\left[\frac{\partial ln\hat{\varphi}_i{}^{\ell}(T,P,\overline{x})}{\partial x_r}\right]_{T,P,y_{k\neq r}}\left[\frac{\partial x_r}{\partial k_j}\right]k_j$$
$$=-\left[\frac{\partial ln\hat{\varphi}_i{}^{\ell}(T,P,\overline{x})}{\partial x_j}\right]_{T,P,y_{k\neq j}}\frac{x_jy_j}{z_j}f \tag{6.21}$$

$$\frac{\partial EQ_i}{\partial lnk_j}=\delta_{ij}+\left[\frac{\partial ln\hat{\varphi}_i{}^{v}(T,P,\overline{y})}{\partial y_j}\right]_{T,P,y_{k\neq j}}\left(y_j-\frac{y_j^2}{z_j}f\right)$$
$$+\left[\frac{\partial ln\hat{\varphi}_i{}^{\ell}(T,P,\overline{x})}{\partial x_j}\right]_{T,P,y_{k\neq j}}\frac{x_jy_j}{z_j}f \tag{6.22}$$

The derivatives of RR are more straightforward. In terms of our variables ($\overline{\alpha}=lnT,lnP,ln\overline{k}$), RR only depends[11] on $ln\overline{k}$ (see Eq. 8.142, with the required derivative derived from Eq. 8.148:

$$\frac{\partial RR}{\partial lnk_j}=\frac{z_jk_j}{[1+f(k_r-1)]^2}=\frac{x_jy_j}{z_j} \tag{6.23}$$

Finally, our *Spec* equation is very simple in terms of α; the pressure is the specified variable:

$$\frac{\partial Spec}{\partial \alpha_j}=\frac{\partial lnP}{\partial \alpha_j}=[0,1,0...0] \tag{6.24}$$

Note that the position of the "1" in Eq. 6.24 depends on where we choose to place P within α (we have chosen a TPn ordering of our variables). As we will see, if we had specified the temperature, there would still be only a single non-zero entry in Eq. 6.24, but it would be at the initial position.

We now look at the calculation of the first point on our phase boundary in an example.

Example 12 (Determination of the Initial Point on the VLE Boundary) With $P_{spec}=1E5$Pa and $\overline{\beta}=\frac{1}{3},\frac{1}{3},\frac{1}{3},1$ we have from Example 11 $T\approx 311.147$K,

[11] If there is confusion here, recall that $lnT,lnP,ln\overline{k}$ are *independent* variables; we only used Wilson k values, which do depend on T and P, as an initial approximation.

$P = 1E5$ Pa, $ln\overline{k} \approx 5.937, 1.276, -0.999$. We now begin to formally assemble the pieces required to solve the system of equations.

$$\begin{aligned}\overline{\alpha} &= lnT, lnP, ln\overline{k}\\ &= 5.740, 11.513, 5.937, 1.276, -0.999...\end{aligned}$$

Note that Wilson k values are used to initialize $ln\overline{k}$ here.

$$\begin{aligned}\overline{\beta} &= \overline{z}, f\\ &= \frac{1}{3}, \frac{1}{3}, \frac{1}{3}, 1.\end{aligned}$$

$$\begin{aligned}\overline{F} &= \overline{EQ}, RR, Spec\\ &= 7.02E-1, 6.05E-2, 4.84E-2, -3.30E-14, 0\end{aligned}$$

Note that $RR = 0$ only within the limits of our Rachford–Rice solver,[12] while $Spec$ is identically zero.

$$\overline{\overline{H}} = \begin{bmatrix} -1.427 & 1.006 & 0.999... & -5.680E-3 & 6.025E-3 \\ -8.208 & 0.972 & -5.374E-5 & 0.999 & 9.704E-4 \\ -11.965 & 0.950 & 5.853E-6 & 9.964E-5 & 0.999... \\ 0 & 0 & 8.803E-4 & 0.093 & 0.906 \\ 0 & 1 & 0 & 0 & 0 \end{bmatrix} \tag{6.25}$$

A numerical test can also be incorporated into the calculation of $\overline{\overline{H}}$ (as detailed in Sect. 2.3). With $\overline{F}$ coded as a function of $\overline{\alpha}$, this simply requires perturbations on $\overline{\alpha}$. If your expression for $\overline{\overline{H}}$ is coded correctly, the array of slopes will be of the form:

$$\begin{bmatrix} 2.0... & 2.0... & 2.0... & 2.0... & 2.0... \\ 2.0... & 2.0... & 2.0... & 2.0... & 2.0... \\ 2.0... & 2.0... & 2.0... & 2.0... & 2.0... \\ X & X & 2.0... & 2.0... & 2.0... \\ X & (small) & X & X & X \end{bmatrix} \tag{6.26}$$

[12] Also, the value shown here is different than the final value in Table 6.2. This results from how the Rachford–Rice solver is coded. After the 8th iteration shown, the next dT step is taken, but the algorithm exits after detecting the error (dT compared to T) is below tolerance. Put another way, the value of RR shown here (Example 12) would have been seen in Example 11 if an additional iteration had been performed.

Table 6.3 Summary of iterations for solving initial point on PT envelope for equimolar methane, n-butane, and n-hexane via SRK EOS

Iter	$\overline{\alpha}$ $lnT, lnP, ln\overline{k}$	$\overline{F}$ $\overline{EQ}, RR, Spec$	$d\overline{\alpha}$
1	5.740, 11.513, 5.937, 1.276, −0.999	0.702, 0.061, 0.048, −3.297E−14, 0	4.317E−3, 0, −0.696, −0.025, 3.258E−3
2	5.745, 11.513, 5.240, 1.251, −0.997	1.742E−4, 1.449E−4, 1.635E−4, −3.074E−4, 0	4.074E−5, 0, −1.183E−4, 1.869E−4, 3.208E−4
3	5.745, 11.513, 5.240, 1.251, −0.996	5.472E−9, 1.128E−8, 1.473E−8, −4.815E−8, 0	5.432E−9, 0, 1.990E−9, 3.296E−8, 4.984E−8

The Xs in the above array might be zeros, NANs, or even errors, depending on the programming language you're using and how you code the test. The important thing to recognize here is that they correspond to terms where $\overline{\overline{H}}$ is identically zero (recall that our test is designed to detect errors in the derivatives of nonlinear functions, not constants). Also note that the *Spec* equation is linear in one element of $\overline{\alpha}$ (in this case the pressure). Therefore, our numerical test produces noise, marked as (small) in the above array, for this trivial derivative. It is of course possible to structure numerical tests to produce a more meaningful test for linear functions as discussed in Sect. 2.3.

Table 6.3 summarizes the first few iterations of the Newton solution to the first point.

Taking the error in the Newton routine to be $Err = \sum_r d\alpha_r^2$, only a few iterations are required to converge. Also note the key features of the output in Table 6.3, such as the relatively minor shift in T (lnT) from the 1st iteration (via Wilson) and the converged result; 311.15K vs. 312.51K. This shows that the Wilson equation is a very good approximation for this system. Finally, the pressure does not change in these iterations ($d\alpha$ for the pressure is always zero) as pressure was the specified variable for this point.

We can now move the pressure slightly upward and solve an additional point on the envelope. To make the following more understandable (and flexible), let us rewrite our specification equation as:

$$Spec = \alpha[spec] - S \tag{6.27}$$

Here "spec" is the index of the variable in α that is being specified[13] (i.e., 2 for the case of pressure in a TPn formalism). For the initial point, we had set $S = ln10^5$. For the next point we could we increase S by increasing the specified pressure by say 10% and repeat the calculation above. We can actually do a bit better, as we can

[13] If you are using a zero-indexed language your index values might be written as 0,1,2...., making $spec = 1$ for the case of pressure.

use the previous solution to guide the next. Suppose we consider the set of $\overline{\alpha}$ and S that give $\overline{F} = 0$.

$$\overline{F}(\overline{\alpha}, S) = \overline{0}$$

If we differentiate this expression with respect to the specification variable S, we get:

$$\sum_r \frac{\partial F_i}{\partial \alpha_r} \frac{\partial \alpha_r}{\partial S} + \frac{\partial F_i}{\partial S} = 0 \tag{6.28}$$

Some clarifications are in order here. First, the set of $\overline{\alpha}$, S are those that give $\overline{F} = \overline{0}$; they are not arbitrary values, nor are they restricted to the values at the point solved in Example 12. As we move through this set, $\overline{F}$ remains zero (as shown on the right side of Eq. 6.28). Also, $\overline{\alpha}$ and S are not independent. This is why the first term on the left is a sum, with a chain-rule (Appendix 8.2) product; α_i depends on S. We also must recognize that this sum of products is the product of a matrix ($\overline{\overline{H}}$ given by Eq. 6.15) and a vector $\frac{\partial \alpha_r}{\partial S}$. We can then rewrite Eq. 6.28 as:

$$\overline{\overline{H}} * \frac{\partial \overline{\alpha}}{\partial S} + \frac{\partial \overline{F}}{\partial S} = 0 \tag{6.29}$$

We have already have the functional form (and the numerical value) of $\overline{\overline{H}}$, we are seeking $\frac{\partial \overline{\alpha}}{\partial S}$, and $\frac{\partial \overline{F}}{\partial S}$ is simply:

$$\frac{\partial \overline{F}}{\partial S} = \begin{bmatrix} 0 \\ \cdots \\ 0 \\ 0 \\ -1 \end{bmatrix} \tag{6.30}$$

The first group (0 ... 0) on the right indicates that $\overline{EQ}$ does not depend[14] on S, the second to last indicates the same about RR, and the final -1 shows how F_{Spec} formally depends on S. We can therefore solve for $\frac{\partial \overline{\alpha}}{\partial S}$ (see Appendix 8.12.2)

$$\frac{\partial \overline{\alpha}}{\partial S} = -\overline{\overline{H}}^{-1} * \frac{\partial \overline{F}}{\partial S} \tag{6.31}$$

[14] Do not confuse this with the question of whether $\overline{EQ}$ depends on P, for example. Of course it does, but here S is not functionally "equal" to the pressure, it is a number with the value $ln P$.

This provides an estimate of how the solution $\overline{\alpha}$ changes in response to S:

$$\overline{\alpha}(S + dS) \approx \overline{\alpha}(S) + \frac{\partial \overline{\alpha}}{\partial S} dS \tag{6.32}$$

We now look at an example in detail how to initialize and solve the second point on the phase boundary.

Example 13 (Determination of the Second Point on the VLE Boundary) From the calculation of the first point (we shall call it point A), we have from Eq. 6.15:

$$\overline{\overline{H}}_A = \begin{bmatrix} -1.39849 & 1.00595 & 0.9993 & -5.923E-3 & 6.623E-3 \\ -8.14885 & 0.972773 & -1.097E-4 & 0.999 & 1.07E-3 \\ -11.8879 & 0.950903 & 1.296E-5 & 1.130E-4 & 0.9999 \\ 0 & 0 & 0.002 & 0.095 & 0.903 \\ 0 & 1 & 0 & 0 & 0 \end{bmatrix}$$

Via Eq. 6.30 and 6.31 we have:

$$\frac{\partial \overline{\alpha}}{\partial S} = 0.083, 1, -0.893, -0.299, 0.033$$

This result is instructive. It approximates how the solution values of $\overline{\alpha}$ change in response to a change in specification S. In the current context, in which the pressure is being increased along the dew-point ($f = 1$) curve, we see that lnT (e.g., T) will have to increase to remain on the curve, lnP will change by exactly dS (this is the $Spec$ variable), and the lnk values adjust, with the stronger effect on the more volatile components. Note that the changes in the k values have different signs. In this way, Eq. 6.31 is an excellent initialization to the next point on the curve, capturing the dynamics and the constraints of the system. With $dS = S_B - S_A = 12.66 - 11.51 = 1.15$ Eq. 6.32 gives:

$$\begin{aligned} \overline{\alpha}_B &\approx \overline{\alpha}_A + \frac{\partial \overline{\alpha}}{\partial S} dS \\ &= 5.84, 12.66, 4.21, 0.91, -0.96... \end{aligned} \tag{6.33}$$

We easily converge Eq. 6.12 to the next point: $\overline{\alpha}_B = 5.85, 12.66, 4.18, 0.90, -0.95$.

As discussed above, we could continue to specify pressures for a while, but ultimately, we would reach and pass the cricondenbar (the maximum pressure on the envelope), after which our equations are unlikely to converge. We instead use Michelsen and Mollerup's approach of specifying the most rapidly changing variable [Michelsen book] . By looking at the elements of $\frac{\partial \overline{\alpha}}{\partial S}$ via Eq. 6.31, we can select that with the largest magnitude.[15] This is the variable that we specify for our next point, particularly so that the other variables do not change excessively. Also, to capture the curvature of the boundary, we may need to do better than a linear extrapolation. Knowing two points on the boundary (represented by $\overline{\alpha}_A$ and $\overline{\alpha}_B$) and their derivatives with respect to the specification variable ($\frac{\partial \overline{\alpha}_A}{\partial S}$ and $\frac{\partial \overline{\alpha}_B}{\partial S}$), we have enough information to approximate $\overline{\alpha}$ as a third-order polynomial in S;

$$\alpha_i = C_{3i} S^3 + C_{2i} S^2 + C_{1i} S + C_{0i} \tag{6.34}$$

$$\frac{\partial \alpha_i}{\partial S} = 3C_{3i} S^2 + 2C_{2i} + C_{1i} \tag{6.35}$$

These four points define the four[16] coefficients C_{ki}:

$$\begin{bmatrix} \alpha_{A1} & \alpha_{A2} & \ldots & \alpha_{AN} \\ \alpha_{B1} & \alpha_{B2} & \ldots & \alpha_{BN} \\ \frac{\partial \alpha_{A1}}{\partial S} & \frac{\partial \alpha_{A2}}{\partial S} & \ldots & \frac{\partial \alpha_{AN}}{\partial S} \\ \frac{\partial \alpha_{B1}}{\partial S} & \frac{\partial \alpha_{B2}}{\partial S} & \ldots & \frac{\partial \alpha_{BN}}{\partial S} \end{bmatrix} = \begin{bmatrix} S_A^3 & S_A^2 & S_A & 1 \\ S_B^3 & S_B^2 & S_B & 1 \\ 3S_A^2 & 2S_A & 1 & 0 \\ 3S_B^2 & 2S_B & 1 & 0 \end{bmatrix} \begin{bmatrix} C_{31} & C_{32} & \ldots & C_{3N} \\ C_{21} & C_{22} & \ldots & C_{2N} \\ C_{11} & C_{12} & \ldots & C_{1N} \\ C_{01} & C_{02} & \ldots & C_{0N} \end{bmatrix} \tag{6.36}$$

Equation 6.36 can be represented in matrix notation as:

$$\overline{\overline{\alpha}}_{4xN} = \overline{\overline{S}}_{4x4} * \overline{\overline{C}}_{4xN} \tag{6.37}$$

Note the dimensions of the matrices given in Eq. 6.37, with $N = C + 2$. We can then solve for the coefficient matrix via:

$$\overline{\overline{C}}_{4xN} = \overline{\overline{S}}_{4x4}^{\;-1} \overline{\overline{\alpha}}_{4xN} \tag{6.38}$$

Then, with S selected, we can estimate $\overline{\alpha}$ by Eq. 6.34. A significant point is that S and therefore $\frac{\partial \overline{\alpha}}{\partial S}$ are based on the *spec* index, the most rapidly changing element of $\frac{\partial \overline{\alpha}}{\partial S}$ from the previous point. This means that if the spec index changes, the S and $\overline{\overline{H}}$ values must be made consistent. Let's look at an example to clarify this.

Example 14 (Subsequent Points on the PT Envelope by Nonlinear Extrapolation) Suppose we have calculated the following points on the boundary, based on the specification of pressure:

$$\overline{\alpha}_A = 5.745, 11.513, 5.240, 1.251, -0.996$$

[15] Even if that element is negative.

[16] Equation 6.34 represents a correlation for each element i within $\overline{\alpha}$, each with its own set of C values. Equation 6.36 makes clear the number of C values.

$$\overline{\overline{H}}_A = \begin{bmatrix} -1.398 & 1.006 & 0.999 & -5.923E-3 & 6.623E-3 \\ -8.149 & 0.973 & -1.097E-4 & 0.999 & 1.070E-3 \\ -11.888 & 0.951 & 1.296E-5 & 1.130E-4 & 0.999 \\ 0 & 0 & 1.766E-3 & 0.095 & 0.903 \\ 0 & 1 & 0 & 0 & 0 \end{bmatrix}$$

$$\frac{\partial \overline{\alpha}_A}{\partial S} = 0.083, 1, -0.893, -0.299, -0.033$$

$$\overline{\alpha}_B = 5.846, 12.664, 4.182, 0.903, -0.948$$

$$\overline{\overline{H}}_B = \begin{bmatrix} -0.747 & 1.014 & 0.997 & -0.013 & 0.016 \\ -6.688 & 0.931 & -4.929E-4 & 0.998 & 2.973E-3 \\ -9.937 & 0.876 & 9.304E-5 & 4.670E-4 & 0.999 \\ 0 & 0 & 5.088E-3 & 0.135 & 0.860 \\ 0 & 1 & 0 & 0 & 0 \end{bmatrix}$$

$$\frac{\partial \overline{\alpha}_B}{\partial S} = 0.094, 1, -0.952, -0.307, 0.054$$

According to $\frac{\partial \overline{\alpha}_B}{\partial S}$ the next specification variable[17] should be pressure-based. The basis of our extrapolation becomes:

$$\overline{\overline{\alpha}} = \begin{bmatrix} 5.745 & 11.513 & 5.240 & 1.251 & -0.996 \\ 5.846 & 12.664 & 4.182 & 0.903 & -0.948 \\ 8.279E-2 & 1 & -0.893 & -0.299 & 3.329E-2 \\ 9.358E-2 & 1 & -0.952 & -0.307 & 5.380E-2 \end{bmatrix}$$

$$\overline{\overline{S}} = \begin{bmatrix} 1526.009 & 132.548 & 11.513 & 1 \\ 2031.118 & 160.382 & 12.664 & 1 \\ 397.642 & 23.0259 & 1 & 0 \\ 481.147 & 25.328 & 1 & 0 \end{bmatrix}$$

$$\overline{\overline{C}} = \begin{bmatrix} 1.182E-4 & 3.664E-15 & -5.019E-3 & -7.703-4 & 1.752E-3 \\ 3.984E-4 & -1.208E-13 & 0.156 & 2.442E-2 & -5.463E-2 \\ 2.659E-2 & 1 & -2.499 & -0.555 & 0.594 \\ 5.205 & 4.093E-12 & 20.935 & 5.575 & -3.273 \end{bmatrix}$$

[17] The largest element of $\frac{\partial \overline{\alpha}_B}{\partial S}$ is that of lnP, and of course this value is 1; we have specified $lnP = S$.

If we make our step $dS = S_B - S_A = 1.151$, we can then adjust S;

$$S = S_B + dS = 13.816$$

estimate $\overline{\alpha}$;

$$\overline{\alpha} \approx 5.960, 13.816, 3.034, 0.543, -0.867$$

and solve:

$$\overline{\alpha} = 5.960, 13.816, 3.024, 0.542, -0.863$$

$$\overline{\overline{H}} = \begin{bmatrix} 0.171 & 1.031 & 0.987 & -0.032 & 0.044 \\ -4.912 & 0.824 & -2.651E-3 & 0.993 & 9.298E-3 \\ -7.616 & 0.688 & 9.112E-4 & 2.282E-3 & 0.997 \\ 0 & 0 & 0.016 & 0.194 & 0.790 \\ 0 & 1 & 0 & 0 & 0 \end{bmatrix}$$

$$\frac{\partial \overline{\alpha}}{\partial S} = 0.103, 1, -1.077, -0.323, 0.101$$

This last result shows us that the specification variable needs to change before proceeding to the next point; the rate of change of lnk_{CH_4} (-1.08) has become higher than that of $ln P$ (1.0). After we "cycle" our points (labeling what is currently point "B" as point "A" and labeling our latest result as point "B"), we have to redefine the $spec$, S and $\overline{\overline{H}}$ values to recompute the $\frac{\partial \overline{\alpha}}{\partial S}$ values due to the change in specification variable.[18] The only changes required in the $\overline{\overline{H}}$ matrices due to the change in specification index are the final row. This suggests that the routine that solves each point (i.e., solve Eq. 6.12) should also return $\overline{\overline{H}}$, as this quantity is used in the extrapolation, and it is inefficient to needlessly recalculate it. We proceed as follows:

$$\overline{\alpha}_A = 5.846, 12.664, 4.182, 0.903, -0.948$$

$$\overline{\overline{H}}_A = \begin{bmatrix} -0.747 & 1.014 & 0.997 & -1.309E-2 & 1.572E-2 \\ -6.688 & 0.9311 & -4.929E-4 & 0.998 & 2.973E-3 \\ -9.937 & 0.876 & 9.304E-5 & 4.670E-4 & 0.999 \\ 0 & 0 & 5.088E-2 & 0.135 & 0.860 \\ 0 & 0 & 1 & 0 & 0 \end{bmatrix}$$

[18] Our two points were calculated by $ln P$ specification, and our extrapolation may fail if we do not remedy this.

$$\frac{\partial \overline{\alpha}_A}{\partial S} = -9.83E-2, -1.051, 1, 0.322, -5.653E-2$$

$$\overline{\alpha}_B = 5.960, 13.816, 3.024, 0.542, -0.863$$

$$\overline{\overline{H}}_B = \begin{bmatrix} 0.171 & 1.031 & 0.987 & -3.172E-2 & 4.444E-2 \\ -4.912 & 0.824 & -2.651E-3 & 0.993 & 9.298E-3 \\ -7.616 & 0.688 & 9.112E-4 & 2.282E-3 & 0.997 \\ 0 & 0 & 1.620E-2 & 0.194 & 0.790 \\ 0 & 0 & 1 & 0 & 0 \end{bmatrix}$$

$$\frac{\partial \overline{\alpha}_B}{\partial S} = -9.592E-2, -0.928, 1, 0.299, -9.399E-2$$

We then extrapolate via:

$$\overline{\overline{\alpha}} = \begin{bmatrix} 5.846 & 12.664 & 4.182 & 0.903 & -0.948 \\ 5.960 & 13.816 & 3.024 & 0.542 & -0.863 \\ -9.833E-2 & -1.051 & 1 & 0.322 & -5.653E-2 \\ -9.592E-2 & -0.928 & 1 & 0.299 & -9.399E-2 \end{bmatrix}$$

$$\overline{\overline{S}} = \begin{bmatrix} 73.158 & 17.492 & 4.182 & 1 \\ 27.663 & 9.147 & 3.024 & 1 \\ 52.476 & 8.365 & 1 & 0 \\ 27.440 & 6.049 & 1 & 0 \end{bmatrix}$$

$$\overline{\overline{C}} = \begin{bmatrix} 1.637E-3 & 6.899E-3 & 9.99E-16 & -1.803E-3 & -3.046E-3 \\ -1.874E-2 & -0.127 & -1.066E-14 & 2.933E-2 & 4.91E-2 \\ -2.751E-2 & -0.347 & 1 & 0.171 & -0.307 \\ 6.169 & 15.840 & -2.132E-14 & -0.195 & -0.298 \end{bmatrix}$$

Making our step $dS = S_B - S_A = -1.158$, we adjust S;

$$S = S_B + dS = 1.866$$

estimate $\overline{\alpha}$ (by Eq. 6.31 and 6.32);

$$\overline{\alpha} \approx 6.063, 14.793, 1.866, 0.215, -0.720$$

and solve:

$$\overline{\alpha} = 6.061, 14.792, 1.866, 0.220, -0.714$$

$$\overline{\overline{H}} = \begin{bmatrix} 1.560 & 1.022 & 0.927 & -7.549E-2 & 0.1488 \\ -2.813 & 0.584 & -1.454E-2 & 0.985 & 2.955E-2 \\ -5.01 & 0.292 & 1.127E-2 & 1.161E-2 & 0.977 \\ 0 & 0 & 5.156E-2 & 0.268 & 0.681 \\ 0 & 0 & 1 & 0 & 0 \end{bmatrix}$$

$$\frac{\partial \overline{\alpha}}{\partial S} = -7.479E-2, -0.749, 1, 0.250, -0.174$$

As indicated, the specification variable remains lnk_{CH_4}.

A final point of discussion concerns stepsize control. When we start computing the boundary, we can specify an initial pressure and even an increase in the pressure (translating these into $S = lnP$. However, as we continue, we need a rational method for determining the step size in the automatically specified variables, which alternate between lnT, lnP, and lnk values. Michelsen and Mollerup [Michelsen book] provide a method for this, restricting the stepsize to keep the number of iterations within a prescribed limit. That is, if more than say 3 iterations are required to converge the current point, the stepsize (dS) is reduced until convergence within 3 iterations takes place. We also require an automatic method of *increasing* the stepsize; otherwise, the computation time for the boundary can become very large, or worse, the algorithm will fail as the stepsize is driven monotonically lower. Here again Michelsen and Mollerup [Michelsen book] provide a solution, whereby a count is made of the number of subsequent successfully converged (i.e., converged within 3 iterations) points. When this count reaches a prescribed number (e.g., 3), the stepsize is increased (i.e., doubled). Also, stepsize control can be applied to ensure that large changes in T and P do not occur, which can affect the appearance of plots (the spacing in T and P in Fig. 6.2 can easily be seen), even if convergence is not problematic. Algorithm 13 summarizes these concepts.

In structuring such an algorithm, there are a few choices for the exit conditions, including limiting the maximum pressure of the boundary (which is useful when dealing with mixtures containing volatiles species like hydrogen, when the routine has returned back below the initial point on the boundary, or when a certain number of points have been determined. There is no universal choice, and this is left to your preference and experience. Also, while quite robust, there are always components (e.g., water or highly volatile species like hydrogen) that can raise challenges in automatically computing such phase boundaries. Also, while the phase fraction f can be taken as a parameter, and it is often possible to compute phase boundaries for $0 \leq f \leq 1$ for simple systems (i.e., alkanes), the initialization on a pressure of 1 bar is not globally robust for $f \neq 1$. For example, mixtures with significant hydrogen show dew-point pressures that are significantly greater than 1 bar across a wide range of temperature.

Algorithm 13 Computing the TP phase boundary

Parameters: $\overline{\beta} = [\overline{z}, f]$, $MaxIter = 5$,$FastIter = 3$, $FastCount = 0$

Initial point: $spec$ for pressure
Estimate T for $P = P_{start}$ (e.g. 1 bar) as in Example 11. Calculate $ln k$ via Wilson (Eq. 8.198)
Estimate $\overline{\alpha}_A \approx lnT, lnP, ln\overline{k}$
Solve for $\overline{\alpha}_A$ in Eq. 6.12 with $S_A = lnP$ as in Example 12
Retain $\overline{\overline{H}}_A$ from solution

Second point: linear extrapolation with $spec$ for pressure
Increase pressure: $dS = 0.1 * S_A$
Compute $\frac{\partial \overline{\alpha}_A}{\partial S}$ via Eq. 6.31
Estimate $\overline{\alpha}_B \approx \overline{\alpha}_A + \frac{\partial \overline{\alpha}_A}{\partial S} * dS$
Solve for $\overline{\alpha}_B$ in Eq. 6.12 with $S_B = S_A + dS$
Retain $\overline{\overline{H}}_B$ and the $Iter$ count from solution

Subsequent points: automatic specification with third order extrapolation
Set $Exit$: number of points, temperature below initial temperature, pressure exceeds a particular P_{max}, etc.
while $Exit == False$ **do**
    Compute $\frac{\partial \overline{\alpha}_B}{\partial S}$ via Eq. 6.31

    Determine $spec$ as index of largest magnitude element of $\frac{\partial \overline{\alpha}_B}{\partial S}$
    if $spec$ changes **then**
        Recompute $S_A = \overline{\alpha}_A[spec]$, $S_B = \overline{\alpha}_B[spec]$
        Recompute $\overline{\overline{H}}_A$ and $\overline{\overline{H}}_B$ (see Example 14)
        Recompute $\frac{\partial \overline{\alpha}_A}{\partial S}$ and $\frac{\partial \overline{\alpha}_B}{\partial S}$ via Eq. 6.31
    end if

    Third order extrapolation
    Compute $\overline{\overline{C}}$ via Eq. 6.38
    Compute step: $dS = S_B - S_A$
    Modify stepsize based on iteration performance
    if $Iter < FastIter$ (previous point rapidly converged) **then**
        $FastCount += 1$
    else(previous point required several iterations
        $FastCount = 0$
    end if
    if $FastCount >= FastIter$ (sequential points that rapidly converge) **then**
        Speed up: $dS* = 2$
        $FastCount = 0$ (reset $FastCount$)
    end if
    Estimate $\overline{\alpha}$ via Eq. 6.34

    Solve for $\overline{\alpha}$ in Eq. 6.12 with $S = S_B + dS$
    Retain $\overline{\overline{H}}$ and the $Iter$ count from solution

    if $Iter >= MaxIter$ **then**
        Cut the stepsize and resolve until $Iter <= MaxIter$ or failure
    end if

    Cycle points: previous point becomes A, current point becomes B

    Check $Exit$ conditions
end while

6.4 Stability Analysis

A stability analysis determines whether a mixture is stable as a single phase in a particular state . Though the stability analysis plays an important role in flash calculations, it is surprisingly not well known in the field of thermodynamics. It is therefore given its own section here. As we shall see, if a mixture is found to be unstable, stability analysis also provides us with an estimate of the composition of the phase that will form, which can be used to initialize a flash calculation. The most common type of stability analysis is at specified temperature, pressure, and composition. Stability at fixed temperature and volume is also useful and is addressed in its own subsection.

6.4.1 TPn Stability Analysis

Suppose we have a multicomponent mixture of known temperature, pressure, and composition. According to Table 1.2, the potential of interest for this specification is the Gibbs free energy, which for a single phase can be written via Eq. 8.50:

$$G^I = G(\overline{n}) = \sum_r \mu_r(\overline{n}) n_r \tag{6.39}$$

Note that for simplicity of nomenclature, we have omitted writing the variables T and P, as they are constant in this context. Now suppose that a small amount of a new phase forms. Being a distinct phase, we might expect it to have different properties (i.e., composition and chemical potentials), and according to the material balance, the formation of the new phase ($\overline{W}$) is at the expense of the mole numbers of the original phase ($\overline{n}$). Therefore, the Gibbs free energy of a corresponding two-phase system can be written by $1st$ order Taylor series (see Appendix 8.19) as:

$$\begin{aligned} G^{II} &= G(\overline{n} - \overline{W}) + G(\overline{W}) \\ &= G(\overline{n}) + \sum_r \left[\frac{\partial G}{\partial n_r}\right]_{n_{k \neq r}} (-W_r) + G(\overline{W}) \\ &= \sum_r \mu_r(\overline{n}) n_r - \sum_r \mu_r(\overline{n}) W_r + \sum_r \mu_r(\overline{W}) W_r \\ &= \sum_r \mu_r(\overline{n}) n_r + \sum_r \left[\mu_r(\overline{W}) - \mu_r(\overline{n})\right] W_r \end{aligned} \tag{6.40}$$

Note that in the above Taylor series, the term $\frac{\partial G}{\partial n_r}_{n_{k \neq r}}$ is evaluated at $\overline{n}$. This results in a difference of chemical potentials between the initial phase ($\overline{n}$) and the emerging phase ($\overline{W}$). A direct comparison of the Gibbs free energies (subtraction of Eq. 6.39 from Eq. 6.40) gives:

$$G^{II} - G^{I} = \sum_r \left[\mu_r(\overline{W}) - \mu_r(\overline{n})\right] W_r \tag{6.41}$$

The quantity on the right-hand side is of primary interest to us. If we can find a phase $\overline{W}$ that makes this quantity negative, while satisfying the material balance, we can conclude that it is possible to reduce the Gibbs free energy of the single phase system by forming (at least a small amount of) a new phase. This would mean that the initial phase is unstable.

An alternate and very illustrative way of developing Eq. 6.41 is to consider the geometric interpretation of phase stability. With the temperature and pressure fixed, we can vary the Gibbs free energy by varying the mole numbers. In this sense, we can think of the Gibbs free energy (at fixed temperature and pressure) to be a surface.[19] The equation of the surface is simply the Gibbs free energy of a phase with mole numbers represented by $\overline{W}$:

$$G_{surf} = G(\overline{W}) = \sum_r \mu_r(\overline{W}) W_r \tag{6.42}$$

Here again we have simplified our notation, omitting the variables T and P, since they are constant in this context, with the independent variables defining the surface clearly stated as $\overline{W}$. In keeping with the material balance, the initial phase ($\overline{n}$) is simply a point on this surface. The equation of the tangent plane to the surface at this point is:

$$G_{tan} = G(\overline{n}) + \sum_r \left[\frac{\partial G}{\partial n_r}\right]_{n_{k \neq r}} (W_r - n_r) \tag{6.43}$$

Of course for the tangent at this point, the derivative is evaluated at $\overline{n}$. Therefore, Eq. 6.43 becomes:

$$G_{tan} = \sum_r \mu_r(\overline{n}) n_r + \sum_r \mu_r(\overline{n})[W_r - n_r] = \sum_r \mu_r(\overline{n}) W_r \tag{6.44}$$

Comparing Eqs. 6.42 and 6.44, we see that Eq. 6.41 is equivalent to the distance between the Gibbs free energy surface and the tangent defined by point $\overline{n}$. More formally, as discussed by Michelsen and Mollerup [Michelsen book], the difference

[19] One can visualize such a surface for the case of two (or possibly three) components. For higher numbers of species, this becomes an abstraction. Also, keep homogeneity (or intensive properties) in mind; variations of the mole numbers without changing the composition (mole fractions) are not of interest to us here.

in Gibbs free energy between a single and a two-phase state (given by Eq. 6.41) is given by the distance *from* the tangent plane *to* the surface:

$$tpd = \sum_{r} \left[\mu_r(\overline{W}) - \mu_r(\overline{n}) \right] W_r \quad (6.45)$$

To solidify this concept, let's look at an example.

Example 15 (TPn Stability for a Binary Mixture) Let us consider binary mixtures of propane and n-hexane at 5 bar as a function of temperature and composition. In this development, we will consider the independent mole numbers to be $\overline{W}$, with corresponding mole fractions $\overline{w}$; the mixture whose stability we are analyzing has mole numbers $\overline{n}$ and mole fractions $\overline{z}$. Rather than plotting in three dimensions (G vs. $W_{C_3H(8)}$ and $W_{nC_6H_{14}}$), we will use the independent mole fraction of propane ($w_{C_3H_8}$) and plot in two dimensions ($g = \frac{G}{W}$ vs. $w_{C_3H_8}$). To develop the correct functional form, consider Eq. 8.50 written for a binary mixture:

$$G(W_1, W_2) = \mu_1(\overline{W})W_1 + \mu_2(\overline{W})W_2$$

The issue that we encounter is that, in order to evaluate μ_i values, we have to compute reference states. If we use Eq. 1.79, then we require the ideal-gas chemical potentials. On the other hand, if we use Eq. 1.80, we need the Gibbs free energy of the pure substances. For the present context, the latter is much more intuitive, as we can relate these reference states to the concept of forming our mixture from the pure component at the temperature and pressure of interest,[20] via:

$$\mu_i - g_i = R_g T \left(lnw_i + ln\hat{\varphi}_i(\overline{w}) - ln\varphi_i \right) \quad (6.46)$$

The utility of this approach is obvious; the right side of the above equation (Eq. 1.80) is entirely calculable from an EOS; no reference state is required. This aside, we will continue to work in mole *numbers*, as it makes the following development clearer. The Gibbs energy of mixing then becomes:

$$G_{mix} = \sum_{r} \mu_r(\overline{W})W_r - \sum_{r} g_r W_r = \sum_{r} \left[\mu_r(\overline{W}) - g_r \right] W_r \quad (6.47)$$

For our binary mixture, factoring out $R_g T$, Eq. 6.47 becomes:

$$\frac{G_{mix}}{R_g T} = \left[\mu_1(\overline{W}) - g_1 \right] W_1 + \left[\mu_2(\overline{W}) - g_2 \right] W_2 \quad (6.48)$$

---

[20] We are not concerned with "heat of mixing" or changes in pressure at fixed volume; we are simply combining the required mole numbers, while maintaining the temperature and pressure.

We now must generate the corresponding tangent (G_{tan}). As shown in Appendix 8.7, gradient of Eq. 6.48 is:

$$\nabla \frac{G_{mix}}{R_g T} = \frac{\partial[(\mu_1(\overline{n}) - g_1)n_1 + (\mu_2(\overline{n}) - g_2)n_2]}{\partial n_1}, \frac{\partial[(\mu_1(\overline{n}) - g_1)n_1 + (\mu_2(\overline{n}) - g_2)n_2]}{\partial n_2}$$
$$= \frac{\partial \mu_1(\overline{n})}{\partial n_1} n_1 + \mu_1(\overline{n}) - g_1 + \frac{\partial \mu_2(\overline{n})}{\partial n_1} n_2, \frac{\partial \mu_1(\overline{n})}{\partial n_2} n_1 + \frac{\partial \mu_2(\overline{n})}{\partial n_2} n_2 + \mu_2(\overline{n}) - g_2$$
$$= [\mu_1(\overline{n}) - g_1], [\mu_2(\overline{n}) - g_2] \tag{6.49}$$

The final result in Eq. 6.49 follows from the Gibbs–Duhem equation (Eq. 8.53). Also note that, as we are developing the tangent to the surface for the phase we are assessing, the terms in Eq. 6.49 are evaluated at $\overline{n}$, not $\overline{W}$. The state $\overline{n}$ is a *point* on that surface.

For a binary system we would rather use mole fraction rather than mole numbers, especially as this will avoid irrelevant results at values of $\overline{n}$, which have equivalent composition. We can of course write Eq. 6.48 as:

$$\frac{G_{mix}}{W} = g_{mix}$$
$$= (\mu_1(\overline{W}) - g_1)\frac{W_1}{W} + (\mu_2(\overline{W}) - g_2)\frac{W_2}{W}$$
$$= (\mu_1(\overline{w}) - g_1)w_1 + (\mu_2(\overline{w}) - g_2)w_2$$
$$= R_g T\left[(lnw_1 + ln\hat{\varphi}_1(\overline{w}) - ln\varphi_1)w_1 + (lnw_2 + ln\hat{\varphi}_2(\overline{w}) - ln\varphi_2)w_2\right] \tag{6.50}$$

Here we have made use of the fact that μ_i is a $0th$ order homogeneous function of mole numbers. We can now simply choose to plot G_{mix} values along the direction $W_1 = w_1, W_2 = 1 - w_1$. As such, our "surface" actually becomes a line. What do we do with Eq. 6.49? What is the "slope" of our G_{surf} line, if Eq. 6.49 is two-dimensional? The answer is to compute the *directional derivative* (see Appendix 8.7) and compute the change in G_{mix} in the direction $\overline{W} - \overline{n}$:

$$G_{mix}^{tan} = G_{mix}(\overline{n}) + \nabla(\frac{G_{mix(\overline{n})}}{R_g T}) * (\overline{W} - \overline{n})$$
$$= \sum_r (\mu_r(\overline{n}) - g_r)n_r + \sum_r (\mu_r(\overline{n}) - g_r)(W_r - n_r)$$
$$= \sum_r (\mu_r(\overline{n}) - g_r)W_r \tag{6.51}$$

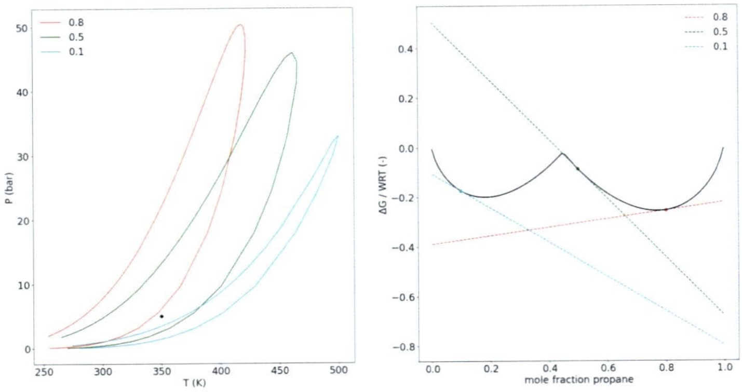

Fig. 6.4 Binary mixtures of propane and n-hexane at 350K and 5 bar computed with the SRK EOS: TPn boundary (left) and stability (right), with the function of feed propane mole fraction (values given in legend)

For a binary system, written on a molar basis, Eq. 6.51 becomes:

$$\begin{aligned} g_{mix}^{tan} &= (\mu_1(\overline{n}) - g_1)\frac{W_1}{W} + (\mu_2(\overline{n}) - g_2)\frac{W_2}{W} \\ &= (\mu_1(\overline{z}) - g_1)w_1 + (\mu_2(\overline{z}) - g_2)w_2 \\ &= R_g T\big[(lnz_1 + ln\hat{\varphi}_1(\overline{z}) - ln\varphi_1)w_1 + (lnz_2 + ln\hat{\varphi}_2(\overline{z}) - ln\varphi_2)w_2\big] \end{aligned} \quad (6.52)$$

Recall from Appendix 8.7 that Eq. 8.63 is the rate of change (i.e., per unit length) of a scalar function in a particular direction. What we want is the *change* in G_{mix} as one moves from the state $\overline{n}$ to $\overline{W}$.

The left side of Fig. 6.4 shows TP phase boundaries for three binary mixtures of propane and n-hexane. As shown, only the mixture with $\overline{z} = 0.5, 0.5$ is expected to be two-phase (i.e., unstable) at $T = 350$K and $P = 5$ bar. The right side of Fig. 6.4 shows the corresponding stability plots for the mixtures. Note that for $z_{C_3H_8} = 0.8$ and 0.1, the distances from the tangent to the Gibbs energy plot are at no point negative, while for $z_{C_3H_8} = 0.5$, negative distances are found below $z_{C_3H_8}$ less than about 0.4. These are instabilities, as they represent opportunities to reduce the Gibbs free energy of our mixture by forming an additional phase. Also note that, as we approach $z_{C_3H_8} = 0.0$ and 1.0 (i.e. pure components), the Gibbs free energy of mixing approaches zero, as it should,[21] yet the tangent plane distances remain negative; pure components are not stable relative to dilute mixtures. Finally, the kink in the $\frac{g_{mix}}{R_g T}$ plot is due to a shift in the most stable root found by the root solver. Therefore, to properly compute this diagram, the root solver must select the most

[21] We cannot compute these points directly, as $ln0$ is undefined.

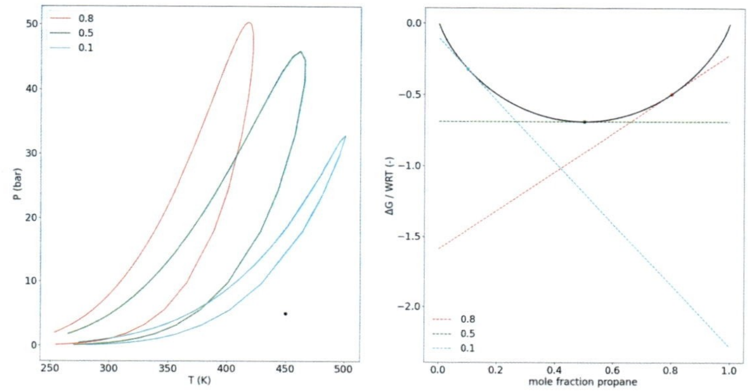

Fig. 6.5 Binary mixtures of propane and n-hexane at 450K and 5 bar computed with the SRK EOS: TPn boundary (left) and stability (right), with the function of feed propane mole fraction (values given in legend)

stable root at each point. If we repeat the calculation at T = 450K, the mixtures are everywhere stable. This is shown in Fig. 6.5.

We now wish to formalize the results in Example 15. The distance *from* the tangent *to* the Gibbs surface is called the *tangent plane distance*, and as stated, negative values indicate that our mixture (defined by T, P, and $\overline{n}$) is unstable. We of course wish to investigate stability for more than two components, cases in which we cannot reduce the problem to a single independent mole fraction. We now return to Eq. 6.45, rewriting μ_i in terms of mole fraction and $ln\hat{\varphi}_i(\overline{z})$ via Eq. 6.46:

$$tpd(\overline{w}) = R_g T \sum_r [lnw_r + ln\hat{\varphi}_r(\overline{w}) - d_r]W_r \tag{6.53}$$

In Eq. 6.53, we have replaced the (constant in this context) term;

$$d_i = lnz_i + ln\hat{\varphi}_i(\overline{z}) \tag{6.54}$$

which is calculated using the most stable root for T, P and $\overline{z}$. By factoring out the term WR_gT, we yield the *reduced* tangent plane distance $rtpd$:

$$rtpd(\overline{w}) = \frac{tpd(\overline{w})}{WR_gT} = \sum_r [lnw_r + ln\hat{\varphi}_2(\overline{w}) - d_r]w_r \tag{6.55}$$

Though not formally calculated in Example 15, the difference between Eqs. 6.50 and 6.52 gives the corresponding result for Eq. 6.55. We seek minimum (especially negative) values of rtpd, but we recognize that mole fractions are not independent of each other. We therefore apply the method of Lagrange multipliers (Appendix 8.10) to seek the minimum value of rtpd while constraining the mole fractions to sum to unity.

$$\mathcal{L}(\overline{w}, \lambda) = rtpd(\overline{w}) - \lambda(\sum_r w_r - 1)$$
$$= \sum_r \big[lnw_r + ln\hat{\varphi}_r(\overline{w}) - d_r\big]w_r - \lambda(\sum_r w_r - 1) \quad (6.56)$$

By virtue of our constraint in Eq. 6.56, we can regard $\overline{w}$ as independent mole numbers that add to unity.[22] Differentiating Eq. 6.56 with respect to the variables $\overline{w}$ and λ, we seek to zero the gradient of $\mathcal{L}(\overline{w}, \lambda)$:

$$\frac{\partial\mathcal{L}}{\partial w_i} = \sum_r \left(\frac{\partial\big[lnw_r + ln\hat{\varphi}_r(\overline{w}) - d_r\big]}{\partial w_i} w_r + \big[lnw_r + ln\hat{\varphi}_r(\overline{w}) - d_r\big]\frac{\partial w_r}{\partial w_i}\right) - \lambda\sum_r \frac{\partial w_r}{\partial w_i}$$
$$= \sum_r \delta_{ri} + \sum_r \frac{\partial ln\hat{\varphi}_r(\overline{w})}{\partial w_i} w_r + \sum_r \big[lnw_r + ln\hat{\varphi}_r(\overline{w}) - d_r\big]\delta_{ri} - \lambda\sum_r \delta_{ri}$$
$$= 1 + [lnw_i + ln\hat{\varphi}_i(\overline{w}) - d_i\big] - \lambda = 0$$
$$\frac{\partial\mathcal{L}}{\partial\lambda} = -\sum_r w_r - 1 = 0 \quad (6.57)$$

Note that in Eq. 6.57, we have used the properties of δ_{ij} (Appendix 8.4) and the Gibbs–Duhem equation (Eq. 8.53). Also, with the constraint that $\sum w_r = 1$, we have written lnw_r rather than $ln\frac{w_r}{\sum_s w_s}$ in Eq. 6.57. It is easily shown that, due to the constraint of $\sum_s w_s = 1$, the solutions (i.e., $\overline{w}$ and $rtpd$) are the same between these two approaches, but the λ values differ by 1. Substituting these results into $\mathcal{L}(\overline{w}, \lambda)$, we have as our solution to the constrained optimization:

$$\mathcal{L}^* = \sum_r (\lambda - 1)w_r - \lambda(0) = \lambda - 1 \quad (6.58)$$

This indicates that negative minima in the reduced tangent plane distance (i.e., instabilities) are characterized by $\lambda < 1$. We can solve Eq. 6.57 via Newton's method (Appendix 8.13), with our target function (for a two-component mixture) being:

$$\begin{bmatrix} 1 + lnw_1 + ln\hat{\varphi}_1(\overline{w}) - d_1 - \lambda \\ 1 + lnw_2 + ln\hat{\varphi}_2(\overline{w}) - d_2 - \lambda \\ w_1 + w_2 - 1 \end{bmatrix} = \begin{matrix} 0 \\ 0 \\ 0 \end{matrix} \quad (6.59)$$

[22] This is the advantage the method of Lagrange multipliers.

By choosing our independent variables to be $\overline{\alpha} = [lnw_1, lnw_2, \lambda]$, we can avoid $w_i \leq 0$. Our Jacobian becomes:

$$\begin{bmatrix} 1 + \frac{\partial ln\hat{\varphi}_1}{\partial w_1} w_1 & \frac{\partial ln\hat{\varphi}_1}{\partial w_2} w_2 & -1 \\ \frac{\partial ln\hat{\varphi}_2}{\partial w_1} w_1 & 1 + \frac{\partial ln\hat{\varphi}_2}{\partial w_2} w_2 & -1 \\ w_1 & w_2 & 0 \end{bmatrix} \tag{6.60}$$

We demonstrate the solution in a clarifying example.

Example 16 (Calculation of the Reduced Tangent Plane Distance for a Mixture) Let us revisit Example 15 and calculate $rtpd$ for binary mixtures of propane and n-hexane at 5 bar and at two different temperatures.

At $T = 350$ K, $P = 5$ bar and $\overline{z} = 0.5, 0.5$, $\overline{d} = -0.728, -0.874$ (see Eq. 6.53). Initialization of a liquid-like phase via Wilson k values (Appendix 8.20) gives:

$$lnw_i = lnz_i - lnk_i$$

$$ln\overline{w} = -2.483, 0.653$$

There is no need to normalize $\overline{w}$, as this is handled by our constraint. Based on Fig. 6.4, we know that an instability is expected (i.e., $\lambda < 1$, so we guess that $\lambda = 0.9$. Table 6.4 summarizes the iterations to solve Eq. 6.59.

The composition corresponding to our solution in Table 6.4 is $\overline{w} = 0.063, 0.937$ in agreement with Fig. 6.4, and $\lambda = 0.430$. Referring to Eq. 6.55, we can calculate the reduced tangent plane distance to be:

$$rtpd(\overline{w}) = \sum_r [lnw_r + ln\hat{\varphi}_r(\overline{w}) - d_r]w_r = -0.570$$

It is interesting to note that Fig. 6.4 indicates that, starting from a feed of $\overline{z} = 0.5, 0.5$, an initial *increase* in Gibbs free energy is required before the lower energy state can be reached. This is a concept we shall explore further in Sects. 6.5 and 6.6.

These results, along with those of other initializations, are summarized in Table 6.5.

In keeping with Fig. 6.4, a liquid-like initialization (corresponding to *lower* concentrations of the more volatile propane) for $\overline{z} = 0.8, 0.2$ yields a non-zero tangent plane distance of $rtpd = 0.157$ at $\overline{w} = 0.210, 0.790$. Vapor-like initializations (*higher* propane mole fractions) of course converge back to $\overline{z}$. The other results in Table 6.5 follow suit. Note that in all cases $rtpd = \lambda - 1$, as in Eq. 6.58. It is also worth noting that some of the cases in Table 6.5 that are "stable" actually produce very small negative values of $rtpd$. This presents a challenge for automated routines seeking $rtpd < 0$ and will be addressed subsequently. All cases in Fig. 6.5 yield $\overline{w} = \overline{z}$, as expected. The $rtpd$ values corresponding to Figs. 6.4 and 6.5 are represented in Fig. 6.6.

Table 6.4 Constrained stability analysis for equimolar propane/n-hexane at 350K and 5 bar via SRK EOS: liquid-like initialization

Iter	$\overline{\alpha}$ $ln\overline{w}, \lambda$	$\overline{F}$ Eq. 6.59	$\overline{\overline{H}}$ Eq. 6.60	$d\overline{\alpha}$
1	−2.483, 0.653, 0.9	−0.194, 0.248, 1.004	0.997, 0.032, −1 1.384E−4, 1.000, −1 8.351E−2, 1.920, 0	−0.076, −0.519, −0.271
2	−2.559, 0.133, 0.629	−3.423E−4, 3.607E−5, 0.220	0.995, 4.83E−3, −1 3.275E−4, 1.000, −1 7.740E−2, 1.142, 0	−0.180, −0.180, −0.180
…	…	…	…	…
5	−2.757, −0.066, 0.430	7.688E−14, 7.555E−14, 1.442E−8	0.995, 4.835E−3, −1 3.28E−4, 1.000, −1 0.063, 0.937, 0	−1.442E−8, −1.442E−8, −1.442E−08

Table 6.5 Summary of solutions to Eq. 6.59 for binary mixtures of propane / n-hexane at 350K and 5 bar via SRK EOS

$\overline{z}$	Initialization	$\overline{w}$	λ	$rtpd$
0.8,0.2	Liquid	0.210, 0.790	1.157	0.157
0.8,0.2	Vapor	0.800, 0.200	0.999..	−9E−17
0.5,0.5	Liquid	0.063, 0.937	0.430	−0.570
0.5,0.5	Vapor	0.500, 0.500	0.999..	−5E−16
0.1,0.9	Liquid	0.100, 0.900	0.999..	−9E−16
0.1,0.9	Vapor	0.624, 0.376	1.333	0.333

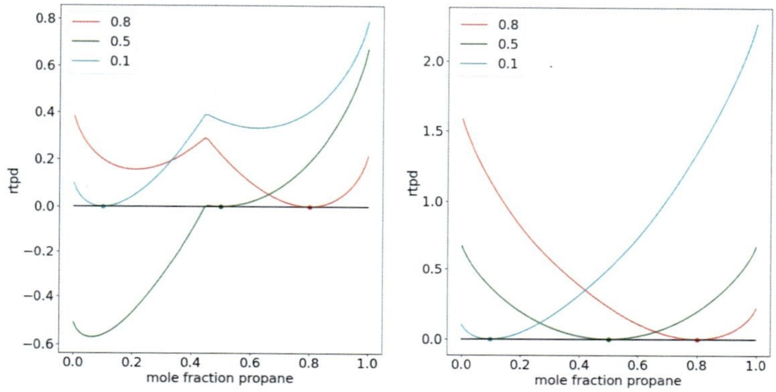

Fig. 6.6 Values of rtpd for mixture of 1 mole of propane and 1 mole of n-hexane at 350K (left) and 450K (right) at 5 bar computed with the SRK EOS. Values in the legend give the mole fraction of propane in the feed

As shown, our solution for $\overline{z} = 0.5, 0.5$ at 350K and 5 bar yields a new phase with propane mole fraction = 0.063 and a *negative* minimum $rtpd$.[23] For all of the other cases, the minimum $rtpd$ values are either positive or zero, consistent with the results given above.

Michelsen and Mollerup [Michelsen book] demonstrate that we can avoid the need to solve for the Lagrange multiplier by converting the constrained optimization to one that is unconstrained. Consider Eq. 6.57, which implies:

$$ln w_i + ln\hat{\varphi}_r(\overline{w}) - d_i = \lambda - 1 \tag{6.61}$$

It would be much more convenient to solve:

$$ln W_i + ln\hat{\varphi}_i(\overline{W}) - d_i = 0 \tag{6.62}$$

[23] Note also that the increase in free energy discussed in Fig. 6.4 is seen in the reduced tangent plane distance plotted in Fig. 6.6.

where W_i are mole *numbers*. To develop this concept, we formally consider Eq. 6.62 to be the gradient (see Appendix 8.7) of a function M:

$$\nabla M_i = lnW_i + ln\hat{\varphi}_i(\overline{W}) - d_i \tag{6.63}$$

Integrating Eq. 6.63 by parts (see Appendix 8.9), we have:

$$M = W_i lnW_i - W_i + \int ln\hat{\varphi}_i dW_i - d_i W_i + N(W_{k\neq i}) \tag{6.64}$$

If we now consider that Eqs. 6.63 and 6.64 must also apply to species other than i, we conclude that:

$$M = \sum_r \big(lnW_r - 1 - d_r\big)W_r + \sum_r \int ln\hat{\varphi}_r dW_r + K \tag{6.65}$$

It is difficult to write a closed form for $\int ln\hat{\varphi}_i dn_i$, particularly at fixed pressure, but we can use integration by parts again to yield:

$$\begin{aligned}\sum_r \int ln\hat{\varphi}_r dW_r &= \sum_r [ln\hat{\varphi}_r W_r - \int W_r dln\hat{\varphi}_r] \\ &= \sum_r [ln\hat{\varphi}_r W_r - \int W_r \sum_k [\frac{\partial ln\hat{\varphi}_r}{\partial W_k}]dW_k] \\ &= \sum_r [ln\hat{\varphi}_r W_r - \int W_r \sum_k [\frac{\partial ln\hat{\varphi}_k}{\partial W_r}]dW_k] \\ &= \sum_r ln\hat{\varphi}_r W_r \end{aligned} \tag{6.66}$$

In the above development, we have also made use of symmetry and the Gibbs–Duhem equation (refer to the development of Eq. 8.56). The resulting form of Eq. 6.65 is:

$$M = \sum_r \big(lnW_r + ln\hat{\varphi}_r - d_r - 1\big)W_r + K \tag{6.67}$$

Now we consider solutions to Eq. 6.62, which yield by Eq. 6.67:

$$M^{Unconstrained} = -\sum_r W_r + K = -W_T + K \tag{6.68}$$

The corresponding solution to the *constrained* problem (Eq. 6.61) yields by Eq. 6.67

$$\begin{aligned} M^{Constrained} &= \sum_r \left(lnw_r + \ln W_T + ln\hat{\varphi}_r - d_r - 1\right)W_r + K \\ &= \sum_r \left(\lambda - 1 + \ln W_T - 1\right)W_r + K \\ &= \left(\lambda - 1 + \ln W_T\right)W_T - W_T + K \end{aligned} \tag{6.69}$$

Consistency between the constrained and unconstrained problems requires:

$$\lambda - 1 + \ln W_T = 0 \tag{6.70}$$

Instabilities in the constrained problem are characterized by $\lambda < 1$ (see Eq. 6.58), meaning $W_T > 1$. Stability ($\lambda \leq 1$) is therefore characterized by $W_T \leq 1$. We can make the values of M consistent with these cases (i.e., we can have $M < 0$ for instabilities, otherwise $M \geq 0$), by choosing $K = 1$ in Eq. 6.68. The final form for M is what Michelsen and Mollerup [Michelsen book] call the *modified* tangent plane distance (t_m);

$$t_m = 1 + \sum_r \left(lnW_r + ln\hat{\varphi}_r - d_r - 1\right)W_r \tag{6.71}$$

with *stationary* values (solutions to Eq. 6.62):

$$t_m^* = 1 - W_T \tag{6.72}$$

We now formalize TPn stability analysis as follows. The target function for our algorithm is Eq. 6.63, which for notational simplicity we will simply write as:

$$F_i = lnW_i + ln\hat{\varphi}_i(\overline{W}) - d_i = 0 \tag{6.73}$$

We can initialize via Wilson k values, generating either a vapor;

$$lnW_i = lnk_{Wi} + lnz_i \tag{6.74}$$

or a liquid;

$$lnW_i = lnz_i - lnk_{Wi} \tag{6.75}$$

These initializations work well for VLE cases that are not too far from ideal-gas/ideal-solution behavior. However, they are not reliable for LLE or other highly non-ideal conditions. Therefore, initialization with phases that are pure (or nearly pure) in one component should also be considered. A simple method to solve Eq. 6.73 for the independent variables $ln\overline{W}$ is Successive Substitution (Appendix 8.18), whereby the values of $ln\overline{W}$ are updated via:

$$ln\overline{W}^{t+1} = \overline{d} - ln\hat{\varphi}_r(\overline{W}^t) \tag{6.76}$$

Note that one must select the root to correspond with the initialization; vapor for Eq. 6.74, liquid for Eq. 6.75 and the most stable for nearly pure initializations. Algorithm 14 outlines the main ideas, while Example 17 provides numerical results.

Algorithm 14 Solving the modified tangent plane distance using successive substitution

```
Calculate d (Eq. 6.54)
Set Iter = 1,,MaxIter = 100, Tol = 1E − 8, Flag = True
while (Flag == True) do

    W̄ = exp(lnW̄) (Extract W̄)
    Calculate lnφ̂(W̄) for the desired root

    Iter+ = 1
    Err = max(abs(lnW̄ + lnφ̂(W̄) − d̄))

    Update lnW̄ = d̄ − lnφ̂(W̄) (Successive substitution)

    if (Iter ≥ MaxIter) or (Err < Tol) then
        Flag = False
    end if
end while

Check convergence
if (Iter ≥ MaxIter) then
    Return Error Message
else
    Return W̄
end if
```

Example 17 (Calculation of t_m for a Mixture) We once again revisit Example 15, considering binary mixtures of propane and n-hexane at 5 bar, using the modified tangent plane formalism.

At $T = 350$ K, $P = 5$ bar and $\overline{z} = 0.5, 0.5$, we have $\overline{d} = -0.728, -0.874$ (see Eq. 6.54). With liquid-like initialization via Wilson k values (Eq. 6.75) we have $ln\overline{w} = -2.483, 0.653$. Table 6.6 summarizes the iterations to solve Eq. 6.73.

Table 6.6 Unconstrained (t_m) stability analysis for 1 mole of propane and 1 mole of n-hexane at 350K and 5 bar via SRK EOS: liquid-like initialization

Iter	$ln\overline{W}$	$\overline{W}$	$ln\hat{\varphi}(\overline{W})$	Err
1	−2.4828, 0.6526	0.084, 1.9204	1.4613, −1.3784	0.1482
2	−2.1890, 0.5043	0.1120, 1.6558	1.4596, −1.3783	9.6955E-5
3	−2.1873, 0.5042	0.1122, 1.6556	1.4596, −1.3783	6.0299E-7
4	−2.1873, 0.5042	0.1122, 1.6556	1.4596, −1.3783	3.1158E-9

For these conditions, we find $t_m = 1 - W_T = -0.768$ and a normalized composition of $\overline{w} = 0.063, 0.937$.[24] For vapor-like initialization at 350K and 5 bar we converge to $t_m = 1 - \sum_r W_r = -4E - 16$. We require a method for deciding whether such t_m values represent very slight instabilities or are zero within our solver tolerance. Taking the latter view, having found $ln W_i$ that satisfies Eq. 6.73 within $\pm Tol$, we have:

$$ln W_i^{high} = ln W_i + Tol$$
$$ln W_i^{low} = ln W_i - Tol \tag{6.77}$$

We can exclude tm values for which:

$$1 - \sum_r W_r^{low} \geq 0$$

We accordingly find the vapor initialization to be stable, as we do for all cases when calculating at 450K (consistent with Example 16).

Pure component initializations can be handled by setting the mole number of a particular component to unity, with all others set to very low numbers.[25] For example, we can search for a propane-rich phase by initializing $ln\overline{W} = ln(1.0, 1E - 14) = 0, -32.236$, which converges to $t_m = 0$ ($\overline{w} = \overline{z}$). We can also seek a phase rich in n-hexane using $ln\overline{W} = ln(10^{-14}, 1.0) = -32.236, 0$, which converges to $tm = -0.768$ and $\overline{w} = 0.063, 0.97$. A general strategy is to conduct $2 + C$ searches, where C is the number of components; liquid-like, vapor-like and a pure initialization for each species. The minimum t_m value is selected, and its accompanying $\overline{W}$ is returned. Conducting multiple searches within the composition (W) space is conservative. Though in the present example pure component initialization yields no new information compared with Wilson k, this is not always the case. For example, the addition of water to the mixture increases the likelihood of forming a water-rich (aqueous) phase in addition to a propane-rich vapor and a liquid rich in n-hexane. If we repeat the above stability analysis for

[24] Compare with Example 16, which indicated instability with $\overline{w} = 0.063, 0.937$ and $\lambda = 0.430$. By Eq. 6.70, $ln W_T = 1 - \lambda = 0.570$, which is consistent with $tm = 1 - W_T = -0.768$.

[25] The use of $ln\overline{W}$ does not permit zero mole numbers.

Table 6.7 Unconstrained (t_m) stability analysis for 1ppm propane and 1ppm n-hexane in water (by moles) at 350K and 5 bar via SRK EOS: various initializations

Initialization	$\overline{W}^{init}$	t_m	$\overline{W}$
Liquid	1.670E−7, 3.841E−6, 10.427	0.000	$\overline{z}$
Vapor	5.987E−6, 2.604E−7, 0.096	−121.325	0.204, 122.050, 0.071
Propane	1.0, 1E−10, 1E−10	−396.526	0.048, 397.420, 0.058
n-Hexane	1E−10, 1.0, 1E−10	−396.526	0.048, 397.420, 0.058
Water	1E−10, 1E−10, 1.0	0.000	$\overline{z}$

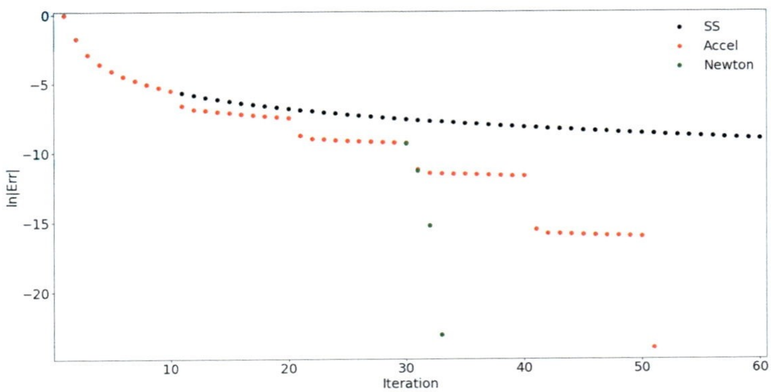

Fig. 6.7 Comparison of solvers in converging Eq. 6.73 for one mole of propane and one mole of n-hexane in the vicinity of the critical point: $T = 463.55$K, $P = 45.60$bar, $T_c = 464.018$K, $P_c = 45.65$bar, with vapor-like initialization. All computations with the SRK EOS

mixtures with traces of propane and n-hexane in water ($z_{propane} = z_{n-hexane} = 1\,ppm$ molar, balance H_2O), we find the results shown in Table 6.7.

The results in Table 6.7 are interesting. Wilson initialization suggests the formation of a vapor (but not a liquid), and fails to predict the larger instability yielded by pure propane or n-hexane initialization. We interpret the detected instability to be a phase with composition $\overline{w} = 1.211E-4, 0.999.., 1.469E-4$. If you are familiar with the behavior of hydrocarbons and water at such conditions, you might expect both an organic and an aqueous liquid. This will be discussed in Sect. 6.7.1.

A final point to consider is that while Successive Substitution is simple and often robust, it is sometime necessary to augment it with more advanced solution techniques. As stated in Michelsen and Mollerup [Michelsen book], and elaborated on in Appendix 8.18.1, a large number of iterations is required to converge a stability analysis near the critical point of the mixture. This is shown in Fig. 6.7.

For the conditions shown, which is within 0.999 of the calculated critical point (see Sect. 6.6), approximately 400 iterations are required to converge (a tolerance of 10^{-8} is approximately -18.4 on the log scale shown) using successive substitution alone. The utility of accelerated successive substitution (See Example 49 in

Appendix 8.18.1) in converging such equations is clear. Also indicated is that Newton's method is superior in terms of convergence rate. However, this comes at a higher computational cost, and as discussed in Appendix 8.13, convergence is not guaranteed. Therefore, we consider combinations of Successive Substitution, Acceleration, and Newton's method. For example, the successive substitution (SS) plot in Fig. 6.7 does not include acceleration or Newton's method. When acceleration (Accel) is used, the dominant eigenvalue method is attempted every tenth iteration of successive substitution. Finally, where Newton's method (Newton) is used, successive substitution is conducted with acceleration attempted every tenth iteration. Once three consecutive accelerations are achieved, Newton's method is applied. A simple treatment of Newton's method considers the target function to be F in Eq. 6.73, with the Jacobian being:

$$H_{ij} = \frac{\partial F_i}{\partial ln W_j} = \delta_{ij} + \left[\frac{\partial ln \hat{\varphi}_i}{\partial W_j}\right]_{T,P,W_{k \neq j}} W_j \tag{6.78}$$

Michelsen and Mollerup [Michelsen book] provide further information on the choice of independence variables with Newton's method.

6.4.2 TVn Stability Analysis

Another useful form of stability analysis is for fixed temperature, volume, and composition; TVn. According to Table 1.2, the Helmholtz free energy is the potential of interest. For a single phase mixture, the Helmholtz free energy is given by Eq. 8.49:

$$A^I = A(T, V, \overline{n}) = -PV + \sum\nolimits_r \mu_r n_r$$

Should a second phase form, it does so at the expense of the first.

$$A^{II} = A(T, V - V^W, \overline{n} - \overline{W}) + A(T, V^W, \overline{W})$$

Via a Taylor series (Sect. 8.19) and Table 1.1 we can write:

$$\begin{aligned} A^{II} &= A(T, V, \overline{n}) + \left(\frac{\partial A}{\partial V}\right)_{T,\overline{n}} (-V^W) + \sum_r \left(\frac{\partial A}{\partial n_r}\right)_{T,V,n_{k\neq r}} (-W_r) + A(T, V^W, \overline{W}) \\ &= A(T, V, \overline{n}) + PV^W - \sum_r \mu_r W_r + -P^W V^W + \sum_r \mu_r^W W_r \\ &= A(T, V, \overline{n}) - (P^W - P)V^W + \sum_r (\mu_r^W - \mu_r) W_r \end{aligned} \tag{6.79}$$

Note that in Eq. 6.79 we have used compact notation. For example, $P^W = P(T, V^W, \overline{W})$. The difference in Helmholtz free energy between the two and the one-phase cases is:

$$A^{II} - A^{I} = -(P^W - P)V^W + \sum_r (\mu_r^W - \mu_r)W_r \tag{6.80}$$

Equation 6.80 is $1st$ order homogeneous[26] (Sect. 8.6) in V^W and $\overline{W}$, so the absolute size of the W phase does not determine the sign of $A^{II} - A^{I}$. We can therefore select a volume for convenience and seek mole numbers that give negative values of the following equation:

$$\Delta A = -(P^W - P)V + \sum_r (\mu_r^W - \mu_r)W_r \tag{6.81}$$

To be clear, we have selected a volume that is numerically equal to that of the feed, and we are looking for mole numbers filling this volume that give negative values of Eq. 6.81. Unlike TPn stability (Sect. 6.4.1), the mole numbers in the present development are independent. We can therefore seek minima of Eq. 6.81 via:

$$\begin{aligned}\left(\frac{\partial \Delta A}{\partial W_i}\right)_{T,V,W_{q\neq i}} &= -\left(\frac{\partial P^W}{\partial W_i}\right)V + \sum_r \left(\frac{\partial \mu_r^W}{\partial W_i}\right)W_r + \sum_r (\mu_r^W - \mu_r)\left(\frac{\partial W_r}{\partial W_i}\right)\\ &= -\left(\frac{\partial P^W}{\partial W_i}\right)V + \sum_r \left(\frac{\partial \mu_r^W}{\partial W_i}\right)W_r + (\mu_i^W - \mu_i)\\ &\to 0\end{aligned} \tag{6.82}$$

Through application of the Gibbs–Duhem equation for fixed TV (Eq. 8.57) Eq. 6.82 becomes:

$$\left(\frac{\partial \Delta A}{\partial W_i}\right)_{T,V,W_{q\neq i}} = \mu_i^W - \mu_i \to 0 \tag{6.83}$$

yielding the extreme value of ΔA:

$$\Delta A^* = -(P^W - P)V \tag{6.84}$$

As stated by Mikyška and Firoozabadi [Firoozabadi paper], unequal pressures imply that the W phase is in chemical but not mechanical equilibrium with the feed. If the pressure of this phase (P^W) is greater than the feed pressure (P), we

[26] P and μ_r are constants in Eq. 6.80.

have found an instability, and the mixture will form an additional phase. It might seem counterintuitive that a *higher* pressure means that another phase will form. This can be understood in a few ways. First, a *higher* pressure suggests a *lower* Helmholtz free energy in Eq. 8.49. Also, two phases at fixed temperature with equal chemical potentials and confined within a fixed volume, but unequal pressures, will exchange energy through boundary work. The higher pressure of the W phase will compress the feed, resulting in changes in chemical potential. Looking at Eq. 8.57 in differential form, we have:

$$\sum_r [d\mu_r]_{T,V,n_{j\neq i}} n_r = V\Big[dP\Big]_{T,V,n_{i\neq j}} \tag{6.85}$$

Though some chemical potentials may increase while others decrease (i.e., the feed phase may become depleted in high-volatility species and enriched in low-volatility species), the sum of the chemical potential changes in the feed phase due to compression ($dP > 0$) will be positive. This leads to material transfer and phase separation.

Let us now formalize an approach to handling TVn stability calculations, starting with our target function (Eq. 6.83). To avoid the embedded reference states in chemical potentials, we rewrite Eq. 1.69 for the feed:

$$\begin{aligned}
\mu_i &= \Big[g_i^{IG}(T,P^o) - R_gTln(P^o)\Big] + R_gTlnz_i + R_gTlnP \\
&\qquad + \left(\frac{\partial A^{Res}}{\partial n_i}\right)_{T,V,n_{k\neq i}} - R_gTlnZ \\
&= \Big[g_i^{IG}(T,P^o) - R_gTln(P^o)\Big] + R_gTln(\frac{n_i}{n}) + R_gTln\frac{ZnR_gT}{V} \\
&\qquad + \left(\frac{\partial A^{Res}}{\partial n_i}\right)_{T,V,n_{k\neq i}} - R_gTlnZ \\
&= \left[g_i^{IG}(T,P^o) - R_gTln(P^o) + R_gTln\frac{R_gT}{V}\right] + R_gTln(n_i) \\
&\qquad + \left(\frac{\partial A^{Res}}{\partial n_i}\right)_{T,V,n_{k\neq i}}
\end{aligned} \tag{6.86}$$

The advantage of this approach is that we have eliminated the lnP terms, which cause numerical issues for early iterations where $P <= 0$, a problem entirely avoided in TPn stability. At fixed T and V, the above equation is written for the W phase as:

$$\mu_i^W = \left[g_i^{IG}(T,P^o) - R_gTln(P^o) + R_gTln\frac{R_gT}{V}\right] + R_gTln(W_i) + \left(\frac{\partial A^{Res}}{\partial W_i}\right)_{T,V,W_{k\neq i}} \tag{6.87}$$

We reemphasize here that the volumes of the feed and the W phases are identical. The difference in chemical potentials called for in Eq. 6.83 gives:

$$\mu_i^W - \mu_i = R_g T ln(W_i) + \left(\frac{\partial A^{Res}}{\partial W_i}\right)_{T,V,W_{k\neq i}} - R_g T ln(n_i) - \left(\frac{\partial A^{Res}}{\partial n_i}\right)_{T,V,n_{k\neq i}} \tag{6.88}$$

We thus make our target function;

$$F_i = \frac{\mu_i^W - \mu_i}{R_g T} = ln(W_i) + \frac{1}{R_g T}\left(\frac{\partial A^{Res}}{\partial W_i}\right)_{T,V,W_{k\neq i}} - b_i \tag{6.89}$$

with:

$$b_i = ln(n_i) + \frac{1}{R_g T}\left(\frac{\partial A^{Res}}{\partial n_i}\right)_{T,V,n_{k\neq i}} \tag{6.90}$$

taking the place of d_i in TPn stability calculations (Eq. 6.54). If we choose our variables to be $ln\overline{W}$, the derivative of our target function is:

$$\begin{aligned} H_{ij} &= \left(\frac{\partial F_i}{\partial ln W_j}\right)_{T,v,ln W_{k\neq j}} \\ &= \delta ij + \frac{1}{R_g T}\left(\frac{\partial^2 A^{Res}}{\partial W_i \partial W_j}\right) W_j \end{aligned} \tag{6.91}$$

These equations allow us to use Newton's method to solve $\overline{F} = \overline{0}$, but we still need an initialization strategy. Stability at fixed $T, V, \overline{n}$ differs from the constant pressure case (Sect. 6.4.1) in an important way. If we consider a feed mixture at stated $T, P, \overline{n}$, the given pressure is the correct pressure of the mixture, regardless of how many phases actually exist at equilibrium. In contrast, if a multiphase feed exists at a stated $T, V, \overline{n}$, it is unlikely that any of the phases have a molar volume of $\frac{V}{n}$, and it is even possible that initially assuming the feed to be single phase can result in a *negative* pressure. This can be seen in Fig. 4.2; there are clearly volumes where $P <= 0$. Negative pressures are simply a mathematically consequence of the EOS, and the mixture will of course have a pressure greater than zero at equilibrium, regardless of the stability or number of phases in the mixture. Yet the possibility of such an initial result prevents us from using the given volume to determine the pressure, which then prevents us from initializing the problem with Wilson approximation k values (Appendix 8.20) to estimate the composition of the W phase.

We therefore require a reliable method to estimate the pressure when the given T, V, and $\overline{n}$ yields a negative number. Some authors [Firoozabadi paper]use pure component vapor pressures for such an estimate. A possible limitation to this approach is when calculations are performed at temperatures above the critical temperatures of several components. A different strategy makes use of the Wilson approximation 8.20 to estimate dew- and bubble-point pressures (see Appendix 8.16.1), initializing the pressure between these values if the pressure of the feed is found to be negative. A possible limitation to this (otherwise reasonable) approach is that it may not predict LLE. It is therefore wise to couple this approach with the use of nearly pure phases (see Sect. 6.4.1).

For each composition estimate, we use our root solver to determine the *molar* volume of our *normalized* composition. We apply this molar volume to determine how many moles of our composition we require to fill our target volume (V). As shown in Sect. 8.18, successive substitution is not robust for such problems, and so we apply Newton's method to converge Eq. 6.89 to zero. Stepsize control is very important in TVn stability, because each EOS has a minimum volume (see Chap. 4). At each Newton iteration, we must ensure the update to $ln\overline{W}$ does not yield a $Vmin$ (calculated from the EOS) that is greater than our volume V. We would also be concerned about avoiding negative mole numbers in our Newton iterations, but the use of $ln\overline{W}$ as independent variables protects against this. Finally, as with any use of Newton's method, we have to guard against large stepsizes that overstep the solution.

Once converged we check the sign of $(P^W - P)$. Negative values mean the feed is unstable as a single phase. Based on our choice of target function, we can interpret the tangent plane distance at fixed volume as;

$$t_V = -(P^W - P)V/(R_g T) \tag{6.92}$$

Example 18 illustrates the main points of the t_V analysis algorithm we have developed here.

Example 18 (Calculation of t_V for a Mixture) Example 17 demonstrated that a mixture of one mole of propane and one mole of n-hexane at 350K and 5 bar is unstable as a single phase. In this example, we will demonstrate the calculation of TVn stability, for which we need a volume. For the purposes of illustration and consistency, we will conduct this example at the volume that this two-phase mixture would occupy at equilibrium, which at this point in our development is unknown to us. The reader is referred ahead to Example 24, which indicates that the volume of such a mixture (according to the SRK EOS) would be $V = 6223.641\text{cm}^3$.

The $\overline{b}$ values (Eq. 6.90) are:

$$\overline{b} = -0.254, -0.494$$

At our volume the pressure is calculated to be 7.620 bar, from which we estimate k values via the Wilson Equation (Eq. 8.198):

$$ln\overline{k} = 1.368, -1.767$$

If we search for a vapor-like phase (interpreting the feed as liquid-like via $ln\overline{W} = ln\overline{k} + ln\overline{z}$), the normalized composition is:

$$w = 0.958, 0.0417$$

If we solve for the vapor-like molar volume of the W phase (at the estimated pressure P^W and $\overline{w}$), we find:

$$v^W = 3501.672\text{cm}^3/mol$$

Then the mole numbers of the W phase required to fill the volume v are:

$$\overline{W} \rightarrow \frac{V}{v^W} * \overline{w} = 1.703, 7.407E - 2\text{moles}$$

With our independent variables being $ln\overline{W}$, Eq. 6.89 is converged with the iterations shown in Table 6.8. As indicated the iterations converge to $W = 1.0, 1.0$, from which we calculate:

$$P^W = 7.620\text{bar}$$

As this pressure corresponds to the pressure we had calculated for the feed, we conclude that no instability was detected. As shown in Fig. 6.6, this is not surprising. At 350K and 5 bar a liquid-like (n-hexane rich) phase is expected to form. If we initialize accordingly (interpreting the feed as vapor-like via $ln\overline{W} = ln\overline{z} - ln\overline{k}$), the normalized composition is:

$$w = 4.167E - 2, 0.958$$

Table 6.8 Iterations for vapor-like TV stability analysis for one mole of propane and one mole of n-hexane at $T = 350$K and $V = 6223.641\text{cm}^3$ using the SRK EOS

Iter	$ln\overline{W}$	$\overline{F}$	$\overline{\overline{H}}$	$dln\overline{W}$	V_{min}^{New} (cm^3)
1	0.533, −2.603	0.625, −2.421	0.854, −0.012 −0.283, 0.976	−0.699, 2.278	142.275
2	−0.167, −0.324	−0.108, −0.211	0.928, −0.119 −0.140, 0.767	0.156, 0.303	182.673
3	−0.011, −0.022	−6.636E−3, −0.013	0.917, −0.159 −0.161, 0.689	0.011, 0.021	185.959
4	−6.596E−5, −1.289E−4	−3.946E−5, −7.726E−5	0.916, −0.163 −0.163, 0.682	6.595E−5, 1.289E−4	185.979
5	−2.423E−9, −4.738E−9	−1.450E−9, −2.839E−9	0.916, −0.163 −0.163, 0.682	2.423E−9, 4.738E−9	185.979

Table 6.9 Iterations for liquid-like TV stability analysis for one mole of propane and one mole of n-hexane at $T = 350$K and $V = 6223.641\text{cm}^3$ using the SRK EOS

Iter	$ln\overline{W}$	$\overline{F}$	$\overline{\overline{H}}$	$dln\overline{W}$	$v^{New}(\text{cm}^3)$
1	0.499, 3.635	−1.386, −0.845	1.165, 6.084 0.265, 10.315	0.879, 0.059	5184.307
2	1.379, 3.694	0.411, 0.716	1.847, 14.671 1.448, 25.605	−6.967E−4, −0.028	5048.297
3	1.378, 3.666	0.055, 0.098	1.662, 10.970 1.113, 18.992	1.811E−3, −5.283E−3	5023.473
4	1.3780, 3.661	1.448E−3, 2.571E−3	1.635, 10.418 1.064, 18.013	3.871E−5, −1.450E−4	5022.791
5	1.3780, 3.661	1.052E−6, 1.867E−6	1.634, 10.404 1.063, 17.987	2.708E−8, −1.054E−7	5022.790

If we solve for the liquid-like molar volume of the W phase at 7.620bar, we find:

$$v^W = 157.395\text{cm}^3/mol$$

Then the mole numbers of the W phase required to fill the volume v is:

$$\overline{W} \rightarrow \frac{V}{v^W} * \overline{W} = 1.648, 37.894\text{moles}$$

Table 6.9 summarizes the iterations to converge Eq. 6.89. In contrast to the vapor-like initialization, the liquid-like initialization converges to a composition different from the feed composition: $W = 3.975, 38.894$ ($w = 9.272E - 2, 0.907$) from which we calculate:

$$P^W = 177.300\text{bar}$$

and via Eq. 6.92

$$t_V = -36.289\text{moles}$$

Recalling that the mole numbers are independent from each other, we can plot the Helmholtz free energy vs. the propane and n-hexane mole numbers to visualize the results in Tables 6.8 and 6.9. Returning to Eq. 6.81, we see that Eq. 6.88 has already provided a useful form for the chemical potentials relative to those of the feed. Now we turn our attention to the $-(P^W - P)V$ term. As stated above, we have found a set of mole *numbers* that minimize ΔA in Eq. 6.81 for a volume V. That is, the feed and the W phase have the same volume (but not necessarily molar volume):

$$(P^W - P)V = R_g T(Z^W W - Zn)$$

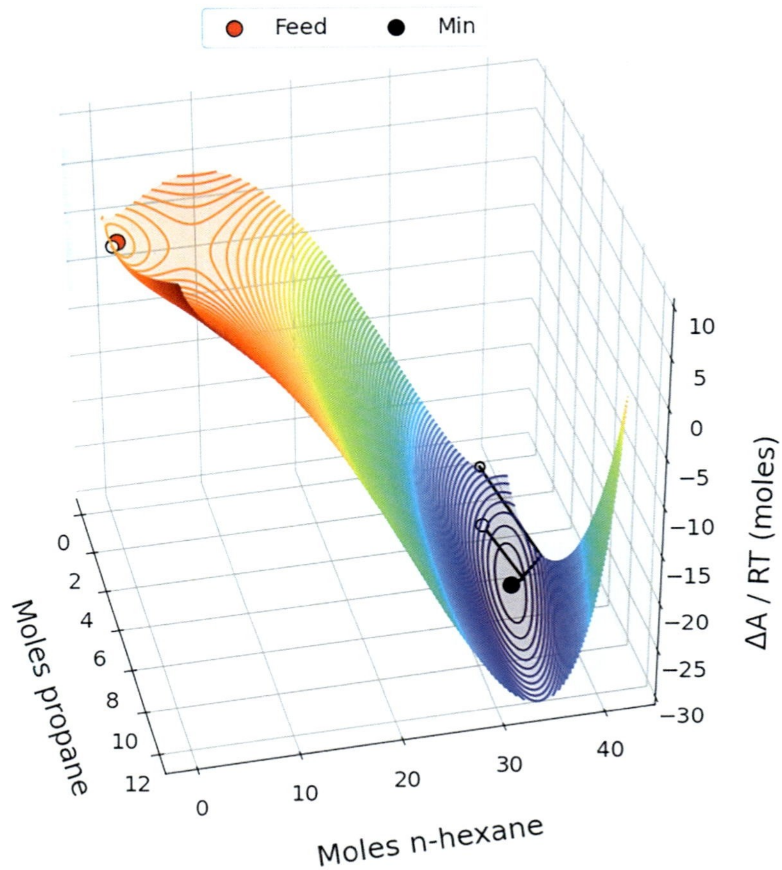

Fig. 6.8 Plot of Eq. 6.81 for a feed of one mole of propane and one mole of n-hexane at 350K and 6223.64cm^3 using the SRK EOS. The vapor-like and nearly pure propane initializations are shown near the Feed and converge to that state. The liquid-like and nearly pure n-hexane initializations are shown at the right and converge to the minimum value of $\frac{\Delta A}{R_g T}$, shown as Min

A plot of Eq. 6.81 is shown in Fig. 6.8, indicating that the feed is unstable as a single phase, yielding a liquid-like (enriched in n-hexane vs. propane) phase.

Note that Fig. 6.8 shows a slight increase in free energy required to pass from the feed to the new phase. It is further interesting to consider the case where more than one local minimum can be detected. Such a case can be found near the critical point of our mixture ($T_c \approx 464$K and $P_c \approx 46$ bar, as in Example 23). Referring to Fig. 6.4 we might select $T = 453$ and $P = 39$bar, where a two-phase flash will indicate a volume of 783.331cm^3. A vapor-like search yields $\overline{W} = 0.838, 0.678$, $P^W = 38.476$bar and $t_V = -2.032E - 3$ moles, while a liquid-like search yields $\overline{W} = 1.342, 2.213$, $P^W = 40.313$bar and $t_V = -0.040$. Based on the more negative

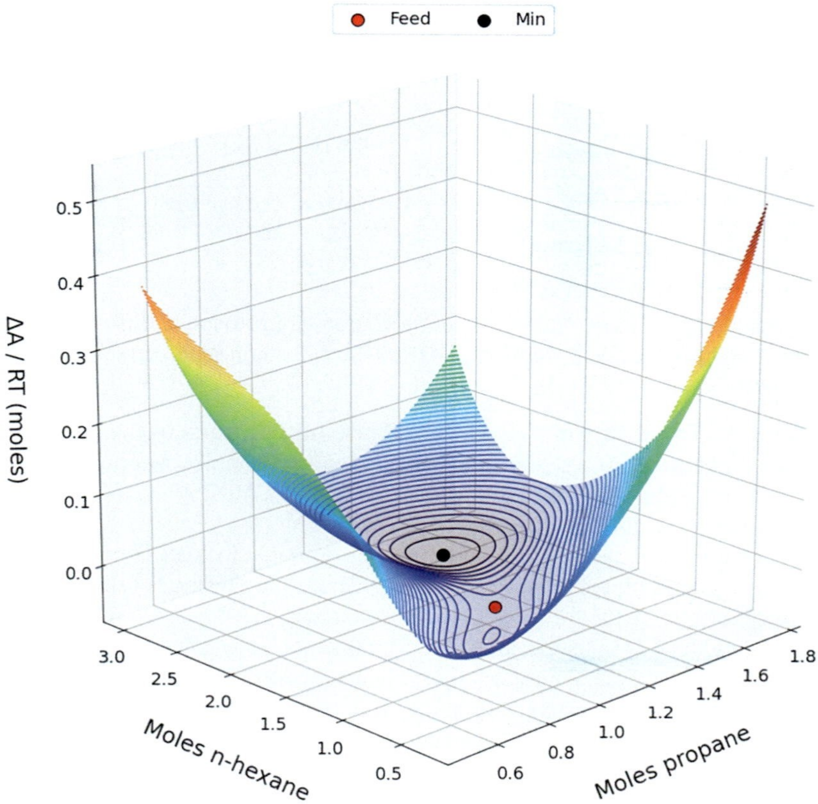

Fig. 6.9 Plot of Eq. 6.81 for a feed of one mole of propane and one mole of n-hexane at 453K and 708.391cm^3 using the SRK EOS. Two minima are detected, and the more negative t_V value is selected as the incipient phase

t_V value, the liquid is the considered to be the incipient phase. The existence of two minima is shown in Fig. 6.9.

An interesting feature of Fig. 6.9 is that there is no thermodynamic barrier (increase in free energy) between the feed state and the minimum detected. This contrasts strongly with Fig. 6.8, and this concept will be discussed in Sect. 6.5.

Finally, note that for the above cases the vapor- and liquid-like initializations both converge without issue, with the composition yielding a V_{min} less than the volume of the feed at each iteration. In the general case, it is advisable to limit stepsizes to maintain this, as volumes less than the $Vmin$ of the EOS lead to numerical issues. Note also that in this example, initializations with (nearly) pure phases provided no additional information. This results from Wilson k values being reliable for such a mixture of hydrocarbons.

Table 6.10 t_V stability analysis for 1 mole of propane, 1 mole of n-hexane and 1 mole of water at 350K and 6000 cm^3 via SRK EOS: various initializations

Initialization	$\overline{W}^{init}$	tV	$\overline{W}$
Vapor	2.501, 0.109, 0.04	0	$\overline{n}$
Liquid	1.110, 25.524, 69.288	Unconverged	Unconverged
Propane	2.618, 2.618E−10, 2.618E−10	0	$\overline{n}$
n-Hexane	3.768E−9, 37.684, 3.768E−9	Unconverged	Unconverged
Water	2.423E−8, 2.423E−8, 242.287	−553.778	6.676E−7, 1.645E−12, 253.243

As stated above, pure component initialization can provide additional information in a stability analysis for liquid–liquid/strongly non-ideal mixtures. This is shown in Example 19.

Example 19 (The Utility of Nearly Pure Phase Initialization in the Calculation of t_V for a Mixture) If we revisit the calculation in Example 18 for an equimolar mixture of propane, n-hexane, and water, with $T = 350$K, $V = 6000$cm^3, and $\overline{n} = 1, 1, 1$, we find the results given Table 6.10:

As shown in Table 6.10, some initializations do not converge. During iterations on the liquid-like root, the composition strays toward pure water, but the V_{min} value becomes larger than the V where we are assessing stability, yielding numerical problems for the EOS. In the n-hexane initialization, the solver also encounters issues with V_{min} and does not converge. It is also worth noting that only the water initialization yields a converged instability (i.e., not via Wilson initialization); the multiple iteration strategies provide a degree of robustness here.

6.5 The Spinodal Boundary

In Sect. 6.4.2, we considered changes in Helmholtz free energy due to the formation of an additional phase. As shown, the possibility of lowering the free energy[27] indicates that a mixture is unstable. This is very much an *equilibrium* thermodynamic argument, making a state-to-state comparison, rather than an along-the-path argument. A further level of analysis relates to changes in potential *between* these states (along the path). Looking back at Fig. 6.8, we observe that while the ultimate formation of the W phase indeed offers a lower energy state, the feed resides within a local free energy well, and formation of phase W (or something close in composition, looking ahead to Fig. 6.16) initially results in an *increase* in free energy. This provides a fascinating connection between thermodynamics and kinetics, as this idea relates to *metastability*, whereby a system that is unstable at equilibrium is nonetheless stable within small perturbations. Though not discussed

[27] The appropriate potential to be used depends on which variables are fixed.

in detail here, this phenomenon affects the mechanism and rate of phase separation. Particularly in systems with low rates of mass transfer (e.g., high-molecular-weight polymer in a solvent), while phase separation may be favored at equilibrium, finite time to separate is required.

Relating our experience with scalar functions, we can recognize that upward curvature (positive second derivative) results in such wells, and changes in the sign of the second derivative characterize phase separation without a thermodynamic barrier. In the case of multicomponent mixtures, this scalar concept breaks down, and we must apply the methods of linear algebra. Taking the formalism of T, V, $\overline{n}$, where our potential of interest is the Helmholtz energy, we write a Taylor series expansion in mole numbers:

$$A(T, V, \overline{n} + d\overline{n}) = A(T, V, \overline{n}) + \sum_r (\frac{\partial A}{\partial n_r})_{T,V} dn_r$$

$$+\frac{1}{2}\sum_r\sum_s (\frac{\partial^2 A}{\partial n_r \partial n_s})_{T,V} dn_r dn_s + \frac{1}{6}\sum_r\sum_s\sum_t (\frac{\partial^3 A}{\partial n_r \partial n_s \partial n_t})_{T,V} dn_r dn_s dn_t + ... \tag{6.93}$$

Equation 6.93 provides the variation in A (not A^{Res}) of a single phase mixture with mole numbers, or put another way, it explores the surface of A around the point $\overline{n}$ at fixed T and V. It provides different information than Fig. 6.8, which looks at *other* compositions (e.g., $\overline{W}$). In developing Eq. 6.93, we aren't so much interested in A itself, but the difference between $A(T, V, \overline{n} + d\overline{n})$ and the tangent plane at $\overline{n}$. With the equation of this tangent plane being given by:

$$A(T, V, \overline{n} + d\overline{n}) = A(T, V, \overline{n}) + \sum_r (\frac{\partial A}{\partial n_r})_{T,V} dn_r$$

we are primarily interested in the quadratic term, particularly when it is zero.

$$\sum_r\sum_s (\frac{\partial^2 A}{\partial n_r \partial n_s})_{T,V} dn_r dn_s = 0 \tag{6.94}$$

Equation 6.94 does not mean the second derivatives *are* zero,[28] but rather that for arbitrary changes in mole numbers, this double sum is zero. Writing Eq. 6.94 in linear algebra notation, we have:

$$d\overline{n} * \left[(\frac{\partial^2 A}{\partial \overline{n}^2})_{T,V} * d\overline{n} \right] = 0 \tag{6.95}$$

[28] Except for a pure substance.

For Eq. 6.95 to be generally true (i.e., for an arbitrary $d\overline{n}$), not only is the term in brackets zero, but the determinant of the matrix with elements $\frac{\partial^2 A}{\partial n_i \partial n_j}$ must be zero;

$$det\left(\frac{\partial^2 A}{\partial \overline{n}^2}\right)_{T,V} = 0 \tag{6.96}$$

otherwise we could solve for $d\overline{n} = \overline{0}$ (see Sects. 8.12.1 and 8.12.1.1).

Equation 6.96 describes the limit of stability [Heidemann critical point paper] or the spinodal.[29] This term may seem in contradiction to the term "spinodal volumes" $(\frac{\partial P}{\partial V} = 0)$ used in Chap. 4, but this is clarified in Appendix 8.17.

With $A = A(T, V, \overline{n})$, application of Eq. 6.96 means only one of T and V is independent for given $\overline{n}$. The expression for the Helmholtz free energy required is:

$$\begin{aligned} A &= A^{IG} + A^{Res} \\ &= -(PV)^{IG} + \sum_r \mu_r^{IG} n_r + A^{Res}(T, V, \overline{n}) \end{aligned}$$

The first derivative with respect to mole numbers (applying Eq. 8.57 to the ideal gas state) is:

$$\left(\frac{\partial A}{\partial n_i}\right)_{T,V,n_{k \neq i}} = \mu_i^{IG} + \left(\frac{\partial A^{Res}}{\partial n_i}\right)_{T,V,n_{k \neq i}}$$

The chemical potential of an ideal gas can be rewritten in terms of T, V and $\overline{n}$ from Eq. 1.43:

$$\mu_i^{IG}(T, P, \overline{n}) = g_i^o + R_g T ln\left(\frac{R_g T}{V P^o}\right) + R_g T ln(n_i)$$

The second derivative with respect to mole numbers of the Helmholtz free energy becomes:

$$\left(\frac{\partial^2 A}{\partial n_i \partial n_j}\right)_{T,V} = R_g T \frac{\delta_{ij}}{n_i} + \left(\frac{\partial^2 A^{Res}}{\partial n_i \partial n_j}\right)_{T,V} \tag{6.97}$$

From the suggestion of Heidemann and Khalil [Heidemann critical point paper], we write Eq. 6.97 in dimensionless form:

$$Q_{ij} = n\left[\frac{\delta_{ij}}{n_i} + \frac{1}{R_g T}\left(\frac{\partial^2 A^{Res}}{\partial n_i \partial n_j}\right)_{T,V}\right] \tag{6.98}$$

[29] The boundary where an additional phase forms (i.e., the PT envelope) is in this context sometimes referred as the "binodal." It is good to know these terms, but their use tends to cause more confusion than clarity in the discussion of thermodynamics.

We also define the determinant of the matrix given by Eq. 6.98 as:

$$M = det\overline{\overline{Q}} \tag{6.99}$$

Example 20 demonstrates the use of Eq. 6.99.

Example 20 (Calculating the Value of Spinodal M) Via the SRK EOS, and the arbitrary conditions $T = 350\text{K}$ and $V = 10320.036\text{cm}^3$, a mixture containing 1 mole each of propane and n-hexane ($\overline{n} = 1, 1$) yields, via Eq. 2.28:

$$(\frac{\partial^2 A^{Res}}{\partial n_i \partial n_j})_{T,V} = \begin{bmatrix} -150.390 & -292.378 \\ -292.378 & -569.772 \end{bmatrix} \quad (\frac{J}{mol^2})$$

Via Eq. 6.98, we have;

$$Q_{ij} = \begin{bmatrix} 1.897 & -0.201 \\ -0.201 & 1.608 \end{bmatrix}$$

and finally via Eq. 6.99:

$$M = det\overline{\overline{Q}} = 3.010$$

Figure 6.10 shows values of M for varying temperature and volume.

Our goal is to find the T and V for a given composition (from which pressure can be calculated) that satisfy $M = 0$ (Eq. 6.99). As stated above, there is one independent variable, meaning that we can compute a *curve* corresponding to the stability limit. As we can see in Fig. 6.10, we can specify the temperature and then solve for the volume that satisfies Eq. 6.96. However, the boundary to be computed does not extend monotonically over the temperature; at some point, the volume may

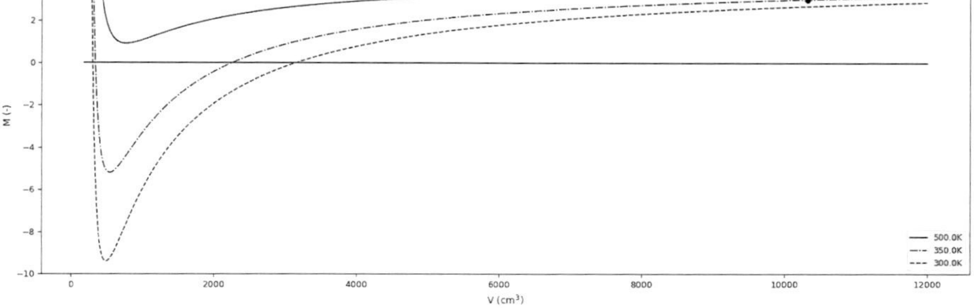

Fig. 6.10 Plot of M (Eq. 6.99) for varying temperature and volume, calculated for a mixture of propane and n-hexane (1 mole of each) using the SRK EOS. The dot corresponds to $T = 350\text{K}$ and $V = 10320.036\text{cm}^3$

be a better choice of specified variable. This can also be seen in Fig. 6.10, which shows that there are no V roots to Eq. 6.96 at 500K. To address this, we will make use of a simplified algorithm of the form used in computing TP phase boundaries (Sect. 6.3). We make our independent variables;

$$\overline{\alpha} = \begin{bmatrix} lnT \\ lnV \end{bmatrix} \tag{6.100}$$

and our target function:

$$\overline{F} = \begin{bmatrix} M \\ \overline{\alpha}[spec] - S \end{bmatrix} \tag{6.101}$$

The second equation within $\overline{F}$ is the analogue of Eq. 6.27 in Sect. 6.3. Here we specify either lnT or lnV using a specification index $spec$, and the value specified is S. Also note that, while we are using the same algorithm as the computation of phase boundaries (Sect. 6.3), this algorithm is significantly simpler, as M is a scalar, regardless of the number of components. With only two unknowns and two equations, the dimension of the system we are solving is in all cases 2. The complication is that we require the temperature and volume derivatives of M, which are difficult to compute analytically with the number of components being arbitrary.[30] To surmount this, we compute the derivatives numerically [Heidemann critical point paper] (see Sect. 8.14). As in Sect. 6.3, the derivative of the specification equation with respect to our independent variables is either [1, 0] or [0, 1], depending on which variable is specified (see the discussion around Eq. 6.24). For example, if the temperature is specified, the Jacobian would be:

$$\overline{\overline{H}} = \begin{bmatrix} \left(\frac{\partial M}{\partial lnT}\right) & \left(\frac{\partial M}{\partial lnV}\right) \\ 1 & 0 \end{bmatrix} \tag{6.102}$$

This allows a Newton's method approach to calculating points on the spinodal boundary, for which we require an initialization. We can generate an initial guess by selecting an initial pressure where we might expect VLE and solve the Rachford–Rice equation with Wilson k values, as in Example 11. Example 21 examines this in detail.

Example 21 (Initialization of the Spinodal Boundary for a Mixture) Consider a mixture of propane and n-hexane (1 mole each). At an initial pressure of 0.1bar, we solve the Rachford–Rice equation for the dew-point temperature using Wilson k values (See Sect. 8.16.1), yielding $T = 265.01$K. Using the SRK EOS, we then solve for the liquid volume at the assumed pressure: $V = 227.50\text{cm}^3$. This allows us to populate alpha (Eq. 6.100):

[30] Note that Eq. 8.86 only applies to two-component mixtures.

Table 6.11 Iterations solving the initial point on spinodal boundary for equimolar propane and n-hexane, initialized at $P = 0.1$bar, using the SRK EOS

Iter	$\overline{\alpha}$	$\overline{F}$	$\overline{\overline{H}}$	$d\overline{\alpha}$
1	5.580, 5.427	76.492, 0.	51.601, −929.128 1, 0	3E−17, 8.233E−2
2	5.580, 5.509	28.758, 0.	50.615, −352.079 1, 0	−2E−17, 8.168E−2
3	5.580, 5.591	8.709, 0.	49.319, −167.809 1, 0	−5E−18, 5.190E−2
4	5.580, 5.643	1.572, 0.	48.389, −111.871 1, 0	3E−18, 1.405E−2
5	5.580, 5.657	0.079, 0.	48.126, −100.833 1, 0	2E−19, 7.851E−4
6	5.580, 5.658	2.268E−4, 0.	48.112,−100.256 1, 0	−6E−22. 2.263E−6

$$\alpha = lnT, lnV = 5.5798, 5.4271$$

The specification is based on temperature:

$$S = lnT = 5.5798$$

The iterations for converging Eq. 6.101 are shown in Table 6.11.

We note in Table 6.11 that the inexact numerical derivatives result in slight changes in the variables to be held fixed (i.e., $dlnT \neq 0$).

After solving the first point, we then compute the slope of the solution with respect to the specification (as in Eq. 6.31):

$$\frac{\partial \overline{\alpha}}{\partial S} = -\overline{\overline{H}}^{-1} \frac{\partial \overline{F}}{\partial S} \tag{6.103}$$

We make a small increase ($dS = 0.01 * S$) to the temperature specification and approximate the next point (B) using the initial point (A) by linear extrapolation (as in Eq. 6.33):

$$\overline{\alpha}_B \approx \overline{\alpha}_A + \frac{\partial \overline{\alpha}}{\partial S} dS \tag{6.104}$$

Once converged, we use automatic specification by selecting the index (i.e., specifying lnT or lnV) of the element of $\frac{\partial \overline{\alpha}}{\partial S}$ with maximal magnitude. We must (as in Sect. 6.3) recalculate values of α and $\overline{\overline{H}}$ to account for a change specification before extrapolating (via Eq. 6.104) to the next point (a linear extrapolation is in many cases adequate). The computation of the boundary can be terminated on a

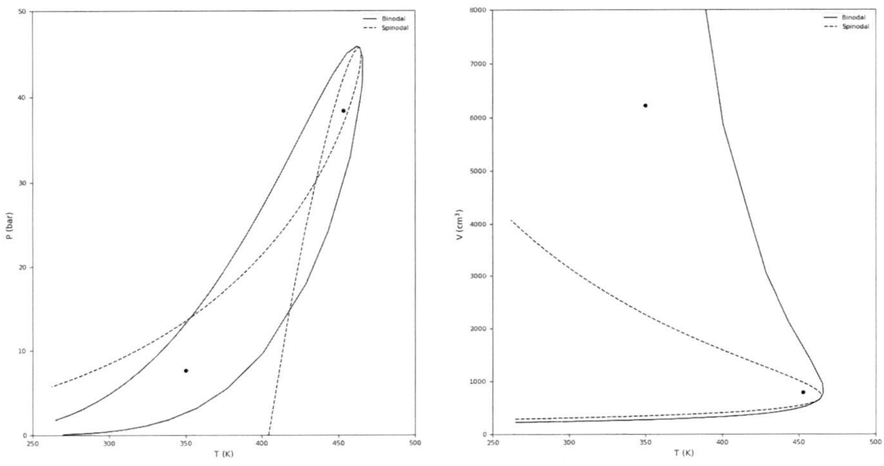

Fig. 6.11 Comparison of the Binodal and Spinodal boundaries of a mixture of one mole of propane and one mole of n-hexane, using the SRK EOS. The dots shown correspond to the states in Figs. 6.8 and 6.9

number of criteria, such as passing below the temperature of the first point. An example is shown in Fig. 6.11.

The left side of Fig. 6.11 is a common way of representing the relationship between the phase boundary (i.e., the binodal) and the limit of stability (i.e., the spinodal). However, plotting in terms of pressure for this case is somewhat confusing because spinodal branches intersect. To clarify, the upper portion of the phase boundary (the bubble point) corresponds to the rapidly descending portion of the spinodal boundary, while the lower portion of the phase boundary (the dewpoint) corresponds to the similarly sloped other branch of the spinodal. The situation is much clearer when plotted in terms of volume, as is shown on the right side of the figure. We then see that the state at 350K and 6223.641cm^3 lies between the spinodal and binodal boundaries. We therefore expect a new phase to form (and we can consider it a liquid, based on the feed's proximity to the dew-point curve, which is *uppermost* when plotted in TV), but we also expect an increase in free energy to achieve the separation. This was made clear in the discussion of Fig. 6.8. A very different result is seen at 453K and 783.331cm^3, as this state lies within the binodal (meaning another phase will form at equilibrium), but also within the spinodal (there is no thermodynamic barrier required to reach the new state). This was also clarified in Fig. 6.9. Finally, note that in both sides of Fig. 6.11, there is a point of intersection between the binodal and spinodal boundaries. This is the critical point of the mixture and is examined in further detail in the next section.

6.6 Determination of Critical Points

The critical point of a substance (pure or a mixture) can be thought of as a limit of stability. In the case of a pure substance, the critical point (T_c, P_c) represents the maximum T and P where the liquid and vapor states can coexist.[31] For a pure substance these are often experimentally measured, and in terms of EOS modeling reproduce:

$$\left(\frac{\partial P}{\partial V}\right)_T = \left(\frac{\partial^2 P}{\partial V^2}\right)_T = 0$$

In fact, these conditions lead to the determination of EOS constants (i.e., the 4.2747 and 8.664 in Sect. 2.1.) in a cubic[32] EOS.

At such a pure component critical point, the properties of the vapor and liquid are identical. For a mixture, a vapor–liquid critical point is the condition where the liquid and vapor states coexist, but in the limiting sense, the two phases are also identical. These ideas are made more concrete by returning the spinodal conditions given in Sect. 6.5, namely Eq. 6.93. In addition to the condition of metastability (Eq. 6.94), we now add the condition that the cubic term is also zero:

$$\sum_r \sum_s \sum_t \left(\frac{\partial^3 A}{\partial n_r \partial n_s \partial n_t}\right)_{T,V} dn_r dn_s dn_t = 0 \tag{6.105}$$

In order to evaluate (let alone make zero) the left side of Eq. 6.105, we require values of $d\overline{n}$. We can determine these from Eq. 6.95:

$$\overline{\overline{Q}} * d\overline{n} = \overline{0} \tag{6.106}$$

According to Sect. 8.12.2, the left side of Eq. 6.106 is written as:

$$\overline{\overline{Q}} * d\overline{n} = \begin{matrix} Q_{11}dn_1 & + Q_{12}dn_2 & + .. + Q_{1(n-1)}dn_{n-1} & + Q_{1n}dn_n \\ Q_{21}dn_1 & + Q_{22}dn_2 & + .. + Q_{2(n-1)}dn_{n-1} & + Q_{2n}dn_n \\ .. & .. \ .. & .. \ .. \ .. \ .. & .. \ .. \\ Q_{(n-1)1}dn_1 & + Q_{(n-1)2}dn_2 & + .. + Q_{(n-1)(n-1)}dn_{n-1} & + Q_{(n-1)n}dn_n \\ Q_{n1}dn_1 & + Q_{n2}dn_2 & + .. + Q_{n(n-1)}dn_{n-1} & + Q_{nn}dn_n \end{matrix} = \begin{matrix} 0 \\ 0 \\ .. \\ 0 \\ 0 \end{matrix}$$

[31] In this sense what is commonly called "the critical point" of a substance is technically the vapor–liquid critical point. Other types of critical points, with different coexisting phases (e.g., liquid–liquid), can also be considered.

[32] Most non-cubic EOSs, such as PC-SAFT, are not parameterized to the critical point and therefore do not reproduce pure component critical points.

Because $det\overline{\overline{Q}} = 0$, $d\overline{n}$ cannot be uniquely determined. However, by arbitrarily assigning one of the elements, we can solve for the others (see [Heidemann critical point paper]). For example, if we designate the final (N) element of $d\overline{n}$: $dn_N = 1$, and remove the corresponding elements from the above set, we have:

$$\begin{matrix} Q_{11}dn_1 & + Q_{12}dn_2 & + .. + Q_{1(N-1)}dn_{N-1} & & -Q_{1N} \\ Q_{21}dn_1 & + Q_{22}dn_2 & + .. + Q_{2(N-1)}dn_{N-1} & = & -Q_{2N} \\ .. & .. .. & .. .. & .. & -.. \\ Q_{(N-1)1}dn_1 & + Q_{(N-1)2}dn_2 & + .. + Q_{(N-1)(N-1)}dn_{N-1} & & -Q_{(N-1)N} \end{matrix}$$

The above equation can be written as;

$$\overline{\overline{r}} * d\overline{v} = \overline{q} \tag{6.107}$$

where for $i, j \neq N$:

$$\begin{aligned} r_{ij} &= Q_{ij} \\ dv_j &= dn_j \\ q_i &= -Q_{iN} \end{aligned} \tag{6.108}$$

We now apply the $d\overline{n}$ to Eq. 6.105, for which we also need $\frac{\partial^3 A}{\partial n_i \partial n_j \partial n_k}$. Differentiating Eq. 6.97, we have:

$$\left(\frac{\partial^3 A}{\partial n_i \partial n_j \partial n_k}\right)_{T,V} = -R_g T \frac{\delta_{ij}\delta_{ik}}{n_i^2} + \left(\frac{\partial^3 A^{Res}}{\partial n_i \partial n_j \partial n_k}\right)_{T,V} \tag{6.109}$$

Just as with the quadratic term in Eq. 6.93, we write Eq. 6.109 in dimensionless form:

$$S_{ijk} = n^2 \left[-\frac{\delta_{ij}\delta_{ik}}{n_i^2} + \frac{1}{R_g T} \left(\frac{\partial^3 A^{Res}}{\partial n_i \partial n_j \partial n_k}\right)_{T,V} \right] \tag{6.110}$$

One could compute the $3rd$ order derivative of A^{Res} analytically, but following the example of Heidemann and Khalil [Heidemann critical point paper], we compute this numerically (see Sect. 8.14). We then compute:

$$N = \sum_r \sum_s \sum_t S_{rst} dn_r dn_s dn_t \tag{6.111}$$

These ideas are demonstrated in Example 22.

Example 22 (Computation of the Spinodal Cubic Term) Returning to Example 20 (at $T = 350$K and $V = 10320.036\text{cm}^3$, with $\overline{n} = 1, 1$) we numerically differentiate $\frac{\partial^2 A^{Res}}{\partial n_i \partial n_j}$ via the SRK EOS. As in Sect. 8.14, we use the five point method, whereby we take:

$$F_{ij}(n_k) = \left(\frac{\partial^2 A^{Res}}{\partial n_i \partial n_j}\right)_{T,V}$$

and compute for $\Delta n_k = \frac{n_k}{10^6} = 10^{-6}$:

$$\left(\frac{\partial^3 A^{Res}}{\partial n_i \partial n_j \partial n_k}\right)_{T,V} = \frac{\partial F_{ij}}{\partial n_k}$$
$$\approx \left(\frac{-F_{ij}(n_k - 2\Delta n_k) + 8F_{ij}(n_k - \Delta n_k) - 8F_{ij}(n_k + \Delta n_k) + F_{ij}(n_k + 2\Delta n_k)}{12\Delta n_k}\right)$$

The required terms are, from (Eq. 2.28) and for k corresponding to propane:

$$F_{ij}(n_{propane} - 2\Delta n_{propane}) = \begin{bmatrix} -150.38988914 & -292.37781456 \\ -292.37781456 & --569.77166545 \end{bmatrix}$$

$$F_{ij}(n_{propane} - \Delta n_{propane}) = \begin{bmatrix} -150.38988713 & -292.37781088 \\ -292.37781088 & --569.7716587 \end{bmatrix}$$

$$F_{ij}(n_{propane} + \Delta n_{propane}) = \begin{bmatrix} -150.38988313 & -292.37780352 \\ -292.37780352 & -569.7716452 \end{bmatrix}$$

$$F_{ij}(n_{propane} + 2\Delta n_{propane}) = \begin{bmatrix} -150.38988113 & -292.37779984 \\ -292.37779984 & -569.77163845 \end{bmatrix}$$

giving:

$$\left(\frac{\partial^3 A^{Res}}{\partial \overline{n}^3}\right)_{T,V} \approx \begin{matrix} \begin{bmatrix} 2.00223514 & 3.68046587 \\ 3.68046574 & 6.7495736 \end{bmatrix} \\ \begin{bmatrix} 3.68046578 & 6.7495736 \\ 6.74957338 & 12.34617079 \end{bmatrix} \end{matrix}$$

$$\overline{\overline{S}} = \begin{matrix} \begin{bmatrix} -3.99724786 & 0.00505893 \\ 0.00505893 & 0.00927753 \end{bmatrix} \\ \begin{bmatrix} 0.00505893 & 0.00927753 \\ 0.00927753 & -3.98302976 \end{bmatrix} \end{matrix}$$

$$N = -3.984664174891974$$

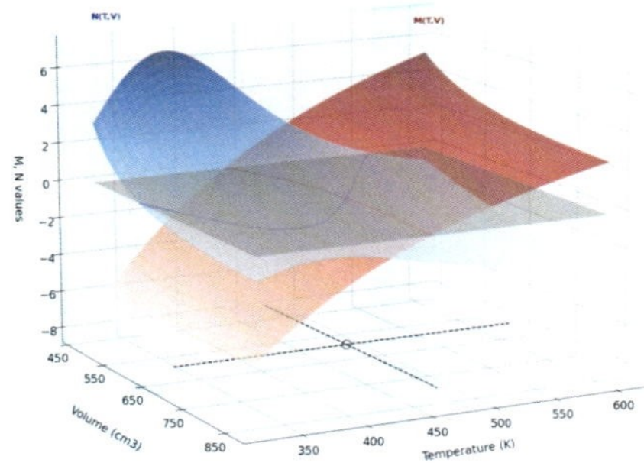

Fig. 6.12 Plot of spinodal M (Eq. 6.99) and N (Eq. 6.111) for a mixture of one mole of propane and one mole of n-hexane, using the SRK EOS. The bold lines show the intersections of the M and N surfaces with the zero plane, and where these lines cross, we find the critical point (464.02K, 664.428cm^3 and 45.65 bar)

Based on M (Example 20) and N (present example), the critical state of the composition considered is not $T = 350$K and $V = 10320.036$cm^3. This is also made clear in Fig. 6.12. We now concern ourselves with determining the critical point ($N = M = 0$).

This we handle in the standard way, defining:

$$\overline{\alpha} = \begin{bmatrix} lnT \\ lnV \end{bmatrix} \tag{6.112}$$

and our target function:

$$\overline{F} = \begin{bmatrix} M \\ N \end{bmatrix} \tag{6.113}$$

We thus require:

$$\overline{\overline{H}} = \frac{\partial \overline{F}}{\partial \overline{\alpha}}$$

$$= \begin{bmatrix} \frac{\partial \overline{M}}{\partial lnT} & \frac{\partial \overline{M}}{\partial lnV} \\ \\ \frac{\partial N}{\partial lnT} & \frac{\partial N}{\partial lnV} \end{bmatrix} \tag{6.114}$$

The derivatives of M and N are difficult to compute analytically, due to the requirements of differentiating the determinant of an arbitrarily sized matrix and the $d\overline{n}$ values. Instead, we determine these numerically as illustrated in Sect. 8.14. In terms of initialization, one has a few options. The simplest is to use Kay's rule:

$$T_c^{Kay} = \sum_r c_r T_{cr} \tag{6.115}$$

$$P_c^{Kay} = \sum_r c_r P_{cr} \tag{6.116}$$

One can then estimate V_c using a root solver. Though Kay's rule will often provide a suitable initialization for the computation of the critical point, another method then can be considered is to compute the spinodal boundary (Sect. 6.5) and select and estimate the point (T, V) with the minimum value of $N^2 + M^2$, from Eqs. 6.99 and 6.111. We will now look at computation of the critical point in an example.

Example 23 (Calculating the Critical Point of a Mixture) We will calculate the critical point for a mixture of propane and n-hexane ($\overline{n} = 1, 1$) as shown in Fig. 6.11. From Table 8.2 we have:

$$T_{ci} = 369.8, 507.5 \text{ K}$$

$$P_{ci} = 42.50, 30.10\text{K bar}$$

Using Kay's rule as an initialization (Eqs. 6.115 and 6.116), we have:

$$T_c^{Kay} = 438.65 \text{ K}$$

$$P_c^{Kay} = 36.30 \text{ bar}$$

Using our root solver, the most stable volume is $V = 448.43\text{cm}^3$, from which we initialize α (Eq. 6.112);

$$\overline{\alpha} = \begin{bmatrix} 6.084 \\ 6.106 \end{bmatrix} \tag{6.117}$$

from which we calculate F (Eq. 6.113);

$$\overline{F} = \begin{bmatrix} 0.649 \\ 5.002 \end{bmatrix} \tag{6.118}$$

and H (Eq. 6.114):

$$\overline{\overline{H}} = \begin{bmatrix} 19.752 & -9.832 \\ -28.479 & -12.572 \end{bmatrix} \tag{6.119}$$

Via Newton's Method 8.13:

$$\overline{\alpha} = \begin{bmatrix} 7.764E-2 \\ 0.222 \end{bmatrix} \tag{6.120}$$

This update yields $T = 474.066$K and $V = 559.879$cm^3. After 5 iterations, the converged result is $T_c = 464.018$K $V_c = 664.428$cm^3, from which, via Eq. 3.9, we calculate $P_c = 45.648$bar. Compare these results to Fig. 6.11.

Example 23 demonstrates that Kay's rule is a somewhat inaccurate estimate of the critical point. It is interesting to note that the true critical temperature (T_c) is often interior to the set of pure component critical temperatures (T_{ci}). This is not always true of the critical pressure, which, like in this example, often exceeds the largest pure component critical pressure. In this sense we might cautiously consider Kay's rule more accurate for estimating T_c than for P_c.

6.7 Flash Calculations

Flash calculations are perhaps the most widely used and well-known application of computational thermodynamics, at least in *chemical* engineering. In general terms, a flash calculation determines the number, composition, and (relative) amount of the phases constituting a mixture at equilibrium. Referring to Table 1.2), there are many permutations we could consider. Two very common types of flash calculations can be appreciated in terms of the problems they solve:

- Fixed $T, P, \overline{n}$ These specifications are very common and intuitive in chemical processing and process conceptualization. The Gibbs free energy is minimized.
- Fixed $T, V, \overline{n}$ A mixture is confined to a rigid container of known volume and temperature. The Helmholtz free energy is minimized.

Other examples are academic, while others are quite are practical. For example when a mixture undergoes a pressure drop at fixed[33] enthalpy, described by a PHn flash, and it is solved by maximizing the entropy. We can even conceive of less *canonical* examples, such as pressurizing a container of fixed volume. If the container initially contains a known quantity of material, how much of a given mixture (in general differing in composition from the initial charge) has been added, and which phases are present, once we reach a particular temperature and pressure in the container?[34] Regardless of the form of flash we are considering, it is helpful to know in advance that the mixture we are considering is unstable as a single phase. This of course is the motivation for stability analysis (Sect. 6.4), which should be conducted in advance. By a wide margin, the flash at fixed temperature, pressure, and mole numbers (often simply called the *isothermal* flash, or the TP flash) is the most common flash type. We therefore examine this important algorithm in detail.

[33] Often the enthalpy remains constant during the change in pressure. This is idealized as throttling across an insulated valve. One can of course consider heat transfer; the requirement is that the final pressure and enthalpy are *known*.

[34] Due to the initial amount of material in the container, this differs from at TPn flash.

6.7.1 Two-Phase TPn Flash Calculations

The two-phase TPn flash calculation determines the composition and relative amounts of two phases coexisting at equilibrium within a mixture of known temperature (T), pressure (P), and overall composition ($\overline{z} = \overline{n}/n$). For clarity of notation, we will formally write our equations assuming[35] that a liquid and vapor coexist, with compositions $\overline{x}$ and $\overline{y}$. Thus, the required equations are those developed in Sect. 6.3 (Eq. 6.7):

$$\overline{F} = \begin{cases} \overline{EQ} = ln\overline{k} + ln\hat{\varphi}^{v}(T, P, \overline{y}) - ln\hat{\varphi}^{\ell}(T, P, \overline{x}) = \overline{0} & \text{C equations} \\ RR = \sum_r \frac{z_r(k_r-1)}{1+f(k_r-1)} = 0 & \text{1 equation} \end{cases} \tag{6.121}$$

With T, P, and $\overline{z}$ given, our unknowns are the lnk values and a single phase fraction (f). In solving flash calculations, it is important that we label our phases in a consistent way, particularly when dealing with mixtures of species with which we have no familiar experience. In this text, when a designation must be made, we will sort our phases from highest to lowest density (mass per volume) using Eq. 1.81. The highest density phase becomes our reference phase, having composition $\overline{x}$ and giving $k_i = \frac{y_i}{x_i}$. Then, referring to the Rachford–Rice equation, the phase fraction becomes that of the less dense phase. The simplest method of solving Eq. 6.121 is Successive Substitution (Sect. 8.18), which we will demonstrate in the following example.

Example 24 (Two-Phase TPn Flash for a Binary Mixture) Let's execute a two-phase flash on the mixture described in Example 17 using Successive Substitution. Using the SRK EOS, at 350 K and 5bar, an equimolar mixture of propane and n-hexane is unstable, with $t_m = -0.768$ and $\overline{w} = 0.063, 0.937$. We therefore have two phases, with mole fraction composition:

$$\overline{\overline{c}} = \begin{bmatrix} 0.500 & 0.063 \\ 0.500 & 0.937 \end{bmatrix}$$

We next compute the volumes (Chap. 4) and then densities (Eq. 1.81) of these phases, as shown in Table 6.12. Based on the densities, we assign the w phase as the reference phase,[36] reorder the phases, and compute the fugacities:

[35] As we will soon see, the algorithm we develop is not predicated on there being a "liquid" and a "vapor," as these are arbitrary labels that we attach to familiar substances. Our goal is that the methods we develop here apply to LLE as well.

[36] In this example, one might say that a liquid phase, rich in n-hexane, forms from a vapor feed, due to familiar experience with propane and n-hexane as substances.

Table 6.12 Initial phase properties from stability analysis of equimolar propane and n-hexane at 350 K and 5bar using the SRK EOS. The volumes are reported in units of cm^3, but at convergence these will of course be molar volumes

Phase	$\overline{c}$	v (cm^3)	$\rho(\frac{g}{cm^3})$
Feed	0.500, 0.500	5160.018	0.013
w	0.063, 0.937	156.523	0.534

Table 6.13 Initialization of $ln\hat{\varphi}$ values for initial phases in equimolar propane and n-hexane at 350 K and 5bar using the SRK EOS

Phase	$\overline{c}$	$ln\hat{\varphi}$
ℓ	0.063, 0.937	1.460, −1.378
ν	0.500, 0.500	−0.035, −0.181

Updating the lnk values by satisfying the equilibria in Eq. 6.121 is the first part of Successive Substitution (Sect. 8.18):

$$lnk_i = ln\hat{\varphi}_i^{\ell} - ln\hat{\varphi}_i^{\nu} \tag{6.122}$$

We then proceed to solve the Rachford–Rice equation (Sect. 8.16.1) for the phase fraction (f).[37] As shown in Example 45 in Appendix 8.16.1, the result is $f = 0.572$. This yields a new set of compositions (through Eqs. 8.139 and 8.140), which allows us to update Table 6.12, sort the phases by density, and then update Table 6.13. At this point we can compute $\overline{EQ}$ in Eq. 6.121 and monitor the convergence against a chosen tolerance. Sample iterations are detailed in Table 6.14. To the limited significant figures, only the EQ values appear to be changing.

Next consider the compositions to which we have converged in Table 6.14. If we perform a stability analysis (Sect. 6.4.1) on these phases, we find a zero tangent plane distance in each case, meaning that the system is stable as a two-phase mixture. Conceptually, we only need to check the stability of one of the phases; when we have solved Eq. 6.121, all coexisting phases have the same values of $d = lnc + ln\hat{\varphi}$. In practice, the d values of the coexisting phases are only equal within a tolerance. A flash that produces phases with significantly different d values is not converged, and a stability analysis will not provide accurate information on whether the mixture is stable. However, even a tightly converged flash can yield phases of which one indicates an instability toward another of the phases detected (with a very small negative tangent plane distance). These issues can be avoided by checking the stability of all phases detected and selecting a tolerance on the tangent plane distance (values within this tolerance are ignored as instabilities). This is an example of where one must choose between speed and robustness.

[37] We know with confidence that our mixture will form (at least) two phases at equilibrium, because we detected a negative tangent plane distance for the feed at the conditions considered.

Table 6.14 Iterations to converge two phase TPn flash of equimolar propane and n-hexane at 350 K and 5bar using the SRK EOS

Iter	$\overline{\overline{c}}$	$v(\text{cm}^3)$	$ln\hat{\varphi}$	lnk	$\overline{EQ}$	f
1	0.0635, 0.5 0.9365, 0.5	156.523, 5160.018	1.460, −0.035 −1.378, −0.181	1.494 −1.197	0.570 0.570	0.572
2	0.168, 0.749 0.832, 0.251	151.419, 5370.302	1.451, −0.045 −1.377, −0.164	1.4958 −1.2135	−1.627E−3 0.016	0.567
3	0.169, 0.753 0.831, 0.247	151.386, 5373.435	1.451 −0.045 −1.377, −0.163	1.496 −1.214	−5.093E−5 3.246E−4	0.567
4	0.169, 0.753 0.831, 0.247	151.386, 5373.497	1.451, −0.045 −1.377, −0.163	1.496 −1.214	−1.269E−6 6.443E−6	0.567
5	0.169, 0.753 0.831, 0.247	151.386, 5373.498	1.451, −0.045 −1.377, −0.163	1.496 −1.214	−2.903E−8 1.276E−7	0.567

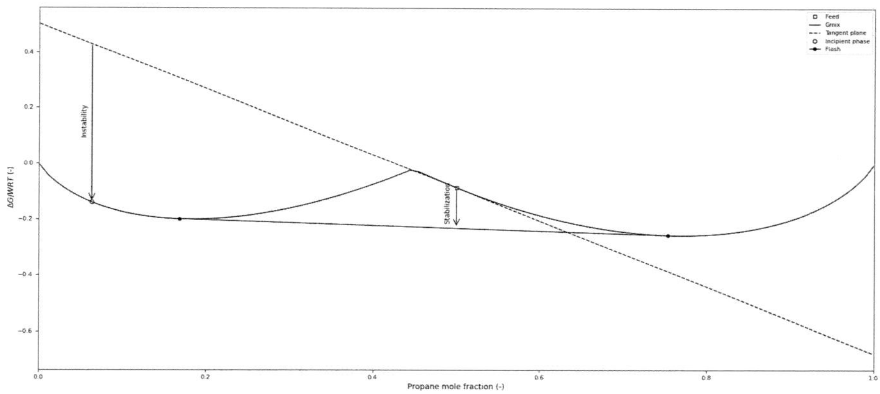

Fig. 6.13 Stability and flash of binary mixture of propane and n-hexane at 350K and 5 bar, using the SRK EOS

The key results of both the stability and flash calculations for this system are shown graphically in Fig. 6.13. Note that the composition indicated by the stability analysis is only an *estimate* for the phases coexisting at equilibrium,[38] and that the separation of the mixture into two phases lowers the Gibbs free energy (in the case of a TPn flash).[39] It is also worth noting that Fig. 6.13 demonstrates that in order to form the two-phase mixture, an initial increase in free energy is required. This corresponds with the results shown in the left side of Fig. 6.6 and in Fig. 6.8.

The molar volumes of the two phases detected are:

$$\overline{v} = 151.39, 5373.50(\text{cm}^3/\text{mol})$$

[38] Figure 6.13 also makes clear that stability analysis is based on matching the tangent slope of the test phase. It does not detect a local minimum in Gibbs free energy.

[39] The Gibbs free energy of the two-phase mixture lies below the Gibbs free energy "surface."

and with the phase fractions found to be:

$$\overline{f} = 0.433, 0.567$$

the molar volume of the two-phase mixture is:

$$v_{mixture} = \sum_{t} v_t f_t = 3111.820 \frac{\text{cm}^3}{\text{mol}}$$

Just as in Stability Analysis (Sect. 6.4.1), flash calculations near critical points require a large number of iterations to converge by successive substitution, and again acceleration by the dominant eigenvalue method is a simple technique to address this. Details can be found in Sect. 8.18.1 and Example 50.

The methods outlined in this section provide a simple framework for a TP flash when two phases are *formally* assumed. However, just because a two-phase flash has converged does not mean that the mixture is stable with only two phases. If we revisit Example 24 with equimolar propane, n-hexane, and water, we again find straightforward convergence to two phases:

$$\overline{\overline{c}} = \begin{bmatrix} 2.139E-6 & 0.462 \\ 3.862E-9 & 0.462 \\ 0.99... & 7.543E-2 \end{bmatrix}$$

indicating a water-rich phase containing only traces of hydrocarbons and a hydrocarbon phase containing significant water[40] However, a stability analysis must be performed on these phases to be confident that *only* two phases exist at equilibrium. This will be discussed in Sect. 6.7.2.

6.7.2 Multiphase TPn Flash Calculations

According to the Gibbs Phase Rule (Eq. 1.63), a three-component mixture can coexists as a three-phase mixture over a range of temperature and pressure. For larger number of components, a higher number of phases is conceptually possible, through in reality it is not the number of components but rather the type of components (highly volatile species like hydrogen, hydrocarbons, polar species, and water) in a mixture that determines how many phases coexist. Regardless, we require methods for solving flash calculations with more than two coexisting phases.

[40] From this information alone, it is difficult to assign identities to these phases; are they "liquids" or "vapors"? One could calculate the densities of these phases and make qualitative assignments. However, for less familiar chemicals, one might need to compute respective phase boundaries to assign these states.

Such problems underscore that labels such as "liquid," "vapor," and "solvent" are constructs with no basis in generality. We instead choose one phase to be the reference (R), which throughout this text is that with the highest mass density. For a C-component, Π-phase mixture, the equilibrium of species i in phase j is written analogously to Eq. 6.6:

$$EQ_i^j = lnk_i^j + ln\hat{\varphi}_i^j - ln\hat{\varphi}_i^R = 0 \tag{6.123}$$

The material balance, in the form of the multiphase Rachford–Rice equation, is given by Eq. 8.160:

$$RR^j = \sum_r \frac{z_r(k_r^j - 1)}{1 + \sum_{s \neq R} f^s(k_r^s - 1)} = 0$$

To solve these equations we consider our unknowns to be the lnk and f values. Successive substitution can be used to solve the equilibria, and this proceeds for the multiphase case just as for the two-phase case (Eq. 6.122), except now we use an index to allow for more than one phase beyond the reference phase:

$$lnk_i^j = ln\hat{\varphi}_i{}^R - ln\hat{\varphi}_i^j \tag{6.124}$$

The Rachford–Rice equations are then solved for the phase fractions by Newton's method (Eq. 8.167 in Sect. 8.16.2). Initialization of these values begin with a stability analysis on the feed (Sect. 6.4.1). If an instability is detected, stability also provides an estimate of the incipient phase composition, and we converge this two-phase flash. Stability is then checked again. If this two-phase mixture is unstable, the stability analysis provides an estimate of the third phase, and the calculations proceed until a converged flash is found to be stable. This was introduced at the end of Sect. 6.7.1, which we now examine in Example 25.

Example 25 (Three-Phase TPn Flash for a Three-Component Mixture) A flash of a mixture containing one mole each of propane, n-hexane, and water at 350K and 5 bar using the SRK EOS yields the following phases and compositions:

$$\bar{\bar{c}} = \begin{bmatrix} 2.139E-6 & 0.462 \\ 3.862E-9 & 0.462 \\ 0.999 & 0.075 \end{bmatrix}$$

with phase fractions:

$$f = 0.721$$

However, this two-phase mixture is unstable, as $t_m = -0.696$ with an incipient phase having the composition $c = 0.0605, 0.905, 0.0345$. This indicates that one[41] more phase will form, and that the two-phase results above represent only a local minimum in the Gibbs free energy of system. To find the global minimum, a multiphase (i.e., more than 2-phase) algorithm must be applied. We then expand the number of phases and the phase fractions:

$$\overline{\overline{c}} = \begin{bmatrix} 2.139E-6 & 0.462 & 0.061 \\ 3.862E-9 & 0.462 & 0.905 \\ 0.999 & 0.075 & 0.035 \end{bmatrix}$$

$$\overline{f} = \begin{bmatrix} 0.721 & 0 \end{bmatrix}$$

We then calculate the densities (Eq. 1.81) of the three phases;

$$\overline{\rho} = \begin{bmatrix} 0.727, 0.012, 0.533 \end{bmatrix}$$

and order the phases accordingly:

$$\overline{\overline{c}} = \begin{bmatrix} 2.139E-6 & 0.061 & 0.462 \\ 3.862E-9 & 0.905 & 0.462 \\ 0.999 & 0.035 & 0.075 \end{bmatrix}$$

$$\overline{f} = \begin{bmatrix} 0 & 0.721 \end{bmatrix}$$

The fugacity coefficients are then computed;

$$ln\overline{\overline{\hat{\varphi}}} = \begin{bmatrix} 12.247 & 1.467 & -0.037 \\ 18.422 & -1.378 & -0.179 \\ -2.621 & 0.214 & -0.037 \end{bmatrix}$$

and then successive substitution (Eq. 6.123) is performed:

$$ln\overline{\overline{k}} = \begin{bmatrix} 10.779 & 12.284 \\ 19.801 & 18.601 \\ -2.831 & -2.585 \end{bmatrix}$$

As shown in Example 46, the compositions and phase fractions are updated by solving the multiphase Rachford–Rice equation (Sect. 8.16.2):

[41] The Gibbs phase rule (Eq. 1.63) indicates that a three-component, three-phase mixture at specified temperature and pressure has zero degrees of freedom. If a fourth phase were to form, we would be relying on coincidence that the stated temperature and pressure correspond to the ternary point.

$$\bar{\bar{c}} = \begin{bmatrix} 3.179E-6 & 0.153 & 0.687 \\ 1.984E-9 & 0.788 & 0.237 \\ 0.999 & 0.059 & 0.075 \end{bmatrix}$$

$$\overline{f} = \begin{bmatrix} 0.297 & 0.419 \end{bmatrix}$$

The alternating process of solving k values from phase fractions (Equilibrium) and phase fractions from k values (Rachford-Rice equation) is continued, yielding:

$$\bar{\bar{c}} = \begin{bmatrix} 3.165E-6 & 0.152 & 0.690 \\ 1.986E-9 & 0.789 & 0.234 \\ 0.999... & 0.059 & 0.076 \end{bmatrix}$$

and the full set of phase fractions;

$$\overline{f} = \begin{bmatrix} 0.284 & 0.299 & 0.417 \end{bmatrix}$$

The calculated molar volumes (cm^3/mol) of the phases are:

$$\overline{v} = \begin{bmatrix} 24.768 & 144.778 & 5382.427 \end{bmatrix}$$

which, using the phase fractions above, and 3 moles of mixture, yield a total volume of:

$$V_{mixture} = n v_{mixture} = n \sum_t v_t f_t = 6885.87 \frac{\mathrm{cm}^3}{\mathrm{mol}}$$

The results of this three-phase flash are shown in Fig. 6.14. The red tie line indicates that, while this false 2-phase state is tangent to the Gibbs surface, it lies

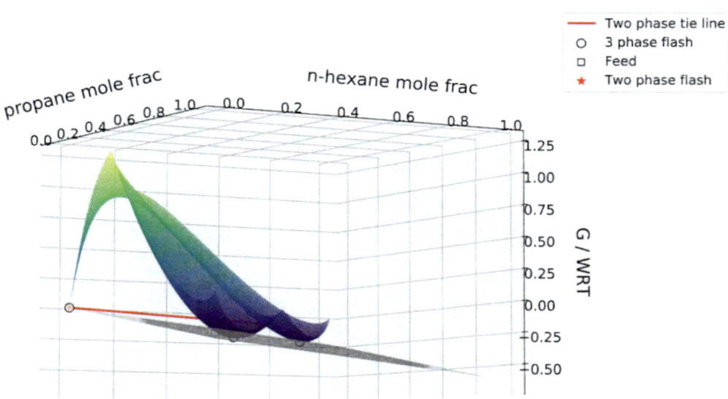

Fig. 6.14 Flash of ternary mixture of propane, n-hexane, and water (1 mole each) at 350 K and 5 bar, yielding a three-phase state. Also shown is the false two-phase state discussed at the beginning of Example 25

above the gray plane of the more stable (lower Gibbs energy) three-phase state. Note that both the two- and three- phase solutions yield a phase that is nearly pure water, and these respective phases are indistinguishable in the figure. This also makes it difficult to see the tangent behavior at this phase. Figure 6.14 also demonstrates how considering any two of the three components will yield phase separation plots such as Fig. 6.13. The successive substitution approach can of course be augmented using acceleration by the dominant eigenvalue method and Newton's method. These are described in Appendices 8.18.2 and 8.13.1, respectively.

6.7.3 Two-Phase TVn Flash Calculations

Fundamentally, the two-phase TVn flash calculation determines the composition and relative amounts of two phases coexisting at equilibrium within a mixture of known temperature (T), volume (V), and composition (n). In contrast to a TPn flash, a TVn flash seeks to minimize the Helmholtz free energy (see Table 1.2). In further contrast with TPn flashes, the pressure in the TVn flash is initially unknown, and regardless of whether one is interested in the value, mechanical equilibrium requires that all phases have the same pressure. Thus, the pressure may be considered an internal variable in a TVn flash algorithm. Just as we did in Sect. 6.7.1, we will formally regard the two phases as "vapor" and "liquid," with the latter being the reference phase. The set of equations to be solved in a C-component two-phase TVn flash could be expressed as:

- Equilibrium between vapor and liquid for each substance (C equations)
- Component balances between vapor and liquid phases through Rachford–Rice (1 equation)
- Equal pressures between the vapor and liquid phases (1 equation)
- The volume of vapor plus liquid equals the feed volume (1 equation)

This gives $C + 3$ equations, which matches the number of unknowns:

- Molar volumes of the vapor and liquid phases (2)
- lnk value for each substance (C)
- Vapor phase fraction f (1)

However, just as the Rachford–Rice equation (Sect. 8.16) allows us to represent x_i and y_i with a single variable k_i, we can apply the same concept to the molar volumes of the phases. The total volume of the two-phase mixture is given by:

$$V = V^{\ell} + V^{\nu}$$

$$nv = n^{\ell} v^{\ell} + n^{\nu} v^{\nu}$$

We can then develop an expression for the molar volume:

$$
\begin{aligned}
v &= f^{\ell} v^{\ell} + f^{\nu} v^{\nu} \\
&= (1 - f) v^{\ell} + f v^{\nu} \\
&= v^{\ell} + f(v^{\nu} - v^{\ell}) \\
&= v^{\ell}(1 + f(\frac{v^{\nu}}{v^{\ell}} - 1) \\
&= v^{\ell}(1 + f(L - 1)
\end{aligned}
\tag{6.125}
$$

Thus, with a single variable $L = \frac{v^{\nu}}{v^{\ell}}$ we can represent the two variables (v^{ℓ} and v^{ν}) via:

$$v^{\ell} = \frac{v}{1 + f(L - 1)} \tag{6.126}$$

and

$$v^{\nu} = \frac{vL}{1 + f(L - 1)} \tag{6.127}$$

This formulation gives the advantage that the volume constraint is automatically satisfied. As we did in the development of stability for TVn specification (Eq. 6.87), we will treat the equilibrium relationship differently than the usual (TPn) formalism used in this text. We begin with the fundamental statement of phase equilibrium (Eq. 1.62)

$$\mu_i^{\nu} = \mu_i^{\ell}$$

By Eq. 1.79, we can write:

$$
\begin{aligned}
\mu_i &= g_i^o + R_g T ln \frac{P}{P^o} + R_g T ln c_i + R_g T ln \hat{\varphi}_i \\
&= \left(g_i^o - R_g T ln P^o\right) + R_g T ln c_i + R_g T ln P + R_g T ln \hat{\varphi}_i \\
&= \left(g_i^o - R_g T ln P^o\right) + R_g T ln c_i + R_g T ln \frac{Z R_g T}{v} + R_g T ln \hat{\varphi}_i \\
&= \left(g_i^o + R_g T ln \frac{R_g T}{P^o}\right) + R_g T ln c_i - R_g T ln v + \left(\frac{\partial A^{Res}}{\partial n_i}\right)_{T,V,n_k \neq i}
\end{aligned}
$$

The last line above results from the definition of fugacity (Eq. 1.71).[42] The equilibrium relationship, with the formalism that the reference phase is the "liquid," then becomes:

$$
\begin{aligned}
EQV_i &= \frac{(\mu_i^v - \mu_i^\ell)}{R_g T} \\
&= \left[lnc_i - lnv + \frac{1}{R_g T}\left(\frac{\partial A^{Res}}{\partial n_i}\right)\right]^v - \left[lnc_i - lnv + \frac{1}{R_g T}\left(\frac{\partial A^{Res}}{\partial n_i}\right)\right]^\ell \\
&= lnk_i - ln\frac{v^v}{v^\ell} + \frac{1}{R_g T}\left[\left(\frac{\partial A^{Res}}{\partial n_i}\right)^v - \left(\frac{\partial A^{Res}}{\partial n_i}\right)^\ell\right] \\
&= lnk_i + \frac{1}{R_g T}\left[\left(\frac{\partial A^{Res}}{\partial n_i}\right)^v - \left(\frac{\partial A^{Res}}{\partial n_i}\right)^\ell\right] - lnL
\end{aligned}
\tag{6.128}
$$

For notational simplicity, we have omitted the statement of fixed variables in the derivatives in Eq. 6.128. For clarification:

$$
\left(\frac{\partial A^{Res}}{\partial n_i}\right)^\ell = \left(\frac{\partial A^{Res,\ell}}{\partial n_i^\ell}\right)_{T,V^\ell,n^\ell_{k\neq i}}
$$

We have also introduced our volume ratio L. The Rachford–Rice equation has already been developed as Eq. 8.142:

$$
RR = \sum_r \frac{z_r(k_r - 1)}{1 + f(k_r - 1)}
$$

Finally, the equality of pressures can be written as:

$$
P^v = P^\ell
$$

$$
\frac{Z^v R_g T}{v^v} = \frac{Z^\ell R_g T}{v^\ell}
$$

$$
\frac{Z^v}{v^v} = \frac{Z^\ell}{v^\ell}
$$

We can then formally write the pressure constraint as:

$$
Press = Z^v - Z^\ell L \tag{6.129}
$$

[42] Mikyška and Firoozabadi [Firoozabadi paper] published a helpful paper on TVn flashes, wherein they introduce a *new* variable Φ. Here we demonstrate that volume-based equilibrium is naturally written in terms of Helmholtz free energy.

These results allow us to assemble our target function for the two-phase TVn flash:

$$\overline{F} = \begin{cases} \overline{EQV} & \text{C equations} \\ RR & \text{1 equation} \\ Press & \text{1 equation} \end{cases} \tag{6.130}$$

We also select as our independent variables $ln\overline{k}$, L and f;

$$\overline{\alpha} = \begin{cases} ln\overline{k} & \text{C values} \\ L & \text{1 value} \\ f & \text{1 value} \end{cases} \tag{6.131}$$

with which we will now derive the elements of the Jacobian of F, beginning with the equilibrium expression.

Formally, $\overline{EQV}$ depends on $\overline{lnk}$ in its first term, and further connections are due to how A^{Res} depends on composition. The first term is trivial, so let's focus on the latter. We've written $\frac{\partial A^{Res}}{\partial n_i}$ in Eq. 6.128, but we are not using mole numbers as our independent variables. However, as A^{Res} is $1st$ order homogeneous in V and $\overline{n}$ (see Sect. 8.6), $\frac{\partial A^{Res}}{\partial n_i}$ is $0th$ order in these variables, and $\frac{\partial^2 A^{Res}}{\partial n_i \partial n_j}$ is $-1th$ (see Eq. 8.48). Mathematically, this means:

$$A^{Res}(T, \frac{1}{n}V, \frac{1}{n}\overline{n}) = (\frac{1}{n})^1 A^{Res}(T, V, \overline{n})$$

$$\frac{\partial A^{Res}}{\partial n_i}(T, \frac{1}{n}V, \frac{1}{n}\overline{n}) = (\frac{1}{n})^0 \frac{\partial A^{Res}}{\partial n_i}(T, V, \overline{n})$$

$$\frac{\partial^2 A^{Res}}{\partial n_i \partial n_j}(T, \frac{1}{n}V, \frac{1}{n}\overline{n}) = (\frac{1}{n})^{-1} \frac{\partial^2 A^{Res}}{\partial n_i \partial n_j}(T, V, \overline{n})$$

Let's be clear on the notation above. $\frac{\partial A^{Res}}{\partial n_i}(T, \frac{1}{n}V, \frac{1}{n}\overline{n})$ is a function whose name is $\frac{\partial A^{Res}}{\partial n_i}$, that depends on T, V and $\overline{n}$, and is being evaluated at T, $\frac{1}{n}V$ and $\frac{1}{n}\overline{n}$. Also, note that the above relations apply to each phase; the V and $\overline{n}$ can be those of the liquid or the vapor phase as required. We can further write;

$$nA^{Res}(T, v, \overline{c}) = A^{Res}(T, V, \overline{n})$$

$$\frac{\partial A^{Res}}{\partial c_i}(T, v, \overline{c}) = \frac{\partial A^{Res}}{\partial n_i}(T, V, \overline{n})$$

$$\frac{1}{n}\frac{\partial^2 A^{Res}}{\partial c_i \partial c_j}(T, v, \overline{c}) = \frac{\partial^2 A^{Res}}{\partial n_i \partial n_j}(T, V, \overline{n})$$

Here we are being even sloppier with our notation, having changed the "names" of the functions on the left-hand side (e.g., $A^{Res}(T, \frac{1}{n}V, \frac{1}{n}\overline{n})$ becomes $A^{Res}(T, v, \overline{c})$). Furthermore, of course we cannot differentiate with respect to mole fraction; $\frac{\partial A^{Res}}{\partial c_i}(T, v, \overline{c})$ means $\frac{\partial A^{Res}}{\partial n_i}$ *evaluated* at $T, \frac{V}{n}, \frac{\overline{n}}{n}$. $\frac{\partial^2 A^{Res}}{\partial c_i \partial c_j}(T, v, \overline{c})$ means $\frac{\partial^2 A^{Res}}{\partial n_i \partial n_j}$ *evaluated* at $T, \frac{V}{n}, \frac{\overline{n}}{n}$.

These relations will allow us to more carefully derive the required Jacobian. We start with $\frac{\partial EQVi}{\partial lnk_j}$. A change in lnk_j causes changes in mole numbers (through composition) in each phase:

$$(\frac{\partial^2 A^{Res}}{\partial n_i \partial lnk_j}) = \sum_r (\frac{\partial^2 A^{Res}}{\partial n_i \partial n_r})\frac{\partial n_r}{\partial lnk_j}$$

$(\frac{\partial^2 A^{Res}}{\partial n_i \partial n_r})$ comes from the particular EOS being used. For example, the form for the SRK EOS is Eq. 2.28. Let us further investigate $\frac{\partial n_r}{\partial lnk_j}$. For the two phases we have:

$$\begin{aligned} n_r^{\nu} &= n^T f^{\nu} y_r = n^T f y_r = n^{\nu} y_r \\ n_r^{\ell} &= n^T f^{\ell} x_r = n^T (1-f) x_r = n^{\ell} x_r \end{aligned} \tag{6.132}$$

Noting that n^T (used here to indicate the *total* mole numbers) and f are constant in the differentiation with respect to lnk_j:[43]

$$\frac{\partial n_r^{\nu}}{\partial lnk_j} = n^{\nu} \frac{\partial y_r}{\partial lnk_j}$$

$$\frac{\partial n_r^{\ell}}{\partial lnk_j} = n^{\ell} \frac{\partial x_r}{\partial lnk_j}$$

The expression for $\frac{\partial x_r}{\partial lnk_j}$ is available in Sect. 8.16, where we also note that the composition of species r in the liquid or the vapor only depends on the lnk_r value (see Eqs. 8.139 and 8.140). Therefore;

$$\frac{\partial c_r}{\partial lnk_j} = \delta_{rj} \frac{\partial c_j}{\partial lnk_j}$$

In the product (for each phase) of $\frac{\partial^2 A^{Res}}{\partial n_i \partial n_j}(T, V, \overline{n})$ and $\frac{\partial n_j}{\partial lnk_j}$, the mole numbers of each phase cancels, we are left with expressions that can be written in terms of v

[43] Be careful here. It is true that the k values affect the flash, including the value of f at the solution, but f is not a *function* of $ln\overline{k}$; they are independent variables that are *coupled* through our set of equations to be solved.

and $\overline{c}$:

$$\frac{\partial EQV_i}{\partial lnk_j} = \delta_{ij} + \frac{1}{R_g T}\left[\left(\frac{\partial^2 A^{Res}}{\partial y_i \partial y_j}\right)^{\nu} \frac{\partial y_j}{\partial lnk_j} - \left(\frac{\partial^2 A^{Res}}{\partial x_i \partial x_j}\right)^{\ell} \frac{\partial x_j}{\partial lnk_j}\right] \tag{6.133}$$

The derivative of $\overline{EQV}$ with respect to L follows the same logic as $\frac{\partial EQV_i}{\partial lnk_j}$, recognizing that changes in L lead to changes in phase volumes by Eqs. 6.126 and 6.127. Formally:[44]

$$\frac{\partial^2 A^{Res}}{\partial n_i \partial L} = \frac{\partial^2 A^{Res}}{\partial V \partial n_i}\frac{\partial V}{\partial L}$$

The volumes of the phases can be written in multiple ways:

$$V^{\nu} = n^T f^{\nu} v^{\nu} = n^T f v^{\nu} = n^{\nu} v^{\nu}$$
$$V^{\ell} = n^T f^{\ell} v^{\ell} = n^T (1-f) v^{\ell} = n^{\ell} v^{\ell} \tag{6.134}$$

but their response to changes in L are:

$$\frac{\partial V^{\nu}}{\partial L} = n^{\nu}\frac{\partial v^{\nu}}{\partial L}$$
$$\frac{\partial V^{\ell}}{\partial L} = n^{\ell}\frac{\partial v^{\ell}}{\partial L}$$

The derivatives of the molar volumes with respect to L are:

$$\frac{\partial v_L}{\partial L} = -\frac{v_L^2 f}{v} \tag{6.135}$$

and

$$\frac{\partial v_V}{\partial L} = v_L + L\frac{\partial v_L}{\partial L} \tag{6.136}$$

These results yield (with a similar argument about the -1 degree homogeneity of $\frac{\partial^2 A^{Res}}{\partial V \partial n_i}$ in V and $\overline{n}$):

$$\frac{\partial EQV_i}{\partial L} = \frac{1}{R_g T}\left[\left(\frac{\partial^2 A^{Res}}{\partial v \partial y_i}\right)^{\nu} \frac{\partial v}{\partial L} - \left(\frac{\partial^2 A^{Res}}{\partial v \partial x_i}\right)^{\ell} \frac{\partial v}{\partial L}\right] - \frac{1}{L} \tag{6.137}$$

[44] We are being somewhat pedantic here by working in the variables TVn. This is an attempt to make things are clear as possible, without guessing around homogeneity.

To complete the differentiation of EQV_i, we consider $\frac{\partial EQV_i}{\partial f}$. Changes in f lead to changes in both the (molar) volumes and the mole numbers (compositions) of the phases:

$$\frac{\partial EQV_i}{\partial f} = \frac{1}{R_g T}\left[\left(\frac{\partial^2 A^{Res}}{\partial n_i \partial f}\right)^{\nu} - \left(\frac{\partial^2 A^{Res}}{\partial n_i \partial f}\right)^{\ell}\right] \tag{6.138}$$

For either phase, we have:

$$\left(\frac{\partial(\frac{\partial A^{Res}}{\partial n_i})}{\partial f}\right) = \left(\frac{\partial^2 A^{Res}}{\partial n_i \partial V}\right)\left(\frac{\partial V}{\partial f}\right) + \sum_r \left(\frac{\partial^2 A^{Res}}{\partial n_i \partial n_r}\right)\left(\frac{\partial n_r}{\partial f}\right) \tag{6.139}$$

To preserve the clarity of the derivation, we have regarded A^{Res} to be a function of T, V, and $\overline{n}$ (not T, v, and $\overline{n}$). The derivatives of A^{Res} depend on the equation of state. Equation 6.134 gives;

$$\frac{\partial V^{\nu}}{\partial f} = \frac{V^{\nu}}{f} + n^{\nu}\frac{\partial v^{\nu}}{\partial f}$$

$$\frac{\partial V^{\ell}}{\partial f} = -\frac{V^{\ell}}{(1-f)} + n^{\ell}\frac{\partial v^{\ell}}{\partial f}$$

while Eq. 6.7.3 gives:

$$\frac{\partial n_r^{\nu}}{\partial f} = \frac{n_r^{\nu}}{f} + n^{\nu}\frac{\partial y_r}{\partial f}$$

$$\frac{\partial n_r^{\ell}}{\partial f} = -\frac{n_r^{\ell}}{(1-f)} + n^{\ell}\frac{\partial x_r}{\partial f}$$

If we substitute the above derivatives into Eq. 6.139 for the vapor phase, we have:

$$\begin{aligned}
\left(\frac{\partial(\frac{\partial A^{Res}}{\partial n_i})}{\partial f}\right)^{\nu} &= \left(\frac{\partial^2 A^{Res}}{\partial n_i \partial V}\right)^{\nu}\left(\frac{\partial V^{\nu}}{\partial f}\right) + \sum_r \left(\frac{\partial^2 A^{Res}}{\partial n_i \partial n_r}\right)^{\nu}\left(\frac{\partial n_r^{\nu}}{\partial f}\right) \\
= \left(\frac{\partial^2 A^{Res}}{\partial n_i \partial V}\right)^{\nu}&\left(\frac{V^{\nu}}{f} + n^{\nu}\frac{\partial v^{\nu}}{\partial f}\right) + \sum_r \left(\frac{\partial^2 A^{Res}}{\partial n_i \partial n_r}\right)^{\nu}\left(\frac{n_r^{\nu}}{f} + n^{\nu}\frac{\partial y_r}{\partial f}\right) \\
&= \frac{1}{f}\left[\left(\frac{\partial^2 A^{Res}}{\partial n_i \partial V}\right)^{\nu} V^{\nu} + \sum_r \left(\frac{\partial^2 A^{Res}}{\partial n_i \partial n_r}\right)^{\nu} n_r^{\nu}\right] \\
&+ n^{\nu}\left(\frac{\partial^2 A^{Res}}{\partial n_i \partial V}\right)^{\nu}\left(\frac{\partial v^{\nu}}{\partial f}\right) + \sum_r n^{\nu}\left(\frac{\partial^2 A^{Res}}{\partial n_i \partial n_r}\right)^{\nu}\frac{\partial y_r}{\partial f}
\end{aligned} \tag{6.140}$$

The term in square brackets is identically zero because A^{Res} is homogeneous to degree one in V and $\overline{n}$ (see Eq. 8.48). The same conclusion follows for the liquid phase. Also, due to homogeneity we can write:

$$n\frac{\partial^2 A^{Res}}{\partial V \partial n_i}(T, V, \overline{n}) = \frac{\partial^2 A^{Res}}{\partial v \partial c_i}(T, v, \overline{c})$$

$$n\frac{\partial^2 A^{Res}}{\partial n_i \partial n_j}(T, V, \overline{n}) = \frac{\partial^2 A^{Res}}{\partial c_i \partial c_j}(T, v, \overline{c})$$

With the molar volume derivatives developed above and the composition derivatives given by Eqs. 8.149 and 8.150, we can write:

$$\begin{aligned}\frac{\partial EQV_i}{\partial f} = \frac{1}{R_g T}\Bigg\{&\left(\frac{\partial^2 A^{Res}}{\partial y_i \partial v}\right)^{v}\left(\frac{\partial v^{v}}{\partial f}\right) - \left(\frac{\partial^2 A^{Res}}{\partial x_i \partial v}\right)^{\ell}\left(\frac{\partial v^{\ell}}{\partial f}\right)\\ &+ \sum_r\left[\left(\frac{\partial^2 A^{Res}}{\partial y_i \partial y_r}\right)^{v}\frac{\partial y_r}{\partial f} - \left(\frac{\partial^2 A^{Res}}{\partial x_i \partial x_r}\right)^{\ell}\frac{\partial x_r}{\partial f}\right]\Bigg\}\end{aligned} \tag{6.141}$$

We now consider the derivatives of RR with respect to our variables. RR does not depend on L, while Eqs. 6.23 and 8.152 provide:

$$\frac{\partial RR}{\partial lnk_j} = \frac{x_j y_j}{z_j}$$

and

$$\frac{\partial RR}{\partial f} = -\sum_r \frac{(y_r - x_r)^2}{z_r}$$

Finally, we derive the derivatives of $Press$, which is $0th$ order in V and $\overline{n}$. Note that in the above development of the derivatives of $\frac{\partial A^{Res}}{\partial n_i}$, we showed clearly that, for this $0th$ order homogeneous function of V and $\overline{n}$, the end results can be naturally written in terms of v and $\overline{c}$. We will forgo this step[45] and work directly in these latter variables when dealing with $Press$.

$$\frac{\partial Press}{\partial lnk_j} = \left(\frac{\partial Z^{v}}{\partial lnk_j}\right) - \left(\frac{\partial Z^{\ell}}{\partial lnk_j}\right)L = \left(\frac{\partial Z^{v}}{\partial y_j}\right)\left(\frac{\partial y_j}{\partial lnk_j}\right) - \left(\frac{\partial Z^{\ell}}{\partial x_j}\right)\left(\frac{\partial x_j}{\partial lnk_j}\right)L \tag{6.142}$$

$$\frac{\partial Press}{\partial L} = \left(\frac{\partial Z^{v}}{\partial L}\right) - \left(\frac{\partial Z^{\ell}}{\partial L}\right)L - Z^{\ell} = \left(\frac{\partial Z}{\partial v}\right)^{v}\left(\frac{\partial v^{v}}{\partial L}\right) - \left(\frac{\partial Z}{\partial v}\right)^{\ell}\left(\frac{\partial v^{\ell}}{\partial L}\right)L - Z^{\ell} \tag{6.143}$$

[45] We will also make use of some of the other intermediate results highlighted above.

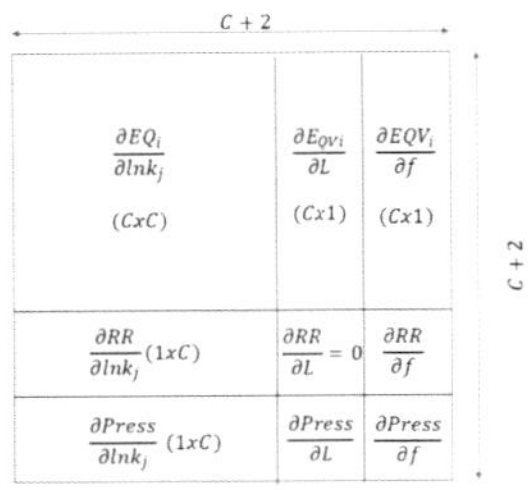

Fig. 6.15 Layout of the Jacobian used in the two-phase TVn Flash

$$\frac{\partial Press}{\partial f} = \left(\frac{\partial Z^{v}}{\partial f}\right) - \left(\frac{\partial Z^{\ell}}{\partial f}\right) L = \left(\frac{\partial Z}{\partial v}\right)^{v} \left(\frac{\partial v^{v}}{\partial f}\right) - \left(\frac{\partial Z}{\partial v}\right)^{\ell} \left(\frac{\partial v^{\ell}}{\partial f}\right) L$$
$$+ \sum_{r} \left[\left(\frac{\partial Z}{\partial y_r}\right)^{v} \left(\frac{\partial y_r}{\partial f}\right) - \left(\frac{\partial Z}{\partial x_r}\right)^{\ell} \left(\frac{\partial x_r}{\partial f}\right) L \right] \tag{6.144}$$

We now assemble the Jacobian of $\overline{F}$ (Eq. 6.130) with respect to our unknowns $\overline{\alpha}$ (Eq. 6.131); $\overline{\overline{H}} = \frac{\partial \overline{F}}{\partial \overline{\alpha}}$. By virtue of the ordering of F and α, our Jacobian takes the form:

$$H = \frac{\partial F}{\partial \alpha} = \left\{ \begin{matrix} \frac{\partial EQV_i}{\partial lnk_j} & \frac{\partial EQV_i}{\partial L} & \frac{\partial EQV_i}{\partial f} \\ \frac{\partial RR}{\partial lnk_j} & \frac{\partial RR}{\partial L} = 0 & \frac{\partial RR}{\partial f} \\ \frac{\partial Press}{\partial lnk_j} & \frac{\partial Press}{\partial L} & \frac{\partial Press}{\partial f} \end{matrix} \right. \tag{6.145}$$

Figure 6.15 clarifies the layout and dimensions of the Jacobian.

We will now consider how to solve a two-phase TVn Flash. Following the detection of an instability (Sect. 6.4.1), the first step is to initialize the flash, which is illustrated by revisiting Example 18 as a seed for our TVn flash example.

Example 26 (Two-Phase TVn Flash for a Binary Mixture) Consider a mixture of 1 mole of propane and 1 mole of n-hexane at 350K and 5.0 bar, modeled with the SRK EOS. As indicated in Example 18, this state corresponds to a two-phase mixture with a volume of $V = 6223.641\text{cm}^3$. Also shown in that example is that an incipient phase $\overline{W} = 3.975, 38.894$ ($\overline{w} = 0.093, 0.907$), with a molar volume of $v^W = \frac{6223.641\text{cm}^3}{3.975+38.894} = 145.180\frac{\text{cm}^3}{\text{mol}}$. It is important to recognize that the TVn stability check produces a phase with the same chemical potential as the phase being tested, but not necessarily the same pressure. In particular, the pressures of the feed (5 bar) and the W phase (177.300 bar) are very different, and we therefore cannot rely on Wilson k values. Instead, consider Eq. 8.138 written for two phases; one being our instability (W) and the other (o) being whatever is leftover from the formation of this phase:

$$z_i = f^w w_i + (1 - f^w) c_i^o$$

We know $\overline{z}$ and $\overline{w}$, but neither f^w nor $\overline{c}^o$. We also know v and v^w, but use of this information simply introduces a new unknown[46] v^o (via rewriting Eq. 6.125):

$$v = f^w v^w + (1 - f^w) v^o$$

What we can say is that the above two equations limit the value of f^w. Consider solving the first equation for c_i^o:

$$c_i^o = \frac{z_i - f^w w_i}{(1 - f^w)} = w_i \left(\frac{\frac{z_i}{w_i} - f^w}{1 - f^w} \right)$$

With $0 < (1 - f^w) < 1$ and $0 < w_i < 1$, we clearly require (with analogy to Eq. 8.154):

$$f^w_{comp,max} = min(\frac{z_i}{w_i}) \tag{6.146}$$

By the same logic, the second equation, when solved for v^o requires:

$$f^w_{vol,max} = \frac{v}{v^w} \tag{6.147}$$

We therefore select:

$$f^w_{max} = min[f^w_{comp,max}, f^w_{vol,max}]$$

With $f^w_{min} = 0$, we initialize f^w via:

$$f^w = \frac{f^w_{min} + f^w_{max}}{2}$$

For the present example, we have:

$$\frac{\overline{z}}{\overline{w}} = 5.393, 0.551$$

$$\frac{v}{v^W} = 21.434$$

We therefore select $f = 0.276$, giving:

$$\overline{w} = 0.093, 0.907$$

[46] We must be cautious that we do not stumble into having to solve the TVn flash here; we are simply trying to *initialize* our solution.

$$\overline{c}^o = 0.655, 0.345$$

$$v^w = 145.180\frac{\text{cm}^3}{\text{mol}}$$

$$v^o = 4240.192\frac{\text{cm}^3}{\text{mol}}$$

With the densities (Eq. 1.81) being $\rho^W = 0.567$ and $\rho^o = 1.38E-2\frac{g}{\text{cm}^3}$, we see that the phases are incorrectly ordered if the "liquid" phase is to be the reference. We thus initialize the independent variables (Eq. 6.131) as:

$$\overline{\alpha} = \begin{cases} ln\overline{k} = ln\frac{\overline{c}^o}{\overline{w}} = 1.955, -0.967 \\ L = \frac{v^o}{v^w} = 29.206 \\ f = 0.724 \end{cases}$$

With $\overline{\alpha}$ initialized, we can compute $\overline{F}$ via Eq. 6.130:

$$\overline{F} = \begin{cases} \overline{EQV} = 0.045, -0.516 \\ RR = 1.110E - 16 \\ Press = -24.946 \end{cases}$$

To facilitate solving the flash via Newton's method (Sect. 8.13), we structure $\overline{\alpha}$ and $\overline{F}$ as:

$$\overline{\alpha} = \begin{bmatrix}1.955 \ -0.967 \ 29.206 \ 0.724\end{bmatrix}$$

$$\overline{F} = \begin{bmatrix}0.045 \ -0.516 \ 1.110E - 16 \ -24.946\end{bmatrix}$$

$$H = \frac{\partial F}{\partial \alpha} = \begin{bmatrix} 1.598 & 5.160 & -0.407 & -25.731 \\ 1.001 & 9.411 & -0.644 & -42.088 \\ 0.121 & 0.626 & 0 & -1.264 \\ 28.658 & 240.642 & -18.175 & -1180.648 \end{bmatrix}$$

The Newton update (Sect. 8.13) to $\overline{\alpha}$ is:

$$d\overline{\alpha} = -\overline{\overline{H}}^{-1} * \overline{F} = \begin{bmatrix}-0.526, \ -0.510, \ 10.731, \ -0.303\end{bmatrix}$$

Stepsize control ($|d\overline{\alpha}| < 0.1|\overline{\alpha}|$) results in:

$$d\overline{\alpha} \rightarrow d\overline{\alpha}/2 = \begin{bmatrix}-0.066 \ -0.064 \ 1.341 \ -0.038\end{bmatrix}$$

The new value of $\overline{\alpha}$ becomes;

$$\overline{\alpha} = 1.889, -1.030, 30.548, 0.687$$

from which we have (according to Eq. 6.131):

$$ln\overline{k} = 1.889, -1.030$$

$$L = 30.548$$

$$f = 0.687$$

We can then recompute the more traditional flash variables:

$$\overline{v} = 146.185, 4465.616\frac{\text{cm}^3}{\text{mol}}$$

$$\overline{\overline{c}} = \begin{bmatrix} 0.103 & 0.681 \\ 0.895 & 0.340 \end{bmatrix}$$

Note that the mole fractions in each phase do not sum to unity; the Rachford–Rice is not yet satisfied. Table 6.15 provides a sample of the iterations in solving the flash.

The final result is:

$$P = 5.000bar$$

Table 6.15 Iterations to converge two-phase TVn flash of equimolar propane and n-hexane at 350 K and 6223.641 cm^3 (5 bar) using the SRK EOS

Iter	$\overline{\alpha}$	$\overline{F}$	$\overline{\overline{H}}$
1	1.955, −0.967, 29.206, 0.724	0.045, −0.516, 1.110E−16, −24.946	1.598, 5.160, −0.407, −25.731 1.001, 9.411, −0.644, −42.088 0.121, 0.626, 0, −1.264 28.658, 240.642, −18.1756, −1180.648
2	1.889,−1.030, 30.548, 0.687	0.079, −0.389, 2.415E−3, −19.785	1.579, 3.913, −0.345, −23.241 0.960, 7.297, −0.536, −37.562 0.140, 0.572, 0, −1.332 29.360, 191.045, −15.736, −1113.474
...	...	...	...
11	1.496, −1.214, 35.495, 0.567	1.076E−8, 1.076E−8, 9.992E−16, 1.009E−12	1.499, 1.271, −0.179, −13.752 0.787, 2.891, −0.253, −20.735 0.254, 0.411, 0, −1.366 31.876, 73.614, −8.310, −767.987

$$\overline{v} = 151.39, 5373.50\text{cm}^3$$

$$\overline{\overline{c}} = \begin{bmatrix} 0.169 & 0.753 \\ 0.831 & 0.247 \end{bmatrix}$$

$$f = 0.567$$

Compare these results, determined without formally *knowing* the equilibrium pressure, with those of Example 24, where the pressure was given. These two examples are consistent with each other. Note also that at each iteration (and at the solution itself), each phase can be plotted on a figure such as Fig. 6.8 by scaling[47] mole numbers according to:

$$\overline{W} = \frac{V}{v^j}\overline{c}^j$$

This is shown in Fig. 6.16, which also demonstrates that the solution is indeed tangent to the Helmholtz surface.

The stability of the solution in Example 26 is indicated by a check of the stability of the two phases. In both cases, using the methods of Sect. 6.4.2, tV is non-negative. Were this not the case, a multiphase flash algorithm must be used. This is developed in Sect. 6.7.4.

6.7.4 Multiphase TVn Flash Calculations

The two-phase TVn flash developed in Sect. 6.7.3 can be easily adapted to the case of π phases. We now use a reference phase (R), which we select as the phase with the highest density. The primary equations to be solved are:

- Equilibrium between phase j and phase R for each substance i $(C(\pi - 1)$ equations)

  We represent this by rewriting Eq. 6.128 for the case of more than two phases:

$$EQV_i^j = lnk_i^j + \frac{1}{R_gT}\left[\left(\frac{\partial A^{Res}}{\partial n_i}\right)^j - \left(\frac{\partial A^{Res}}{\partial n_i}\right)^R\right] - lnL^j \tag{6.148}$$

[47] The coexisting phases do not each have the volume of the mixture, but we can determine how many moles of each of the coexisting phases, given their molar volumes and compositions, would be required to fill that volume.

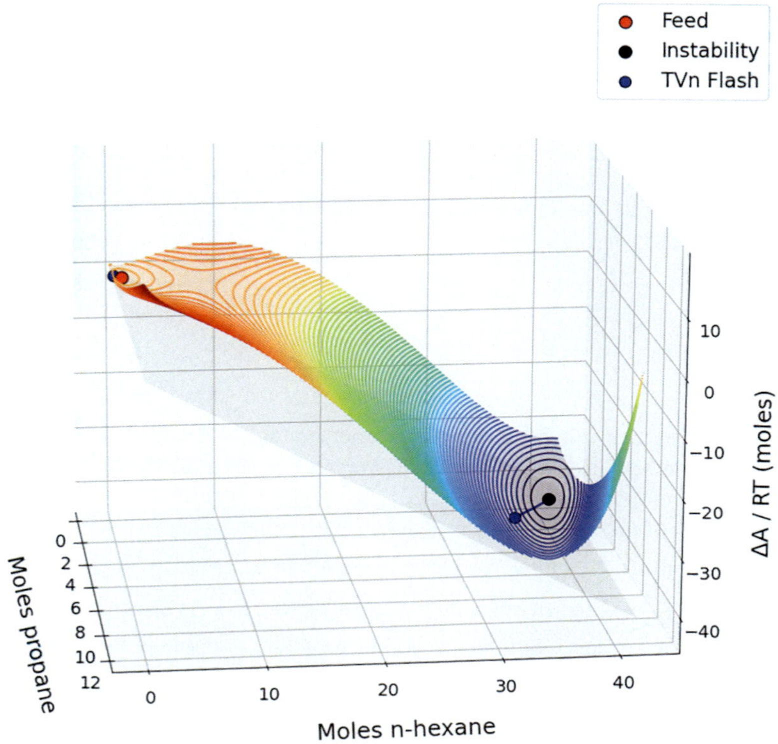

Fig. 6.16 Solution of TVn flash for a mixture of one mole of propane and one mole of n-hexane at 350K and 6223.64 cm3 (5 bar), using the SRK EOS

- Component balances between phase j and R ($\pi - 1$ equations)
  This is described by the multiphase Rachford–Rice equation (Eq. 8.160):

$$RR^j = \sum_r \frac{z_r(k_r^j - 1)}{1 + \sum_{s \neq R} f^s(k_r^s - 1)}$$

- Equal pressures between phase j and phase R ($\pi - 1$ equations)
  We represent this by rewriting Eq. 6.129 as:

$$Press^j = Z^j - Z^R L^j \tag{6.149}$$

We can then assemble the objective function:

$$F = \begin{cases} EQV_i^j & C(\pi - 1) \text{ equations} \\ RR^j & \pi - 1 \text{ equations} \\ Press^j & \pi - 1 \text{ equations} \end{cases} \tag{6.150}$$

We also select as our independent variables lnk, L and f, which are for the multiphase TVn flash:

$$\alpha = \begin{cases} lnk_i^j & C(\pi - 1) \text{ values} \\ L^j & \pi - 1 \text{ values} \\ f^j & \pi - 1 \text{ values} \end{cases} \tag{6.151}$$

We recognize that F and α are two-dimensional, which complicates the use of Newton's method in solving the flash. We will resolve this using the mapping concept we developed in Appendix 8.13.1, but first we develop the functional form of the elements of the Jacobian.

We begin with $(\frac{\partial EQV_i^j}{\partial lnk_q^t})$; the response of EQV_i^j to a change in lnk_q^t with all other independent variables held constant.[48] Returning to Eq. 6.148, we recognize that $(\frac{\partial A^{Res}}{\partial n_i})$ is not an explicit function of lnk, but we addressed this in the development of Eq. 6.133. Following the same logic we have:

$$\begin{aligned} \frac{\partial}{\partial lnk_q^t}(\frac{\partial A^{Res}}{\partial n_i})^j &= \sum_r (\frac{\partial^2 A^{Res}}{\partial n_i \partial n_r})^j \frac{\partial n_r^j}{\partial lnk_q^t} \\ &= \sum_r \frac{1}{n^j}(\frac{\partial^2 A^{Res}}{\partial c_i \partial c_r})^j \frac{\partial c_r^j}{\partial lnk_q^t} n^j \\ &= \sum_r (\frac{\partial^2 A^{Res}}{\partial c_i \partial c_r})^j \frac{\partial c_r^j}{\partial lnk_q^t} n^j \\ &= \sum_r (\frac{\partial^2 A^{Res}}{\partial c_i \partial c_r})^j \delta_{rq} \frac{\partial c_q^j}{\partial lnk_q^t} \\ &= (\frac{\partial^2 A^{Res}}{\partial c_i \partial c_q})^j \frac{\partial c_q^j}{\partial lnk_q^t} \end{aligned}$$

We have arrived at this result by making use of the homogeneity of A^{Res} (Sect. 8.6) and the fact that c_r^j does not depend on the lnk values of species other

[48] In the derivative indicated, only lnk_q^t is changing; all the other lnk values are fixed, as are the L and f values.

than r. This dependency is given by Eq. 8.162. The same logic can be applied to the reference phase, giving:

$$\frac{\partial EQV_i^j}{\partial lnk_q^t} = \delta_{iq}\delta_{jt} + \frac{1}{R_g T}\left[\left(\frac{\partial^2 A^{Res}}{\partial c_i \partial c_q}\right)^j \frac{\partial c_q^j}{\partial lnk_q^t} - \left(\frac{\partial^2 A^{Res}}{\partial c_i \partial c_q}\right)^R \frac{\partial c_q^R}{\partial lnk_q^t}\right] \tag{6.152}$$

Next we develop $(\frac{\partial EQV_i^j}{\partial L^t})$, which requires:

$$\begin{aligned}
\frac{\partial}{\partial L^t}(\frac{\partial A^{Res}}{\partial n_i})^j &= (\frac{\partial^2 A^{Res}}{\partial V \partial n_i})^j \frac{\partial V^j}{\partial L^t} \\
&= \frac{1}{n^j}(\frac{\partial^2 A^{Res}}{\partial v \partial c_i})^j \frac{\partial v^j}{\partial L^t} n^j \\
&= (\frac{\partial^2 A^{Res}}{\partial v \partial c_i})^j \frac{\partial v^j}{\partial L^t}
\end{aligned}$$

In order to progress, we require the multiphase analogue of Eq. 6.125:

$$\begin{aligned}
V = \sum_s V^s &= V^R + \sum_{s \neq R} V^s \\
nv = \sum_s n^s v^s &= n^R v^R + \sum_{s \neq R} n^s v^s \\
v = \sum_s f^s v^s = f^R v^R + \sum_{s \neq R} f^s v^s &= v^R + \sum_{s \neq R} f^s (v^s - v^R) \\
&= v^R \left[1 + \sum_{s \neq R} f^s (L^s - 1)\right]
\end{aligned} \tag{6.153}$$

where we have introduced the multiphase $L^j = \frac{v^j}{v^R}$. This allows us to write the multiphase analogues of Eqs. 6.126 and 6.127:

$$v^R = \frac{v}{1 + \sum_{s \neq R} f^s (L^s - 1)} \tag{6.154}$$

$$v^j = L^j v^R = \frac{v L^j}{1 + \sum_{s \neq R} f^s (L^s - 1)} \tag{6.155}$$

The use of these equations reduces the number of unknowns and automatically satisfies the volume constraint. The Jacobian requires the derivative of these expressions, which we develop analogously to Eqs. 6.135 and 6.136.

$$\frac{\partial v^R}{\partial L^t} = -\frac{(v^R)^2 f^t}{v} \tag{6.156}$$

$$\begin{aligned}\frac{\partial v^j}{\partial L^t} &= \delta_{jt} v^R + L^j \frac{\partial v^R}{\partial L^t} \\ &= \delta_{jt} v^R - \frac{v^R v^j f^t}{v}\end{aligned} \tag{6.157}$$

We can now write $(\frac{\partial EQV_i^j}{\partial L^t})$ as:

$$\frac{\partial EQV_i^j}{\partial L^t} = \frac{1}{R_g T}\left[(\frac{\partial^2 A^{Res}}{\partial v \partial c_i})^j \frac{\partial v^j}{\partial L^t} - (\frac{\partial^2 A^{Res}}{\partial v \partial c_i})^R \frac{\partial v^R}{\partial L^t} \right] - \frac{\delta_{jt}}{L^j} \tag{6.158}$$

The final derivative of EQV required is that with respect to f^t, which raises the question of how $(\frac{\partial A^{Res}}{\partial n_i})^j$ depends on f^t. We proceed as before, but note that changes in f values affect *both* volume *and* composition:

$$\begin{aligned}
\frac{\partial}{\partial f^t}(\frac{\partial A^{Res}}{\partial n_i})^j &= (\frac{\partial^2 A^{Res}}{\partial V \partial n_i})^j (\frac{\partial V^j}{\partial f^t}) + \sum_r (\frac{\partial^2 A^{Res}}{\partial n_i \partial n_r})^j (\frac{\partial n_r^j}{\partial f^t}) \\
&= \frac{1}{n^j}(\frac{\partial^2 A^{Res}}{\partial v \partial c_i})^j (\frac{\partial [nfv]^j}{\partial f^t}) + \frac{1}{n^j}\sum_r (\frac{\partial^2 A^{Res}}{\partial c_i \partial c_r})^j (\frac{\partial [nfc_r]^j}{\partial f^t}) \\
&= \frac{n}{n^j}(\frac{\partial^2 A^{Res}}{\partial v \partial c_i})^j (\delta_{jt} v^j + f^j \frac{\partial v^j}{\partial f^t}) + \frac{n}{n^j}\sum_r (\frac{\partial^2 A^{Res}}{\partial c_i \partial c_r})^j (\delta_{jt} c_r^j + f^j \frac{\partial c_r^j}{\partial f^t}) \\
&= \frac{\delta_{jt}}{f^j}(\frac{\partial^2 A^{Res}}{\partial v \partial c_i})^j v^j + (\frac{\partial^2 A^{Res}}{\partial v \partial c_i})^j (\frac{\partial v^j}{\partial f^t}) + \frac{\delta_{jt}}{f^j}\sum_r (\frac{\partial^2 A^{Res}}{\partial c_i \partial c_r})^j c_r^j + \sum_r (\frac{\partial^2 A^{Res}}{\partial c_i \partial c_r})^j (\frac{\partial c_r^j}{\partial f^t}) \\
&= \frac{\delta_{jt}}{f^j}\left[(\frac{\partial^2 A^{Res}}{\partial v \partial c_i})^j v^j + \sum_r (\frac{\partial^2 A^{Res}}{\partial c_i \partial c_r})^j c_r^j \right] + (\frac{\partial^2 A^{Res}}{\partial v \partial c_i})^j (\frac{\partial v^j}{\partial f^t}) + \sum_r (\frac{\partial^2 A^{Res}}{\partial c_i \partial c_r})^j (\frac{\partial c_r^j}{\partial f^t}) \\
&= (\frac{\partial^2 A^{Res}}{\partial v \partial c_i})^j (\frac{\partial v^j}{\partial f^t}) + \sum_r (\frac{\partial^2 A^{Res}}{\partial c_i \partial c_r})^j (\frac{\partial c_r^j}{\partial f^t})
\end{aligned}$$

The term in the square brackets is identically zero due to $(\frac{\partial A^{Res}}{\partial n_i})$ being a $0th$ order homogeneous function of V and $\overline{n}$. See the development of Eq. 6.140 for more details. The analogous expression can be written for the reference phase:

$$\frac{\partial}{\partial f^t}(\frac{\partial A^{Res}}{\partial n_i})^R = (\frac{\partial^2 A^{Res}}{\partial v \partial c_i})^R(\frac{\partial v^R}{\partial f^t}) + \sum_r (\frac{\partial^2 A^{Res}}{\partial c_i \partial c_r})^R(\frac{\partial c_r^R}{\partial f^t})$$

The f derivatives of the volume expressions (Eqs. 6.154 and 6.155) are:

$$\frac{\partial v^R}{\partial f^t} = -\frac{(v^R)^2}{v}\sum_{s \neq R}\delta_{st}(L^s - 1) = -\frac{(v^R)^2}{v}(L^t - 1) \tag{6.159}$$

$$\frac{\partial v^j}{\partial f^t} = L^j\frac{\partial v^R}{\partial f^t} = -L^j\frac{(v^R)^2}{v}(L^t - 1) = -\frac{v^R v^j}{v}(L^t - 1) \tag{6.160}$$

The composition derivatives with respect to f are given by Eqs. 8.163 and 8.164. We can therefore write:

$$\frac{\partial EQV_i^j}{\partial f^t} = \frac{1}{R_g T}\Bigg[\left(\frac{\partial^2 A^{Res}}{\partial v \partial c_i}\right)^j\left(\frac{\partial v^j}{\partial f^t}\right) - \left(\frac{\partial^2 A^{Res}}{\partial v \partial c_i}\right)^R\left(\frac{\partial v^R}{\partial f^t}\right)$$
$$+\sum_r\Bigg[\left(\frac{\partial^2 A^{Res}}{\partial c_i \partial c_r}\right)^j\left(\frac{\partial c_r^j}{\partial f^t}\right) - \left(\frac{\partial^2 A^{Res}}{\partial c_i \partial c_r}\right)^R\left(\frac{\partial c_r^R}{\partial f^t}\right)\Bigg]\Bigg] \tag{6.161}$$

The derivatives of RR with respect to lnk and f have already been developed (Eqs. 8.166 and 8.167), while $\frac{\partial RR^j}{\partial L^t} = 0$ (see Eq. 8.160).

Finally, we develop the derivatives of $Press$ (Eq. 6.149) with respect to α (Eq. 6.151). Starting with lnk. By the same logic used in the development of Eq. 6.152, we can write:

$$\frac{\partial Press^j}{\partial lnk_q^t} = \left(\frac{\partial Z}{\partial c_q}\right)^j\frac{\partial c_q^j}{\partial lnk_q^t} - \left(\frac{\partial Z}{\partial c_q}\right)^R\frac{\partial c_q^R}{\partial lnk_q^t}L^t \tag{6.162}$$

Similarly, the derivative of $Press^j$ with respect to L^t is:

$$\frac{\partial Press^j}{\partial L^t} = \left(\frac{\partial Z}{\partial v}\right)^j\frac{\partial v^j}{\partial L^t} - \left(\frac{\partial Z}{\partial v}\right)^R\frac{\partial v^R}{\partial L^t}L^t - Z^R\delta_{jt} \tag{6.163}$$

Finally, the derivative of $Press^j$ with respect to f^t is:

$$\frac{\partial Press^j}{\partial f^t} = \left(\frac{\partial Z}{\partial v}\right)^j \frac{\partial v^j}{\partial f^t} - \left(\frac{\partial Z}{\partial v}\right)^R \frac{\partial v^R}{\partial f^t} L^t + \sum_r \left[\left(\frac{\partial Z}{\partial c_r}\right)^j \frac{\partial c_r^j}{\partial f^t} - \left(\frac{\partial Z}{\partial c_r}\right)^R \frac{\partial c_r^R}{\partial f^t} L^j\right] \tag{6.164}$$

At this point, we recognize a challenge associated with how we have structured our equations. For example, both our objective function and our set of unknowns are each technically (C + 2) by ($\pi - 1$) in dimension. This can be seen by reviewing Eqs. 6.150 and 6.151, and understood by considering each phase (excluding the reference phase) as having C (the components) plus two (the Rachford Rice and the pressure constraint) elements. As a result, our Jacobian is technically (C+2) by ($\pi - 1$) by (C+2) by ($\pi - 1$). This makes inverting the Jacobian for the purposes of Newton's method very difficult. Instead, we adapt the mapping concept developed in Appendix 8.13.1. For a three-component three-phase mixture, we would then have:

$$\alpha_{short} = \begin{matrix} lnk_1^1 & lnk_1^2 \\ lnk_2^1 & lnk_2^2 \\ lnk_3^1 & lnk_3^2 \\ L^1 & L^2 \\ f^1 & f^2 \end{matrix} \tag{6.165}$$

$$F_{short} = \begin{matrix} EQV_1^1 & EQV_1^2 \\ EQV_2^1 & EQV_2^2 \\ EQV_3^1 & EQV_3^2 \\ RR^1 & RR^2 \\ Press^1 & Press^2 \end{matrix} \tag{6.166}$$

The mapping is then:

$$Map = \begin{matrix} 1 & 6 \\ 2 & 7 \\ 3 & 8 \\ 4 & 9 \\ 5 & 10 \end{matrix} \tag{6.167}$$

The number of rows in $\overline{\overline{F}}_{short}$ is two more than the number of components in the mixture, while the number of columns is one less than the number of phases. Algorithm 15 outlines the required mapping;

Algorithm 15 Multiphase TVn Flash Mapping

```
Initialize idx
for each row in F_short do
    for each col in F_short do
        Map[row, col] = idx
        idx+ = 1
    end for
end for
```

providing a definition for the long form in terms of the short:

$$Long[Map[row, col]] = Short[row, col]$$

In terms of F, this yields:

$$F_{long} = \begin{matrix} EQ_1^1 \\ EQ_2^1 \\ EQ_3^1 \\ RR^1 \\ Press^1 \\ EQ_1^2 \\ EQ_2^2 \\ EQ_3^2 \\ RR^2 \\ Press^2 \end{matrix} \tag{6.168}$$

The same mapping provides a definition for α_{long}:

$$\alpha_{long} = \begin{matrix} lnk_1^1 \\ lnk_2^1 \\ lnk_3^1 \\ L^1 \\ f^1 \\ lnk_1^2 \\ lnk_2^2 \\ lnk_3^2 \\ LL^2 \\ f^2 \end{matrix} \tag{6.169}$$

This mapping also defines a long form structure for H for a C-component, π-phase system, as outlined in Algorithm 16:

Algorithm 16 Multiphase TVn Jacobian Mapping

Create H_{long} of dimension $(C+2)*(\pi-1)$ by $(C+2)*(\pi-1)$

EQV with lnk
for $species_i$ **do**
  for $phase_j$ **do** (not including reference phase)
    for $species_q$ **do**
      for $phase_t$ **do** (not including reference phase)
        $idx_1 = Map[species_i, phase_j]$
        $idx_2 = Map[species_q, phase_t]$
        $H_{long}[idx_1, idx_2] = \frac{\partial \overline{\overline{EQV}}}{\partial \overline{\overline{lnk}}}[species_i, phase_j, species_q, phase_t]$
      end for
    end for
  end for
end for

EQV with L
Set idx2 to second last column of H_{long}
for $phase_t$ **do** (not including reference phase)
  $idx2+ = C + 2$ (This assumes your language supports zero indexing and index arithmetic)
  for $species_i$ **do**
    for $phase_j$ (not including reference phases) **do**
      idx1 = $Map[species_i, phase_j]$
      $H[idx1, idx2] = \frac{\partial \overline{\overline{EQV}}}{\partial \overline{L}}[species_i, phase_j, phase_t]$
    end for
  end for
end for

EQV with f
Set idx2 to last column of H_{long}
for $phase_t$ **do** (not including reference phase)
  $idx2+ = C + 2$ (This assumes your language supports zero indexing and index arithmetic)
  for $species_i$ **do**
    for $phase_j$ (not including reference phases) **do**
      idx1 = $Map[species_i, phase_j]$
      $H[idx1, idx2] = \frac{\partial \overline{\overline{EQV}}}{\partial \overline{f}}[species_i, phase_j, phase_t]$
    end for
  end for
end for

RR with lnk
Set idx1 to last second last row of H_{long}
for $phase_j$ **do** (not including reference phase)
  $idx1+ = C + 2$ (This assumes your language supports zero indexing and index arithmetic)
  for $species_q$ **do**
    for $phase_t$ (not including reference phases) **do**
      idx2 = $Map[species_q, phase_t]$
      $H[idx1, idx2] = \frac{\partial \overline{RR}}{\partial \overline{\overline{lnk}}}[phase_j, species_q, phase_t]$
    end for
  end for
end for

Algorithm 16 Continued.

RR with $L = 0$

RR with f
Set idx1 to last second last row of H_{long}
for $phase_j$ **do** (not including reference phase)
  $idx1+ = C + 2$ (This assumes your language supports zero indexing and index arithmetic)
  Set idx2 to last column of H_long
  for $phase_t$ **do** (not including reference phase)
    $idx2+ = C + 2$ (This assumes your language supports zero indexing and index arithmetic)
    $H[idx1, idx2] = \frac{\partial \overline{RR}}{\partial f}[phase_j, phase_t]$
  end for
end for

$Press$ with lnk
Set idx1 to last row of H_{long}
for $phase_j$ **do** (not including reference phase)
  $idx1+ = C + 2$ (This assumes your language supports zero indexing and index arithmetic)
  for $species_q$ **do**
    for $phase_t$ **do** (not including reference phase)
      $idx2 = Map[species_q, phase_t]$
      $H[idx1, idx2] = \frac{\partial \overline{Press}}{\partial ln\overline{\overline{k}}}[phase_j, comp_q, phase_t]$
    end for
  end for
end for

$Press$ with L
Set idx1 to last row of H_{long}
for $phase_j$ **do** (not including reference phase)
  $idx1+ = C + 2$ (This assumes your language supports zero indexing and index arithmetic)
  Set idx2 to second last column of H_{long}
  for $phase_t$ **do** (not including reference phase)
    $idx2+ = C + 2$ (This assumes your language supports zero indexing and index arithmetic)
    $H_{long}[idx1, idx2] = \frac{\partial \overline{Press}}{\partial \overline{L}}[phase_j, phase_t]$
  end for
end for

$Press$ with f
Set idx1 to last row of H_{long}
for $phase_j$ **do** (not including reference phase)
  $idx1+ = C + 2$ (This assumes your language supports zero indexing and index arithmetic)
  Set idx2 to last column of H_{long}
  for $phase_t$ **do** (not including reference phase)
    $idx2+ = C + 2$ (This assumes your language supports zero indexing and index arithmetic)
    $H_{long}[idx1, idx2] = \frac{\partial \overline{Press}}{\partial \overline{f}}[phase_j, phase_t]$
  end for
end for

$$H_{long} = \begin{matrix} \frac{\partial EQ_1^1}{\partial lnk_1^1} & \frac{\partial EQ_1^1}{\partial lnk_2^1} & \frac{\partial EQ_1^1}{\partial lnk_3^1} & \frac{\partial EQ_1^1}{\partial L^1} & \frac{\partial EQ_1^1}{\partial f^1} & \frac{\partial EQ_1^1}{\partial lnk_1^2} & \frac{\partial EQ_1^1}{\partial lnk_2^2} & \frac{\partial EQ_1^1}{\partial lnk_3^2} & \frac{\partial EQ_1^1}{\partial L^2} & \frac{\partial EQ_1^1}{\partial f^2} \\ \frac{\partial EQ_2^1}{\partial lnk_1^1} & \frac{\partial EQ_2^1}{\partial lnk_2^1} & \frac{\partial EQ_2^1}{\partial lnk_3^1} & \frac{\partial EQ_2^1}{\partial L^1} & \frac{\partial EQ_2^1}{\partial f^1} & \frac{\partial EQ_2^1}{\partial lnk_1^2} & \frac{\partial EQ_2^1}{\partial lnk_2^2} & \frac{\partial EQ_2^1}{\partial lnk_3^2} & \frac{\partial EQ_2^1}{\partial L^2} & \frac{\partial EQ_2^1}{\partial f^2} \\ \frac{\partial EQ_3^1}{\partial lnk_1^1} & \frac{\partial EQ_3^1}{\partial lnk_2^1} & \frac{\partial EQ_3^1}{\partial lnk_3^1} & \frac{\partial EQ_3^1}{\partial L^1} & \frac{\partial EQ_3^1}{\partial f^1} & \frac{\partial EQ_3^1}{\partial lnk_1^2} & \frac{\partial EQ_3^1}{\partial lnk_2^2} & \frac{\partial EQ_3^1}{\partial lnk_3^2} & \frac{\partial EQ_3^1}{\partial L^2} & \frac{\partial EQ_3^1}{\partial f^2} \\ \frac{\partial RR^1}{\partial lnk_1^1} & \frac{\partial RR^1}{\partial lnk_2^1} & \frac{\partial RR^1}{\partial lnk_3^1} & \frac{\partial RR^1}{\partial L^1} & \frac{\partial RR^1}{\partial f^1} & \frac{\partial RR^1}{\partial lnk_1^2} & \frac{\partial RR^1}{\partial lnk_2^2} & \frac{\partial RR^1}{\partial lnk_3^2} & \frac{\partial RR^1}{\partial L^2} & \frac{\partial RR^1}{\partial f^2} \\ \frac{\partial Press^1}{\partial lnk_1^1} & \frac{\partial Press^1}{\partial lnk_2^1} & \frac{\partial Press^1}{\partial lnk_3^1} & \frac{\partial Press^1}{\partial L^1} & \frac{\partial Press^1}{\partial f^1} & \frac{\partial Press^1}{\partial lnk_1^2} & \frac{\partial Press^1}{\partial lnk_2^2} & \frac{\partial Press^1}{\partial lnk_3^2} & \frac{\partial Press^1}{\partial L^2} & \frac{\partial Press^1}{\partial f^2} \\ \frac{\partial EQ_1^2}{\partial lnk_1^1} & \frac{\partial EQ_1^2}{\partial lnk_2^1} & \frac{\partial EQ_1^2}{\partial lnk_3^1} & \frac{\partial EQ_1^2}{\partial L^1} & \frac{\partial EQ_1^2}{\partial f^1} & \frac{\partial EQ_1^2}{\partial lnk_1^2} & \frac{\partial EQ_1^2}{\partial lnk_2^2} & \frac{\partial EQ_1^2}{\partial lnk_3^2} & \frac{\partial EQ_1^2}{\partial L^2} & \frac{\partial EQ_1^2}{\partial f^2} \\ \frac{\partial EQ_2^2}{\partial lnk_1^1} & \frac{\partial EQ_2^2}{\partial lnk_2^1} & \frac{\partial EQ_2^2}{\partial lnk_3^1} & \frac{\partial EQ_2^2}{\partial L^1} & \frac{\partial EQ_2^2}{\partial f^1} & \frac{\partial EQ_2^2}{\partial lnk_1^2} & \frac{\partial EQ_2^2}{\partial lnk_2^2} & \frac{\partial EQ_2^2}{\partial lnk_3^2} & \frac{\partial EQ_2^2}{\partial L^2} & \frac{\partial EQ_2^2}{\partial f^2} \\ \frac{\partial EQ_3^2}{\partial lnk_1^1} & \frac{\partial EQ_3^2}{\partial lnk_2^1} & \frac{\partial EQ_3^2}{\partial lnk_3^1} & \frac{\partial EQ_3^2}{\partial L^1} & \frac{\partial EQ_3^2}{\partial f^1} & \frac{\partial EQ_3^2}{\partial lnk_1^2} & \frac{\partial EQ_3^2}{\partial lnk_2^2} & \frac{\partial EQ_3^2}{\partial lnk_3^2} & \frac{\partial EQ_3^2}{\partial L^2} & \frac{\partial EQ_3^2}{\partial f^2} \\ \frac{\partial RR^2}{\partial lnk_1^1} & \frac{\partial RR^2}{\partial lnk_2^1} & \frac{\partial RR^2}{\partial lnk_3^1} & \frac{\partial RR^2}{\partial L^1} & \frac{\partial RR^2}{\partial f^1} & \frac{\partial RR^2}{\partial lnk_1^2} & \frac{\partial RR^2}{\partial lnk_2^2} & \frac{\partial RR^2}{\partial lnk_3^2} & \frac{\partial RR^2}{\partial L^2} & \frac{\partial RR^2}{\partial f^2} \\ \frac{\partial Press^2}{\partial lnk_1^1} & \frac{\partial Press^2}{\partial lnk_2^1} & \frac{\partial Press^2}{\partial lnk_3^1} & \frac{\partial Press^2}{\partial L^1} & \frac{\partial Press^2}{\partial f^1} & \frac{\partial Press^2}{\partial lnk_1^2} & \frac{\partial Press^2}{\partial lnk_2^2} & \frac{\partial Press^2}{\partial lnk_3^2} & \frac{\partial Press^2}{\partial L^2} & \frac{\partial Press^2}{\partial f^2} \end{matrix} \tag{6.170}$$

This allows an update to α_{long} via the usual application of Newton's method:

$$d\overline{\alpha}_{long} = -\overline{\overline{H}}_{long}^{-1} * \overline{F}_{long} \tag{6.171}$$

We can then reverse the mapping to extract the short form of the updated α_{long} via Algorithm 17, which provides all the elements required to recalculate F. This is clarified, along with a method of initialization, in Example 27.

Algorithm 17 Multiphase TVn Long to Short

```
Initialize idx
for each row in α_short do
    for each col in α_short do
        α_short[row, col] = α_long[idx]
        idx+ = 1
    end for
end for
```

Example 27 (Three-Phase TVn Flash) Consider an equimolar mixture of propane, n-hexane, and water ($\overline{n} = 1, 1, 1$) at 350K and a volume $V = 6885.87\text{cm}^3$, modeled with the SRK EOS. Note that this example has been structured to correspond with Example 25. A stability analysis (Sect. 6.4.2) yields:

$$W = 9.287E - 7, 3.09E - 12, 290.253(moles)$$

$$t_V = -608.811(moles)$$

$$P_V = 2583.067(bar)$$

The methods of Sect. 6.7.3[49] yield:

$$\overline{v} = 24.767, 3254.199(\frac{\text{cm}^3}{\text{mol}})$$

$$\overline{\overline{c}} = \begin{bmatrix} 3.182E-6 & 0.474 \\ 5.249E-9 & 0.474 \\ 0.999.. & 5.178E-2 \end{bmatrix}$$

$$\overline{f} = 0.297, 0.703$$

It is worth noting that the stability analysis detected a water-rich phase and that the flash has indeed converged accordingly. However, the compositions of propane and n-hexane in the coexisting phase are identical to the significant digits shown. Although our intention is to develop automatic routines that do not rely entirely on our chemical insight, this suggests that our two-phase flash is not the global minimum of our mixture's (Helmholtz free energy) potential; so far we have simply found a state where propane and n-hexane are (nearly) equally unhappy to be mixed with water, despite the fact that these chemicals could further separate from each other. Put another way, it might be said that the cost of mixing with each other is lower than the penalty of mixing with water. To be sure that no additional phases could form, we check the stability of our two-phase mixture, but which volume and composition do we use? Neither of the coexisting phases occupy the volume V entirely. We could use the molar volume and composition of one[50] of the phases, but we want our algorithm to handle two or more phases in a consistent way. We therefore *scale* our test phase to the total volume of the mixture:

$$v \rightarrow V$$

$$c \rightarrow \frac{V}{v}c$$

In words, we are testing the stability of our *container* with volume V, assuming it was *entirely* filled with a mixture of composition c. Such a stability check (Sect. 6.4.2) yields:

$$W = 4.215, 42.536, 2.383(moles)$$

$$t_V = -39.172(moles)$$

[49] The methods of the present section should be implemented to handle two or more phases. Section 6.7.3 is referenced for those attempting to develop the present section.

[50] Besides questions of numerical tolerance, the chemical potentials and pressures of both phases are identical in our current two-phase state, so our stability test should give consistent information, regardless of the phase we check.

$$P_V = 172.951(bar)$$

We therefore generate the following incipient phase and molar volume:

$$\overline{w} = \frac{\overline{W}}{W} = 0.086, 0.866, 0.048$$

$$v = 140.147(\frac{\text{cm}^3}{\text{mol}})$$

Regarding initialization, we must consider how much of the w phase we can possibly form. For the sake of generality, we must recognize that *all* the phases contribute to the formation of the incipient phase, despite the above formalism that "the organic phase has yet to split." Looking carefully at the interim (two-phase) results, we see that the aqueous phase has only traces of propane and hexane. Looking specifically at n-hexane concentrations, we see that the aqueous phase, with a hexane fraction of $5.25E-9$, needs to contribute to a w phase with a hexane fraction of 0.86. Clearly the aqueous phase can contribute only a small amount of material to form w. The key to resolving this is to consider both mole fractions (c) and phase fractions (f) simultaneously, as was explained in Example 26. The volumes and mole numbers in each phase before the formation of the incipient phase are:

$$V^s = nf^s v^s$$
$$n_i^s = nf^s c_i^s$$

Upon the formation of the incipient (w) phase, we have:

$$\begin{aligned} V^{s'} &= n(f^s v^s - f_{w,v}^s v^w) \\ n_i^{s'} &= n(f^s c_i^s - f_{w,i}^s w_i) \end{aligned} \tag{6.172}$$

As indicated, each phase s makes a contribution f_w^s to the w phase, and this amount is limited by the roots of Eq. 6.172:

$$(f^s v^s - f_{w,v}^s v^w) = 0$$
$$(f^s c_i^s - f_{w,i}^s w_i) = 0$$

These roots are:

$$\begin{aligned} f_{w,v}^s &= \frac{f^s v^s}{v^w} \\ f_{w,i}^s &= \frac{f^s c_i^s}{w_i} \end{aligned} \tag{6.173}$$

We then have $C + 1$ values of f_w for each phase s. Note that Eqs. 6.173 are consistent[51] with Eqs. 6.146 and 6.147, and in order to satisfy the volume and material balance constraints, we take half[52] of the lesser of these two values:

$$f^s_{w,max} = \frac{min(f^s_{w,v}, f^s_{w,i})}{2}$$

We then have contributions from *each* phase to the incipient w phase, and these are debited from the volumes, mole numbers, and phase fractions of the existing phases via:

$$V^{s'} = n(f^s v^s - f^s_{w,max} v^w)$$

$$n_i^{s'} = n(f^s c_i^s - f^s_{w,max} w_i)$$

$$f^{s'} = f^s - f^s_{w,max}$$

We finally create the incipient phase via:

$$V^{w'} = n(\sum_s f^s_{w,max}) v^w$$

$$n^{w'} = n(\sum_s f^s_{w,max}) w_i$$

$$f^{w'} = \sum_s f^s_{w,max}$$

and then normalize the volumes and composition:

$$\overline{v}' = \frac{\overline{V}'}{n'}$$

$$\overline{c}' = \frac{\overline{n}'}{n'}$$

[51] When stability is performed at the beginning of the two-phase flash, the "initial" phase is the feed with a phase fraction of unity.

[52] We must avoid negative volumes and mole fractions in the phases we have already identified.

To illustrate, let's consider the results of the two-phase TVn flash converged at the beginning of the present example. Equation 6.173 gives:

$$f^s_{w,v} = 0.05, 16.325$$

$$f^s_{w,i} = \begin{bmatrix} 1.101E-5 & 3.886 \\ 1.800E-9 & 0.385 \\ 6.123 & 0.751 \end{bmatrix}$$

The appropriate contributions are then $f^s_w = 9.000E-10, 0.193$. The normalized volumes, compositions and phase fractions become:

$$\overline{v} = 24.767, 4428.435, 147.145$$

$$\overline{\overline{c}} = \begin{bmatrix} 3.182E-6 & 0.621 & 0.086 \\ 2.62E-9 & 0.326 & 0.866 \\ 0.999... & 0.053 & 0.048 \end{bmatrix}$$

$$\overline{f} = 0.297, 0.510, 0.193$$

The phase with 62% propane is likely a "vapor", and when densities are compared ($\overline{\rho} = 0.727, 0.013, 0.566 \frac{g}{\text{cm}^3}$), we recognize that the phases need to be reordered. The initialized variables of interest (Equation 6.151) for the three-phase flash become:

$$\alpha_{short} = \begin{bmatrix} 10.202 & 12.181 \\ 19.614 & 18.639 \\ -3.026 & -2.937 \\ 5.659 & 178.807 \\ 0.193 & 0.511 \end{bmatrix}$$

The initial values within F (Eq. 6.150) are:

$$F_{short} = \begin{bmatrix} 8.142E-5 & 0.040 \\ 0.693 & 0.166 \\ 1.209E-8 & -0.206 \\ -6.661E-16 & 0 \\ 0.797 & -0.231 \end{bmatrix}$$

The *long* form of the Jacobian (Eq. 6.170) is:

$$H_{long}$$

$$= \left[\begin{matrix} 1.575 & 5.129 & 3.323 & -2.331 & -360.114 \\ 0.966 & 9.397 & 6.303 & -3.773 & -677.861 \\ 0.176 & 1.621 & 1.914 & -0.839 & -106.583 \\ 0.082 & 0.433 & 0.075 & 0 & -4.987 \\ 0.908 & 7.860 & 5.682 & -3.358 & -611.699 \\ 3.7484E-3 & 0.038 & 3.222 & -0.239 & -333.975 \\ 7.285E-3 & 0.074 & 6.127 & -0.455 & -635.034 \\ 3.819E-3 & 0.038 & 0.947 & -0.071 & -98.000 \\ -0.031 & -0.163 & 0.027 & 0 & -3.711 \\ 4.907E-3 & 0.033 & 174.252 & -12.925 & -18073.1 \end{matrix}\right.$$

$$\left.\begin{matrix} -0.574 & -5.129 & 9.334 & -0.573 & -537.364 \\ -0.964 & -8.397 & 17.750 & -1.106 & -1024.42 \\ -0.175 & -1.621 & 2.748 & -0.168 & -157.989 \\ -0.082 & -0.433 & 0.077 & 0 & -3.711 \\ -0.906 & -7.86 & 15.977 & -0.992 & -921.523 \\ 0.997 & -0.038 & 9.337 & -0.639 & -547.638 \\ -5.369E-3 & 0.926 & 17.754 & -1.211 & -1041.32 \\ -3.454E-3 & -0.038 & 3.739 & -0.192 & -160.991 \\ 0.031 & 0.163 & 0.130 & 0 & -4.165 \\ 0.061 & -0.033 & 505.299 & -34.283 & -29613.1 \end{matrix}\right]$$

We now convert F to the long form (Eq. 6.168) and then apply Newton's method (Sect. 8.13) via Eq. 6.171. This gives the following update to α_{long}:

$$d\overline{\alpha}_{long} = -\overline{\overline{H}}_{long}^{-1} * \overline{F}_{long}$$

$$= 0.551, 0.178, 0.203, 0.105, 0.113, 0.102, -0.070, 0.353, 34.212, -0.101$$

With the update rather large in magnitude, stepsize control is applied to restrict the fractional change in α_{long} to less than 0.1 in magnitude. After the update, we convert to α_{short} and extract the flash variables;

$$ln\overline{k} = \begin{bmatrix} 10.271 & 12.194 \\ 19.637 & 18.630 \\ -3.001 & -2.893 \end{bmatrix}$$

$$\overline{L} = 5.672, 183.084$$

$$\overline{f} = 0.207, 0.498$$

The converged 3-phase flash gives;

$$\overline{v} = 24.768, 144.778, 5382.427 \frac{cm3}{\mathrm{mol}}$$

$$\overline{\overline{c}} = \begin{bmatrix} 3.165E-6 & 0.152 & 0.690 \\ 1.986E-9 & 0.789 & 0.234 \\ 0.999... & 0.059 & 0.076 \end{bmatrix}$$

$$\overline{f} = 0.284, 0.299, 0.417$$

and stability indicates no further tendency to separate. Notice also that the results of this flash are entirely consistent with the results in Example 24.

Chapter 7
Glossary

Canonical
As a word, canonical means following a rule or agreement. In thermodynamics, it is often used to describe the natural or preferred variables to describe variation. For example, the canononical variables of enthalpy (i.e., those following directly from the Thermodynamic Laws) are S, P and $\overline{n}$. At fixed mole numbers, the canonical representation of enthalpy is:

$$\begin{aligned} dH &= \left(\frac{\partial H}{\partial S}\right)_S dS + \left(\frac{\partial H}{\partial P}\right)_P dP \\ &= TdS + VdP \end{aligned} \tag{7.1}$$

If instead one wishes to describe enthalpy at fixed composition in terms of temperature and pressure, we must now write:

$$\begin{aligned} dH &= \left(\frac{\partial H}{\partial T}\right)_P dT + \left(\frac{\partial H}{\partial P}\right)_P dP \\ &= CpdT + \left[V - T\left(\frac{\partial V}{\partial T}\right)_P\right] dP \end{aligned} \tag{7.2}$$

Both of these representations of enthalpy (at fixed mole numbers) are equally valid. The simplicity of the former over the latter illustrates the significance of the term canonical.

Extensive and Intensive Properties
An extensive property is one that depends on the amount of material in the system. To avoid confusion, we use the word *amount* rather than *size* of the system. Size might be interpreted as the volume of the system, and one could argue that certain changes in volume lead to changes in temperature, leading to the false conclusion

B. R. Greenhalgh, *An Introduction to Computational Thermodynamics*,
https://doi.org/10.1007/978-3-032-30317-2_7

that temperature is an extensive property. What we really mean by amount is mass or moles. Examples of extensive properties include volume (in cm^3) and enthalpy (in J). An intensive property is one that does not depend on the amount of material in the system. Examples of intensive properties include temperature, pressure, and density, as well as specific properties such as specific enthalpy (in $\frac{J}{mol}$).

Path and State Functions
The change in value of a path function depends on the path connecting the initial and final states. This is in contrast to a state function, for which this change depends only on the initial and final states. Work and heat are examples of path functions, while internal energy and fugacity are examples of state functions.

System, Surroundings, and Boundary
The material on which we focus is called the System, while everything else in the Universe is called the Surroundings. This naturally leads to a division between the two, which we call the Boundary. The nature of the Boundary further characterizes the System. If material can cross the Boundary, we refer to the System as *open*, while in the opposite case we call it *closed*. Other terms used to describe the boundary are *adiabatic* (no heat flow across the Boundary) and *rigid* (no movement of the Boundary). If the Boundary prevents material and energy exchange, the System is termed *isolated*. Many other classifications exist.

Chapter 8
Appendices

8.1 The Association Contribution to A^{Res}

Cubic EOS extend the ideal gas model, from assuming no interactions between particles to both attractive and repulsive ones. These combined give rise to phase behavior with a continuous function. Association extends this idea, assigning to some molecule sites of high and low electron densities (donors and acceptors, representing hydrogen bonding or other such associations), which affect the Helmholtz free energy beyond the simple cubic models. For example, water can be modeled as having two donor sites (due to the lone pairs on the oxygen atom) and two acceptor sites (due to the two protons) as an attempt to explain phenomena such as the high boiling point of water despite its low molecular weight. Further discussion can be found in the published work of Marshall [Marshall paper].

As shown by Michelsen and Hendriks [Michelsen and Hendriks paper], the association contribution to A^{Res} is given by:

$$A^{assoc} = R_g T \sum_r n_r \sum_{Ar} (lnX_{Ar} - \frac{1}{2} X_{Ar} + \frac{1}{2}) \tag{8.1}$$

In Eq. 8.1, X_{Ai} is the fraction of sites A on a molecule i that are *not bonded* to other sites (note that with all $X = 1$, we have $A^{assoc} = 0$). Of course, some species have no bonding sites (e.g., n-alkanes), some may be modeled as having only donors (e.g., CO_2) and while as above water is often modeled as having both donors and acceptors. We also must contend with mixtures of such species. Our task is then to determine the extent of bonding of all sites (i.e., to determine X_{Ai}). Michelsen and

B. R. Greenhalgh, *An Introduction to Computational Thermodynamics*,
https://doi.org/10.1007/978-3-032-30317-2_8

Hendriks [Michelsen and Hendriks paper] show that this can be represented as an optimization problem, wherein one minimizes the following function:[1]

$$Q(T, V, \overline{n}, \overline{X}) = \sum_r n_r \Bigg[\sum_{Ar} (lnX_{Ar} - X_{Ar} + 1) - \frac{N_{av}}{2E24V} \sum_s n_s \sum_{Ar} X_{Ar} \sum_{Bs} X_{Bs} \Delta^{ArBs} \Bigg] \tag{8.2}$$

The function Q has been structured [Michelsen and Hendriks paper] such that, at its minimum with respect to X values:

$$Q(T, V, \overline{n}, \overline{X}^*) = \sum_r n_r \sum_{Ar} (lnX^*_{Ar} - \frac{1}{2} X^*_{Ar} + \frac{1}{2}) \tag{8.3}$$

That is, the minimum value of the function $R_g T Q$ is *equal* to A^{assoc} at the solution:

$$A^{assoc}(T, V, \overline{n}) = R_g T Q(T, V, \overline{n}, \overline{X^*}) \tag{8.4}$$

Differentiation of Eq. 8.2 with respect to X_{Ck} gives:

$$\begin{aligned}
\left(\frac{\partial Q}{\partial X_{Ck}}\right)_{T,V,\overline{n},X_{Bs \neq Ck}} &= \sum_r n_r \Bigg[\sum_{Ar} \left(\frac{\delta_{AC}\delta_{rk}}{X_{Ar}} - \delta_{AC}\delta_{rk} \right) \\
&- \frac{N_{av}}{2E24V} \sum_s n_s \sum_{Ar} \left(\delta_{AC}\delta_{rk} \sum_{Bs} X_{Bs} \Delta^{ArBs} + X_{Ar} \sum_{Bs} \delta_{BC}\delta_{sk} \Delta^{ArBs} \right) \Bigg] \\
&= n_k \left[\left(\frac{1}{X_{Ck}} - 1 \right) - \frac{N_{av}}{2E24V} \sum_s n_s \sum_{Bs} X_{Bs} \Delta^{CkBs} \right] \\
&- \sum_r n_r \left[\frac{N_{av}}{2E24V} n_k \sum_{Ar} X_{Ar} \Delta^{ArCk} \right] \\
&= n_k \left[\frac{1}{X_{Ck}} - 1 - \frac{N_{av}}{1E24V} \sum_s n_s \sum_{Bs} X_{Bs} \Delta^{CkBs} \right]
\end{aligned} \tag{8.5}$$

Setting the derivative to zero;

$$\left(\frac{\partial Q}{\partial X_{Ck}}\right)_{T,V,\overline{n},X_{Bs \neq Ck}} = 0 \tag{8.6}$$

[1] Here we have modified the expression to account for our chosen units.

noting that $n_k \neq 0$, and recognizing that the choice of index is quite arbitrary, we have:

$$X_{Ai} = \frac{1}{1 + \frac{N_{av}}{1E24V} \sum_s n_s \sum_{Bs} X_{Bs} \Delta^{AiBs}} \quad (8.7)$$

Equation 8.7 must be satisfied[2] for all association sites in the mixture. We now turn our attention to the calculation of Δ^{AiBj}. The basic idea is that an associating molecule has a particular energy[3] (ϵ_i^{AB}) and fractional volume (κ_i^{AB}). We now consider the interaction of two sites (either from like or unlike molecules):

$$\nabla^{AiBj} = \kappa_{ij}^{AB} \sigma_{ij}^3 g_{ij}^{hs} \left[exp(\frac{\epsilon_{ij}^{AB}}{T}) - 1\right] \quad (8.8)$$

The following mixing rules (in addition to Eqs. 2.38 and 2.45) are assumed:

$$\epsilon_{ij}^{AB} = \frac{1}{2}\left[\epsilon_i^{AB} + \epsilon_j^{AB}\right] \quad (8.9)$$

$$\kappa_{ij}^{AB} = \frac{\sqrt{\sigma_i^3 \kappa_i^{AB} \sigma_j^3 \kappa_j^{AB}}}{\sigma_{ij}^3} \quad (8.10)$$

We now further consider that only sites of opposite type form bonds; there is no association between donors or between acceptors. We handle this by translating the above interaction matrix (Eq. 8.8) into an association matrix:

$$\Delta^{AiBj} = \begin{cases} \nabla^{AiBj} & Type(Ai) \neq Type(Bj) \\ 0 & Type(Ai) = Type(Bj) \end{cases} \quad (8.11)$$

We therefore need to create site mappings, one connecting the sites with the species, and another stating the type of each site. This is best illustrated by the following example.

Example 28 (Association Site Mapping for the PC-SAFT EOS) This is an extension of Example 2 in Sect. 2.2. We will now generate the site mappings for a mixture of 1 mole of propane, 1 mole of n-hexane and 1 mole of water. We first

[2] For ease of notation, we will no longer write X^*; it will be understood that when calculating EOS properties, we use X values that satisfy Eq. 8.7.

[3] The dimension of this parameter is technically a temperature. Some implementations write this parameter as following the form of $\frac{\epsilon_i}{k}$ as in Sect. 2.2.

declare the number of sites in all species, as a list of the number of donors and acceptors:

$$\overline{n}_d = 0, 0, 2$$
$$\overline{n}_a = 0, 0, 2$$

We have four sites, all in this case belonging to water. We handle this computationally by creating a species-for-site array (using zero indexing);

$$SFS = [2, 2, 2, 2]$$

and a type-for-site array:

$$TFS = [-1, -1, 1, 1]$$

In our TFS array, we have (arbitrarily) represented donor sites with -1 and acceptors with 1. As will be seen, we simply require values that can be compared. Algorithm 18 outlines pseudocode for automatically populating these arrays.

This process will produce TFS and SFS, each of length equal to the total number of sites. It is also worth noting that the frame of reference for mapping the sites is species-based; Algorithm 18 goes species by species, examining each donor and each acceptor site thereon, and captures the type and species name of each site.

Next we turn our attention to the computation of the interaction matrix (Δ^{AiBj}). For illustration, we continue with an example.

Example 29 (Computation of the PC-SAFT Association Matrix) This is an extension of Example 2 in Sect. 2.2. We first compute the interaction matrix for the mixture of 1 mole of propane, 1 mole of n-hexane and one moles of water at 350K and at $V = 116.38\,\text{cm}^3$. The relevant PC-SAFT parameters for this mixture are shown in Table 8.1 (excerpted from Table 8.3).

The mixture association parameters are (Eq. 8.9):

$$\overline{\overline{\epsilon}}^{AB} = \begin{bmatrix} 0 & 0 & 858 \\ 0 & 0 & 858 \\ 858 & 858 & 1716 \end{bmatrix}$$

Table 8.1 PC-SAFT association parameters used for propane, n-hexane, and water

Species	$\frac{\epsilon^{AB}}{k}$ [K]	κ^{AB} [–]	Donors/acceptors [–/–]
Propane	0	0	0/0
n-Hexane	0	0	0/0
Water	1716	0.12	2/2

Algorithm 18 Association site mapping

```
Calculate sites from donors (n_d) and acceptors (n_a)
Initialize sfs and tfs, both length sites
Initialize site

Mapping
for i (species) do
    for the number of d given by n_d[i] do
        sfs[site] = i
        tfs[site] = −1
        site+ = 1
    end for
    for the number of a given by n_a[i] do
        sfs[site] = i
        tfs[site] = 1
        site+ = 1
    end for
end for

Association Matrix
Initialize Δ (sites x sites)
for s1 (sites) do
    Type1 = tfs[s1]
    for s2 (sites) do
        Type2 = tfs[s2]
        if Type1 == Type2 then
            Δ[s1, s2] = 0
        else
            spec1 = sfs[s1]
            spec2 = sfs[s2]
            Δ[s1, s2] = ∇[spec1, spec2]
        end if
    end for
end for
```

and (Eq. 8.10)

$$\overline{\overline{K}}^{AB} = \begin{bmatrix} 0 & 0 & 0 \\ 0 & 0 & 0 \\ 0 & 0 & 0.120 \end{bmatrix}$$

According to Eq. 8.8, the interaction matrix is:

$$\overline{\overline{\nabla}}^{AB} = \begin{bmatrix} 0 & 0 & 0 \\ 0 & 0 & 0 \\ 0 & 0 & 5379.913 \end{bmatrix}$$

With TFS and SFS determined in Algorithm 18 as:

$$SFS = [2, 2, 2, 2]$$

$$TFS = [-1, -1, 1, 1]$$

Algorithm 18 outlines how to determine Δ^{AiBj}, which connects the site- and species-based frames of reference; having mapped the sites into the species (SFS) and type (TFS) arrays we can simply go site by site, and extract the corresponding species and type. In the present example, the association matrix is:

$$\overline{\overline{\Delta}}^{AB} = \begin{bmatrix} 0 & 0 & 5379.914 & 5379.914 \\ 0 & 0 & 5379.914 & 5379.914 \\ 5379.914 & 5379.914 & 0 & 0 \\ 5379.914 & 5379.914 & 0 & 0 \end{bmatrix}$$

In this simple example, there is only one unique site-to-site interaction, but this will not always be the case. This approach can be used to handle general mixtures of associating species.

The next step is to compute the site fractions ($\overline{X}$) which are *governed* by Eq. 8.7, but as it is implicit, it cannot be used to calculate them directly. Instead, we must *solve* Eq. 8.7, for which we develop the following objective function:

$$F_{Ai} = 1 + \frac{Nav}{1E24V} \sum_{s} n_s \sum_{Bs} X_{Bs} \Delta^{AiBs} - \frac{1}{X_{Ai}} \tag{8.12}$$

We also need the Jacobian of this function, with respect to site fraction:

$$H_{AiCk} = \left(\frac{\partial F_{Ai}}{\partial X_{Ck}}\right)_{T,V,\overline{n},X_{Dq\neq Ck}} \tag{8.13}$$

Note that, despite the complexity of the indices, the derivative required in Eq. 8.13 involves only one changing site fraction (X_{Ck}); all other variables, including the other site fractions,[4] are held constant. This means:

$$H_{AiCk} = \frac{Nav}{1E24V} \sum_{s} n_s \sum_{Bs} \left(\frac{\partial X_{Bs}}{\partial X_{Ck}}\right) \Delta^{AiBs} + \frac{1}{X_{Ai}^2} \left(\frac{\partial X_{Ai}}{\partial X_{Ck}}\right)$$

[4] Note that the site fractions are independent; they do not sum to unity. Also note that we must use indices different from Ai and Bs for the derivative in Eq. 8.13 to capture the change in an arbitrary site fraction.

$$= \frac{Nav}{1E24V}\sum_s n_s \sum_{Bs} \delta_{BC}\delta_{sk}\Delta^{AiBs} + \frac{\delta_{AC}\delta_{ik}}{X_{Ai}^2}$$

$$= \frac{Nav}{1E24V} n_k \Delta^{AiCk} + \frac{\delta_{AC}\delta_{ik}}{X_{Ai}^2} \tag{8.14}$$

The site fractions in Eq. 8.12 are initialized by assuming the X values are all unity (unbonded). Newton's method is applied, with a stepsize control that limits the magnitudes of $\frac{d\overline{X}}{\overline{X}}$. This is illustrated in Example 30.

Example 30 (Solving the X^{assoc} Site Fractions for the PC-SAFT EOS) This is an extension of Example 2 in Sect. 2.2. With the association matrix calculated in Example 29, we compute the site fractions for 1 mole of propane, 1 mole of n-hexane, and 1 mole water at $T = 350$K and $V = 116.38\,\text{cm}^3$.

We initialize with:

$$\overline{X} = 1, 1, 1, 1$$

and then compute:

Equation 8.12 gives:

$$\overline{F} = 55.676, 55.676, 55.676, 55.676$$

while Eq. 8.14 produces a Jacobian of:

Equation 8.12 gives:

$$\overline{\overline{H}} = \begin{bmatrix} 1 & 0 & 27.838 & 27.838 \\ 0 & 1 & 27.838 & 27.838 \\ 27.838 & 27.838 & 1 & 0 \\ 27.838 & 27.838 & 0 & 1 \end{bmatrix}$$

Newton's method (Sect. 8.13) gives:

$$d\overline{X} = -0.982, -0.982, -0.982, -0.982$$

After a few iterations, the site fractions converge to $X = 0.125, 0.125, 0.125, 0.125$. As suggested in this example, you may wish to make your code more sophisticated, identifying the *unique* sites. This would significantly economize such calculations, in this case reducing a four dimensional system to one.

With the site fractions determined, the next step is to compute the value of Q given by Eq. 8.2.[5] This is shown in Example 31.

[5] It is computationally more efficient to use Eq. 8.3 for this calculation. Doing so assumes that the underlying code has been correctly structured, particularly that X minimizes Q as given in Eq. 8.2.

Example 31 (Calculation of A^{assoc} for the PC-SAFT EOS) This is an extension of Example 2 in Sect. 2.2. The values of $\overline{\overline{\Delta^{AB}}}$ and $\overline{X}$ were calculated in Examples 29 and 30 for a mixture of 1 mole of propane, 1 mole of n-hexane, and 1 mole of water at 350K and $V = 116.38\,\text{cm}^3$. We then use Algorithm 19 to compute Q according to Eq. 8.2:

Algorithm 19 Association Q function

```
Initialize Q = 0
for Ar (sites) do
    r = SFS[Ar]
    Q+ = n[r](lnX[Ar] − X[Ar] + 1)
    for Bs (sites) do
        s = SFS[Bs]
        Q− = Nav/(2E24V) n[r]n[s]X[Ar]X[Bs]Δ[Ar, Bs]
    end for
end for
```

$$Q = -6.558 \text{ (moles)}$$

giving (by Eq. 8.4):

$$A^{assoc} = -19083.116 \text{ (J)}$$

Combining this result with those of Example 2 in Sect. 2.2, we have from Eq. 2.29:

$$A^{Res} = 163324.577 \text{ (J)}$$

Hopefully the examples in this section have clarified the basic ideas behind the association contribution to equations of state. The advantage to the approach of Michelsen and Hendriks [Michelsen and Hendriks paper] (the use of the function Q) becomes clear when computing derivatives of the A^{assoc} contribution. For example, consider the temperature derivative of A^{assoc}. The temperature affects the value (solution for) $\overline{X}$, which in turn affects Q. Therefore A^{assoc} depends on T. However, the situation is a bit more complicated than that. The X values cannot freely vary with T; they must minimize Q. So what do we mean by the temperature derivative of A^{assoc}? Which variables are being held constant? The answer is, we want to vary the temperature holding V and $\overline{n}$ constant, while retaining the idea

For this reason, the use of Q in Eq. 8.2 is perhaps a more useful (and, as is often the case, less efficient) approach, as it allows more thorough numerical testing.

that $\overline{X}$ solves Eq. 8.6, which means that if T changes in the differentiation, $\overline{X}$ will naturally have to change as well:

$$\left(\frac{\partial A^{assoc}}{\partial T}\right)_{V,\overline{n}} = \frac{D(R_g T Q)}{DT} \tag{8.15}$$

The nomenclature for the derivative on the right side means, as the temperature is changing, the X values change to maintain Q at its minimum.[6] Looking at Eq. 8.4, we have:

$$\left(\frac{\partial A^{assoc}}{\partial T}\right)_{V,\overline{n}} = \frac{A^{assoc}}{T} + R_g T \frac{DQ}{DT} \tag{8.16}$$

Recalling that Q is a function of T, V, $\overline{n}$ and $\overline{X}$, we write the final term on the right side of Eq. 8.16 as:

$$\frac{DQ}{DT} = \left(\frac{\partial Q}{\partial T}\right)_{V,\overline{n},\overline{X}} + \sum_r \sum_{Ar} \left(\frac{\partial Q}{\partial X_{Ar}}\right)_{T,V,n,X_{Bs \neq Ar}} \left(\frac{\partial X_{Ar}}{\partial T}\right)_{V,\overline{n}} \tag{8.17}$$

The double sum in Eq. 8.17 is meant to convey summation over all sites on all species. This equation also suggests that we require $\left(\frac{\partial X_{Ar}}{\partial T}\right)_{V,\overline{n}}$ in order to determine $\left(\frac{\partial A^{assoc}}{\partial T}\right)_{V,\overline{n}}$. However, the definition of Q is such that, at equilibrium:

$$\left(\frac{\partial Q}{\partial X_{Ar}}\right)_{T,V,n,X_{Bs \neq Ar}} = 0$$

Therefore, we can rewrite Eq. 8.16 as:

$$\left(\frac{\partial A^{assoc}}{\partial T}\right)_{V,\overline{n}} = \frac{A^{assoc}}{T} + R_g T \left(\frac{\partial Q}{\partial T}\right)_{V,\overline{n},\overline{X}} \tag{8.18}$$

We can easily determine the derivative required on the right side of Eq. 8.18. Referring back to Eq. 8.2, reminding ourselves that the differentiations are at fixed V, $\overline{n}$ and $\overline{X}$, we write:

$$\left(\frac{\partial Q}{\partial T}\right)_{V,\overline{n},\overline{X}} = -\frac{N_{av}}{2E24V} \sum_r n_r \sum_s n_s \sum_{Ar} X_{Ar} \sum_{Bs} X_{Bs} \left(\frac{\partial \Delta^{ArBs}}{\partial T}\right)_{V,\overline{n}} \tag{8.19}$$

[6] The X values can be thought of as variables internal to Q.

The derivative of the association matrix follows from that of the interaction matrix. Based on Eq. 8.8, we have:

$$\left(\frac{\partial \nabla^{AiBj}}{\partial T}\right)_{V,\overline{n}} = \kappa_{ij}^{AB}\sigma_{ij}^{3}\left[\left(\frac{\partial g_{ij}^{hs}}{\partial T}\right)_{V,\overline{n}}\left[exp(\frac{\epsilon_{ij}^{AB}}{T})-1\right] - g_{ij}^{hs}exp(\frac{\epsilon_{ij}^{AB}}{T})\frac{\epsilon_{ij}^{AB}}{T^2}\right] \tag{8.20}$$

In terms of site association, we have:

$$\left(\frac{\partial \Delta^{AiBj}}{\partial T}\right)_{V,\overline{n}} = \begin{cases} \left(\frac{\partial \nabla^{AiBj}}{\partial T}\right)_{V,\overline{n}} & type(Ai) \neq type(Bj) \\ 0 & Type(Ai) = Type(Bj) \end{cases} \tag{8.21}$$

The algorithms used for Δ^{AiBj} and Q are used analogously to populate $\left(\frac{\partial \Delta^{AiBj}}{\partial T}\right)_{V,\overline{n}}$ and $\left(\frac{\partial Q}{\partial T}\right)_{V,\overline{n},\overline{X}}$.

Example 32 (Calculation of $\left(\frac{\partial A^{assoc}}{\partial T}\right)_{V,\overline{n}}$ for the PC-SAFT EOS) This is an extension of Example 3 in Sect. 2.2 and makes use of the results of the examples presented earlier in this section for a mixture of 1 mole of propane, 1 mole of n-hexane, and 1 mole of water at $T = 350\text{K}$ and $V = 116.38\,\text{cm}^3$. Starting with Eq. 8.20:

$$\left(\frac{\partial \overline{\overline{\nabla}}^{AB}}{\partial T}\right)_{V,\overline{n}} = \begin{bmatrix} 0 & 0 & 0 \\ 0 & 0 & 0 \\ 0 & 0 & -87.664 \end{bmatrix}$$

The derivative of the association matrix becomes (via Eq. 8.21):

$$\left(\frac{\partial \overline{\overline{\Delta}}^{AB}}{\partial T}\right)_{V,\overline{n}} = \begin{bmatrix} 0 & 0 & -87.664 & -87.664 \\ 0 & 0 & -87.664 & -87.664 \\ -87.664 & -87.664 & 0 & 0 \\ -87.66 & -87.664 & 0 & 0 \end{bmatrix}$$

Then, from the previous example;

$$\overline{X} = 0.125, 0.125, 0.125, 0.125,$$

which allows us to compute via Eq. 8.19;

$$\left(\frac{\partial Q}{\partial T}\right)_{V,\overline{n},\overline{X}} = 0.029\left(\frac{1}{mol K}\right)$$

and finally:

$$(\frac{\partial A^{assoc}}{\partial T})_{V,\overline{n}} = 28.427\frac{J}{K}$$

Combining this result with those of Example 3 in Sect. 2.2, we have from Eq. 8.18:

$$(\frac{\partial A^{Res}}{\partial T})_{V,\overline{n}} = 525.152\frac{J}{K}$$

A key conclusion is that first-order derivatives of A^{assoc} do not require derivatives of the X values. Starting with Eq. 8.4:

$$(\frac{\partial A^{assoc}}{\partial V})_{T,\overline{n}} = R_g T\left(\frac{\partial Q}{\partial V}\right)_{T,\overline{n}} + \sum_r \sum_{Ar}\left(\frac{\partial Q}{\partial X_{Ar}}\right)_{T,V,n,X_{Bs\neq Ar}}\left(\frac{\partial X_{Ar}}{\partial V}\right)_{T,\overline{n}} \tag{8.22}$$

As with Eq. 8.18, the summation term in Eq. 8.22 is zero because $\left(\frac{\partial Q}{\partial X_{Ai}}\right)_{T,V,n,X_{Bs\neq Ai}} = 0$. Therefore:

$$(\frac{\partial A^{assoc}}{\partial V})_{T,\overline{n}} = R_g T\left(\frac{\partial Q}{\partial V}\right)_{T,\overline{n}} \tag{8.23}$$

The mole number derivative of A^{assoc} is written as simply as that for T and V:

$$(\frac{\partial A^{assoc}}{\partial n_i})_{T,V,n_{q\neq i}} = R_g T\left(\frac{\partial Q}{\partial n_i}\right)_{T,V,n_{k\neq i}} \tag{8.24}$$

However, we do need to pay further attention to the computation of $(\frac{\partial Q}{\partial n_i})_{T,V,n_{k\neq i}}$. The computation of Q and its derivatives with respect to T and V are all handled from a *site* frame of reference (See Algorithm 19). The mole number derivatives of Q introduce a species frame of reference, regardless of whether a given species has any association sites. This can be understood by differentiating Eq. 8.2 with respect to mole numbers:

$$\left(\frac{\partial Q}{\partial n_i}\right)_{T,V,n_{k\neq i},\overline{X}} = \sum_{Ai}\left[ln X_{Ai} - X_{Ai} + 1 - X_{Ai}\frac{Nav}{1E24V}\sum_s n_s \sum_{Bs} X_{Bs}\Delta^{AiBs}\right]$$
$$-\frac{Nav}{2E24V}\sum_r n_r \sum_s n_s \sum_{Ar} X_{Ar} \sum_{Bs} X_{Bs}(\frac{\partial \Delta^{ArBs}}{\partial n_i})_{T,V,n_{k\neq i}} \tag{8.25}$$

The summation over Ai in Eq. 8.25 takes a *species* frame of reference; what are X_{Ai} for n-hexane in Example 29? This sum is zero in such an example. On the other hand, despite n-hexane being modeled as having no sites, the moles of this species do affect the value of Q through final sum in Eq. 8.25, specifically in $\left(\frac{\partial \Delta^{ArBs}}{\partial n_i}\right)_{T,V,n_{k\neq i}}$. Algorithm 20 demonstrates how to compute $\left(\frac{\partial Q}{\partial n_i}\right)_{T,V,n_{k\neq i}}$.

Algorithm 20 Association $\left(\frac{\partial Q}{\partial n_i}\right)_{T,V,n_{k\neq i}}$ function

---

Initialize $\frac{\partial Q}{\partial n}$ (length = species)
for i (species) **do**
    for Ar (sites) **do**
        if $sfs[Ar] == i$ **then**
            $\frac{\partial Q}{\partial n}[i] += ln(X[Ar]) - X[Ar] + 1$
            for Bs (sites) **do**
                $s = sfs[Bs]$
                $\frac{\partial Q}{\partial n}[i] -= X[Ar](\frac{Nav}{1E24V})n[s]X[Bs]\Delta[Ar, Bs]$
            end for
        end if
        $r = sfs[Ar]$
        for Bs (sites) **do**
            $s = sfs[Bs]$
            $\frac{\partial Q}{\partial n}[i] -= (\frac{Nav}{1E24V})n[r]n[s]X[Ar]X[Bs]\frac{\partial \Delta}{\partial n}[Ar, Bs, i]$
        end for
    end for
end for

Note how Algorithm 20 combines both site and species frames of reference, identifying the species of a given site using our site (sfs) mapping, and the sites of a given species with an *if* statement. The required mole number derivative of the association matrix follows from that of the interaction matrix. Based on Eq. 8.8, we have:

$$\left(\frac{\partial \nabla^{AiBj}}{\partial n_k}\right)_{T,V,n_{q\neq k}} = \kappa_{ij}^{AB}\sigma_{ij}^{3}\left(\frac{\partial g_{ij}^{hs}}{\partial n_k}\right)_{T,V,n_{q\neq k}}\left[exp(\frac{\epsilon_{ij}^{AB}}{T}) - 1\right] \tag{8.26}$$

In terms of site assocation, we have;

$$\left(\frac{\partial \Delta^{AiBj}}{\partial n_k}\right)_{T.V,n_{q\neq k}} = \begin{cases} \left(\frac{\partial \nabla^{AiBj}}{\partial n_k}\right)_{T,V,n_{q\neq k}} & type(Ai) \neq type(Bj) \\ 0 & Type(Ai) = Type(Bj) \end{cases} \tag{8.27}$$

Example 33 looks at this concept in detail.

Example 33 (Calculation of $\left(\frac{\partial A^{assoc}}{\partial \overline{n}}\right)_{T,V}$ for the PC-SAFT EOS) This is an extension of Example 4 in Sect. 2.2 and makes use of the results of the examples presented earlier in this section for a mixture of 1 mole of propane, 1 mole of n-hexane, and 1 mole of water at $T = 350$K and $V = 116.38\,\text{cm}^3$. Starting with Eq. 8.26:

$$\left(\frac{\partial \overline{\overline{\nabla}}^{AB}}{\partial \overline{n}}\right)_{T,V} = \begin{matrix} \begin{bmatrix} 0 & 0 & 0 \\ 0 & 0 & 0 \\ 0 & 0 & 0 \end{bmatrix} \\ \begin{bmatrix} 0 & 0 & 0 \\ 0 & 0 & 0 \\ 0 & 0 & 0 \end{bmatrix} \\ \begin{bmatrix} 0 & 0 & 0 \\ 0 & 0 & 0 \\ 12811.174 & 22752.318 & 3756.424 \end{bmatrix} \end{matrix}$$

While $\left(\frac{\partial \nabla^{AiBj}}{\partial n_k}\right)_{T,V,n_{q \neq k}}$ is 3 by 3 by 3, owing to three components in the mixture, the derivative of the association matrix (via Eq. 8.27) is 4 by 4 by 3, as there are 4 total (all of which belong to H_2O):

$$\left(\frac{\partial \overline{\overline{\Delta}}^{AB}}{\partial \overline{n}}\right)_{T,V} = \begin{matrix} \begin{bmatrix} 0 & 0 & 0 \\ 0 & 0 & 0 \\ 12811.174 & 22752.318 & 3756.424 \\ 12811.174 & 22752.318 & 3756.424 \end{bmatrix} \\ \begin{bmatrix} 0 & 0 & 0 \\ 0 & 0 & 0 \\ 12811.174 & 22752.318 & 3756.424 \\ 12811.174 & 22752.318 & 3756.424 \end{bmatrix} \\ \begin{bmatrix} 12811.174 & 22752.318 & 3756.424 \\ 12811.174 & 22752.318 & 3756.424 \\ 0 & 0 & 0 \\ 0 & 0 & 0 \end{bmatrix} \\ \begin{bmatrix} 12811.174 & 22752.318 & 3756.424 \\ 12811.174 & 22752.318 & 3756.424 \\ 0 & 0 & 0 \\ 0 & 0 & 0 \end{bmatrix} \end{matrix}$$

Then, from Example 30:

$$\overline{X} = 0.125, 0.125, 0.125, 0.125$$

we compute via Eq. 8.25;

$$(\frac{\partial Q}{\partial \overline{n}})_{T,V,n_{q\neq k},\overline{X}} = -4.166, -7.398, -9.528\frac{1}{\text{mol}}$$

and finally:

$$(\frac{\partial A^{assoc}}{\partial \overline{n}})_{T,V} = -12122.367, -21529.014, -27728.228\frac{\text{J}}{\text{mol}}$$

From the results of Example 4, we conclude:

$$(\frac{\partial A^{Res}}{\partial \overline{n}})_{T,V} = 485312.308, 829710.166, 136975.315\frac{\text{J}}{\text{mol}}$$

The derivatives of the site fractions are required for higher-order derivatives. Consider $(\frac{\partial^2 A^{assoc}}{\partial V^2})_{T,\overline{n}}$.

If we now differentiate equation 8.23 with respect to volume, we have:

$$(\frac{\partial^2 A^{assoc}}{\partial V^2})_{T,\overline{n}} = R_g T\frac{D(\frac{\partial Q}{\partial V})_{T,\overline{n},\overline{X}}}{DV} \tag{8.28}$$

The derivative on the right side is:

$$\frac{D(\frac{\partial Q}{\partial V})_{T,\overline{n}}}{DV} = \left(\frac{\partial^2 Q}{\partial V^2}\right)_{T,\overline{n}} + \sum_r \sum_{Ar} \left(\frac{\partial^2 Q}{\partial V \partial X_{Ar}}\right)\left(\frac{X_{Ar}}{\partial V}\right)_{T,\overline{n}} \tag{8.29}$$

We now make a few remarks regarding Eq. 8.29. First, we can easily derive $\left(\frac{\partial^2 Q}{\partial V^2}\right)_{T,\overline{n}}$, which ultimately involves volume derivatives of g^{hs}. Second, $\left(\frac{\partial^2 Q}{\partial V \partial X_{Ai}}\right)$ is not identically zero and must be calculated. True:

$$\left(\frac{\partial Q}{\partial X_{Ai}}\right)_{T,V,\overline{n},X_{Bs\neq Ai}} = 0$$

at our particular T, V and $\overline{n}$, but this does not mean that the value of $\left(\frac{\partial Q}{\partial X_{Ai}}\right)_{T,V,\overline{n},X_{Bs\neq Ai}}$ doesn't depend[7] on V. It is the product of the changes in Eq. 8.29, summed over all sites, that maintains $\left(\frac{\partial Q}{\partial X_{Ai}}\right)_{T,V,\overline{n},X_{Bs\neq Ai}} = 0$ as we move

[7] Just because a function equals zero at a solution does not mean its derivatives are zero at that solution. Think about some of the other optimizations considered in this text, or even the trivial case of a straight line with non-zero slope passing through zero.

from one equilibrium state to the next by changing the volume. If the summation term were neglected, the system would no longer be in *internal* equilibrium with regard to site fractions. This brings us to our third point; in contrast to first derivatives, second-order derivatives require the derivatives of the site fractions. In the present context, this means we require $(\frac{X_{Ai}}{\partial V})_{T,\overline{n}}$.

We begin with $\left(\frac{\partial^2 Q}{\partial V \partial X_{Ai}}\right)_{T,V,\overline{n},X_{Bs\neq Ai}}$, which is derived from Eq. 8.5: =

$$\left(\frac{\partial^2 Q}{\partial V \partial X_{Ai}}\right) = -\frac{N_{av}}{1E24V}\sum_s n_s \sum_{Bs} X_{Bs}\left[(\frac{\partial \Delta^{AiBs}}{\partial V})_{T,\overline{n},\overline{X}} - \frac{\Delta^{AiBs}}{V}\right] \tag{8.30}$$

Note that in Eq. 8.30, we are differentiating with respect to V with all other variables fixed, including $\overline{X}$. Just as differentiation with respect to V in Eq. 8.30 considers $\overline{n}$ to be constant, no volume derivatives of $\overline{X}$ are generated. These derivatives are nonetheless required in Eq. 8.29, which follow from Eq. 8.7:

$$\begin{aligned}\left(\frac{\partial X_{Ai}}{\partial V}\right)_{T,\overline{n}} &= -\frac{1}{\left[1+\frac{N_{av}}{1E24V}\sum_s n_s \sum_{Bs} X_{Bs}\Delta^{AiBs}\right]^2} \\ &\times \frac{\partial\left[1+\frac{N_{av}}{1E24V}\sum_s n_s \sum_{Bs} X_{Bs}\Delta^{AiBs}\right]}{\partial V} \\ &= -X_{Ai}^2\left[-\frac{1}{V}\left[\frac{1}{X_{Ai}}-1\right]+\frac{N_{av}}{1E24V}\sum_s n_s \sum_{Bs}\left[(\frac{\partial X_{Bs}}{\partial V})_{T,\overline{n}}\Delta^{AiBs}\right.\right. \\ &\left.\left.+ X_{Bs}(\frac{\partial \Delta^{AiBs}}{\partial V})_{T,\overline{n}}\right]\right]\end{aligned} \tag{8.31}$$

This implicit expression for $\left(\frac{\partial X_{Ai}}{\partial V}\right)_{T,\overline{n}}$ can be rewritten as an objective function:

$$\begin{aligned}F_{Ai} &= \frac{\left(\frac{\partial X_{Ai}}{\partial V}\right)_{T,\overline{n}}}{X_{Ai}^2} - \frac{1}{V}\left[\frac{1}{X_{Ai}}-1\right]+\frac{N_{av}}{1E24V} \\ &\times \sum_s n_s \sum_{Bs}\left[(\frac{\partial X_{Bs}}{\partial V})_{T,\overline{n}}\Delta^{AiBs} + X_{Bs}(\frac{\partial \Delta^{AiBs}}{\partial V})_{T,\overline{n}}\right]\end{aligned} \tag{8.32}$$

The unknowns in this objective function are not the site fractions (these are already known in this context), but rather their volume derivatives. The derivative of this objective function with respect to $\left(\frac{\partial X_{Ai}}{\partial V}\right)_{T,\overline{n}}$ provides the Jacobian:

$$\left(\frac{\partial F_{Ai}}{\partial\left(\frac{\partial X_{Ck}}{\partial V}\right)_{T,\overline{n}}}\right)_M = \frac{\delta_{AC}\delta_{ik}}{X_{Ai}^2} + \frac{N_{av}}{1E24V}\sum_s n_s \sum_{Bs}\left[\delta_{BC}\delta_{sk}\Delta^{AiBs}\right]$$
$$= \frac{\delta_{AC}\delta_{ik}}{X_{Ai}^2} + \frac{N_{av}}{1E24V} n_k \Delta^{AiCk} \tag{8.33}$$

An important observation here is that, despite the abstract complexity, our objective function (Eq. 8.32) is linear in our unknowns $\left(\frac{\partial X_{Ai}}{\partial V}\right)_{T,\overline{n}}$. This makes the solution very straightforward by Newton's method (Sect. 8.13), with an initial guess of $\left(\frac{\partial X_{Ai}}{\partial V}\right)_{T,\overline{n}} = 1$ converging in a single iteration.

Other higher-order derivatives can be handled as demonstrated here. For example, consider $\left(\frac{\partial^2 A^{assoc}}{\partial V \partial n_i}\right)$:

$$\left(\frac{\partial^2 A^{assoc}}{\partial V \partial n_i}\right) = \left(\frac{\partial\left(\frac{\partial A^{assoc}}{\partial n_i}\right)_{T,V,n_{k\neq i}}}{\partial V}\right)_{T,\overline{n}} = \left(\frac{\partial\left[R_g T\left(\frac{\partial Q}{\partial n_i}\right)\right]}{\partial V}\right)_{T,\overline{n}}$$
$$= R_g T\left[\left(\frac{\partial^2 Q}{\partial V \partial n_i}\right) + \sum_r \sum_{Ar}\left(\frac{\partial^2 Q}{\partial n_i \partial X_{Ar}}\right)\left(\frac{\partial X_{Ar}}{\partial V}\right)_{T,\overline{n}}\right] \tag{8.34}$$

This indicates, through the symmetry of mixed partial derivatives, that we can choose for convenience to calculate in terms of volume (rather than mole number) derivatives of the site fractions.

In the second derivative with respect to mole numbers, the mole number derivatives of the site fractions cannot be avoided:

$$\left(\frac{\partial^2 A^{assoc}}{\partial n_i \partial n_j}\right) = \left(\frac{\partial\left(\frac{\partial A^{assoc}}{\partial n_i}\right)_{T,V,n_{k\neq i}}}{\partial n_j}\right)_{T,V,n_{k\neq j}} = \left(\frac{\partial\left[R_g T\left(\frac{\partial Q}{\partial n_i}\right)_{T,V,n_{k\neq i}}\right]}{\partial n_j}\right)_{T,V,n_{k\neq j}}$$
$$= R_g T\left[\left(\frac{\partial^2 Q}{\partial n_i \partial n_j}\right) + \sum_r \sum_{Ar}\left(\frac{\partial^2 Q}{\partial n_i \partial X_{Ar}}\right)\left(\frac{\partial X_{Ar}}{\partial n_j}\right)_{T,V,n_{k\neq j}}\right] \tag{8.35}$$

The first term in the brackets on the right side of Eq. 8.35 follows from Eq. 8.25:

$$\left(\frac{\partial^2 Q}{\partial n_i \partial n_j}\right) = -\frac{Nav}{1E^{24}V}\left[\sum_{Ai} X_{Ai}\sum_{Bj} X_{Bj}\Delta^{AiBj}\right.$$
$$+ \sum_{Ai} X_{Ai}\sum_s n_s \sum_{Bs} X_{Bs}\left(\frac{\partial \Delta^{AiBs}}{\partial n_j}\right)_{T,V,n_{k\neq j}}$$

$$+\sum_{Aj} X_{Aj} \sum_{s} n_s \sum_{Bs} X_{Bs} (\frac{\partial \Delta^{AjBs}}{\partial n_i})_{T,V,n_{k\neq i}}$$

$$+\frac{1}{2}\sum_{r} n_r \sum_{s} n_s \sum_{Ar} X_{Ar} \sum_{Bs} X_{Bs} (\frac{\partial^2 \Delta^{ArBs}}{\partial n_i \partial n_j})\Bigg] \tag{8.36}$$

The second mole number derivative of the interaction matrix follows from Eq. 8.26:

$$\left(\frac{\partial^2 \nabla^{AiBj}}{\partial n_k \partial n_q}\right)_{T,V} = \kappa_{ij}^{AB} \sigma_{ij}^3 \left(\frac{\partial^2 g_{ij}^{hs}}{\partial n_k \partial n_q}\right)_{T,V} [exp(\frac{\epsilon_{ij}^{AB}}{T}) - 1] \tag{8.37}$$

In terms of site assocation, we have:

$$\left(\frac{\partial^2 \Delta^{AiBj}}{\partial n_k \partial n_q}\right)_{T,V} = \begin{cases} \left(\frac{\partial^2 \nabla^{AiBj}}{\partial n_k \partial n_q}\right)_{T,V} & type(Ai) \neq type(Bj) \\ 0 & Type(Ai) = Type(Bj) \end{cases} \tag{8.38}$$

The pseudocode in Algorithm 21 demonstrates how to compute $(\frac{\partial^2 Q}{\partial n_i \partial n_j})$.

Algorithm 21 Association $(\frac{\partial^2 Q}{\partial n_i \partial n_j})$ function

Initialize $(\frac{\partial^2 Q}{\partial \bar{n} \partial \bar{n}})$ (comp x comp)
for i (species) **do**
  for j (species) **do**
    for Ar (sites) **do**
      for Bs (sites) **do**
        $r = SFS[Ar]$
        $s = SFS[Bs]$
        if $sfs[Ar]) == i$ **then**
          if $SFS[Bs]) == j$ **then**
            $(\frac{\partial^2 Q}{\partial n^2})[i, j] -= (\frac{Nav}{1E24V}) X[Ar] X[Bs] \Delta[Ar, Bs]$
          end if
          $(\frac{\partial^2 Q}{\partial n^2})[i, j] -= (\frac{Nav}{1E24V}) X[Ar] n[s] X[Bs] (\frac{\partial \Delta}{\partial n})[Ar, Bs, j]$
          if $SFS[Ar] == j$ **then**
            $(\frac{\partial^2 Q}{\partial n^2})[i, j] -= (\frac{Nav}{1E24V}) X[Ar] n[s] X[Bs] (\frac{\partial \Delta}{\partial n})[Ar, Bs, i]$
          end if
          $(\frac{\partial^2 Q}{\partial n^2})[i, j] -= (\frac{Nav}{2E24V}) n[r] n[s] X[Ar] X[Bs] (\frac{\partial^2 \Delta}{\partial n^2})[Ar, Bs, i, j]$
        end if
      end for
    end for
  end for
end for

We also require the $\left(\frac{\partial^2 Q}{\partial n_i \partial X_{Ar}}\right)$ and $\left(\frac{\partial X_{Ar}}{\partial n_j}\right)_{T,V,n_{k\neq j}}$. The former follows from Eq. 8.5, taking $\overline{n}$ and $\overline{X}$ to be independent:

$$\left(\frac{\partial^2 Q}{\partial n_i \partial X_{Aj}}\right) = \frac{\delta_{ij}}{n_j}\left(\frac{\partial Q}{\partial X_{Aj}}\right)_{T,V,\overline{n},X_{Bs\neq Aj}} - n_j \frac{N_{av}}{1E24}\Bigg[\sum_{Bi} X_{Bi}\Delta^{AjBi} + \sum_s n_s \sum_{Bs} X_{Bs}\left(\frac{\partial \Delta^{AjBs}}{\partial n_i}\right)_{T,V,n_{k\neq i}}\Bigg] \tag{8.39}$$

Note further that this expression is intended to be generally applicable, not simply at the *solution* (where $\left(\frac{\partial Q}{\partial X_{Aj}}\right)_{T,V,\overline{n},X_{Bs\neq Aj}} = 0$). Algorithm 22 demonstrates how to populate $\left(\frac{\partial^2 Q}{\partial n_i \partial X_{Aj}}\right)$.

Algorithm 22 Association $\left(\frac{\partial^2 Q}{\partial n_i \partial X_{Aj}}\right)$ function

Initialize $\left(\frac{\partial^2 Q}{\partial \overline{n} \partial \overline{X}}\right)$ (species by sites)
for i (species) **do**
  for A_j sites **do**
    $j = SFS[Aj]$
    if $(i == j)$ **then**
      $\left(\frac{\partial^2 Q}{\partial \overline{n} \partial \overline{X}}\right)[i, Aj] = \frac{1}{n[j]}\frac{\partial Q}{\partial X}[Aj]$
    end if
    for Bs (sites) **do**
      if $sfs[Bs] == i$ **then**
        $\left(\frac{\partial^2 Q}{\partial \overline{n} \partial \overline{X}}\right)[i, Aj] = -n[j]\left(\frac{Nav}{1E24V}\right)X[Bs]\Delta[Aj, Bs]$
      end if
      $s = sfs[Bs]$
      $\left(\frac{\partial^2 Q}{\partial \overline{n} \partial \overline{X}}\right)[i, Aj] = -n[j]\left(\frac{Nav}{1E24V}\right)n[s]X[Bs]\left(\frac{\partial \Delta}{\partial n}\right)[Aj, Bs, i]$
    end for
  end for
end for

The latter cannot be written explicitly, but must be solved numerically. Differentiating equation 8.7 with respect to mole numbers we have:

$$\left(\frac{\partial X_{Ai}}{\partial n_j}\right)_{T,V,n_{k\neq j}} = -\frac{N_{av}}{1E24V} \frac{\sum_{Bj} X_{Bj}\Delta^{AiBj} + \sum_s n_s \sum_{Bs}\left[\left(\frac{\partial X_{Bs}}{\partial n_j}\right)_{T,V,n_{k\neq j}}\Delta^{AiBs} + X_{Bs}\left(\frac{\partial \Delta^{AiBs}}{\partial n_j}\right)_{T,V,n_{k\neq j}}\right]}{\left(1 + \frac{N_{av}}{1E24V}\sum_s n_s \sum_{Bs} X_{Bs}\Delta^{AiBs}\right)^2}$$

$$= -X_{Ai}^2 \frac{N_{av}}{1E24V}\left(\sum_{Bj} X_{Bj}\Delta^{AiBj} + \sum_s n_s \sum_{Bs}\left[\left(\frac{\partial X_{Bs}}{\partial n_j}\right)_{T,V,n_{k\neq j}}\Delta^{AiBs}\right.\right.$$

$$\left.\left. + X_{Bs}\left(\frac{\partial \Delta^{AiBs}}{\partial n_j}\right)_{T,V,n_{k\neq j}}\right]\right) \tag{8.40}$$

Rewriting Eq. 8.40

$$F_{A_i,n_j} = \frac{N_{av}}{1E24V}\left(\sum_s n_s \sum_{Bs}\left[\left(\frac{\partial X_{Bs}}{\partial n_j}\right)_{T,V,n_{k\neq j}}\Delta^{AiBs} + X_{Bs}\left(\frac{\partial \Delta^{AiBs}}{\partial n_j}\right)_{T,V,n_{k\neq j}}\right]\right.$$

$$\left. + \sum_{Bj} X_{Bj}\Delta^{AiBj}\right)$$

$$+ \frac{1}{X_{Ai}^2}\left(\frac{\partial X_{Ai}}{\partial n_j}\right)_{T,V,n_{k\neq j}} \to 0 \tag{8.41}$$

Solving Eq. 8.41 is not as difficult as it would seem, since only the $\left(\frac{\partial \overline{X}}{\partial \overline{n}}\right)$ are unknown, and the Jacobian needed to solve Eq. 8.41 is evaluated at fixed T, V, $\overline{n}$, and $\overline{X}$;

$$H_{A_i n_j C_p n_q} = \left(\frac{\partial F_{A_i,n_j}}{\partial\left(\frac{\partial X_{Cp}}{\partial n_q}\right)_{T,V,n_{k\neq q}}}\right)_* \tag{8.42}$$

where $*$ is a shorthand for the conditions of the derivative stated above. As a result, differentiating yields terms like:

$$\frac{\partial\left(\frac{\partial X_{Ai}}{\partial n_j}\right)_{T,V,n_{k\neq j}}}{\partial\left(\frac{\partial X_{Cp}}{\partial n_q}\right)_{T,V,n_{k\neq q}}} = \delta_{AC}\delta_{ip}\delta_{jq} \tag{8.43}$$

This is simply a statement that the $\left(\frac{\partial \overline{X}}{\partial \overline{n}}\right)$ are independent.

$$H_{A_i n_j C_p n_q} = \frac{N_{av}}{1E24V}\left(\sum_s n_s \sum_{Bs}\delta_{BC}\delta_{sp}\delta_{jq}\Delta^{AiBs}\right) + \frac{1}{X_{Ai}^2}\delta_{AC}\delta_{ip}\delta_{jq}$$

$$= \delta_{jq}\left(\frac{N_{av}}{1E24V}n_p\Delta^{AiCp} + \frac{1}{X_{Ai}^2}\delta_{AC}\delta_{ip}\right) \tag{8.44}$$

Note that $\overline{F}$ is linear in $\left(\frac{\partial \overline{X}}{\partial \overline{n}}\right)$ (H does not depend on those values), meaning that only a single iteration by Newton's method is required to converge Eq. 8.41. This aside, the high dimensionality of Eqs. 8.41 and 8.44 presents a challenge in terms of populating the required structures. This we resolve using a similar mapping between short and long forms as we did in Appendix 8.13.1. The procedures for populating the short forms are given in Algorithms 23 and 24.

Algorithm 23 Populating F_{short} (Eq. 8.41)

```
for Ai (sites) do
    for j (species) do
        for Bs (sites) do
            s = SFS[Bs]
            F[Ai, j]+ = (Nav/(1E24V)) n[s] [ (∂X/∂n)[Bs, j] Δ[Ai, Bs] + X[Bs] (∂Δ/∂n)[Ai, Bs, j] ]
            if s == j then
                F[Ai, j]+ = (Nav/(1E24V)) X[Bs] Δ[Ai, Bs]
            end if
        end for
        F[Ai, j]+ = (1/X[Ai]^2) (∂X/∂n)[Ai, j]
    end for
end for
```

Algorithm 24 Populating H_{short} (Eq. 8.44)

```
Initialize H_short as zeros (sites x species x sites x species)
for Ai (sites) do
    i = SFS[Ai]
    for j (species) do
        for Cp (sites) do
            p = SFS[Cp]
            for q (species) do
                if (j == q) then
                    H_short[Ai, j, Cp, q] = ( (Nav/(1E24V)) n[p] Δ[Ai, Cp] )
                    if (Ai == Cp) and (i == p) then
                        H_short[Ai, j, Cp, q]+ = 1/X[Ai]^2
                    end if
                end if
            end for
        end for
    end for
end for
```

The *long* forms of $\overline{\overline{F}}$ and $\left(\frac{\partial \overline{X}}{\partial \overline{n}}\right)$, both of which we represent here as *short* with size sites by species, can be generated using Algorithm 25.

The generation of the long form of H in Eq. 8.44 is outlined in Algorithm 26.

Algorithm 25 Generating the long form of F in Eq. 8.41 and values of $\left(\frac{\partial X}{\partial n}\right)$

```
Initialize iter
for q (species) do
    for Ap (sites) do
        long[iter] = short[Ap, q]
        iter+ = 1
    end for
end for
```

Algorithm 26 Generating the long form of H in Eq. 8.44

```
Initialize row
Initialize col
for q (species) do
    for C_p (sites) do
        for j (species) do
            for A_i (sites) do
                H_long[row, col] = H_short[A_i, j, C_p, q]
                row+ = 1
            end for
        end for
        Reinitialize row
        col+ = 1
    end for
end for
```

The given algorithms allow for direct use of Newton's method, after which the *long* form of $\left(\frac{\partial \overline{X}}{\partial \overline{n}}\right)$ can be updated. Unpacking these values is facilitated by converting back to the short form, which is detailed in Algorithm 27.

Algorithm 27 Generating the short form of $\frac{\partial X}{\partial n}$ from the long form

```
Initialize iter
for i (species) do
    for Aj (sites) do
        short[Aj, i] = long[iter]
        iter+ = 1
    end for
end for
```

Example 34 outlines the computation of $\left(\frac{\partial \overline{X}}{\partial \overline{n}}\right)$.

Example 34 (Calculation of $\left(\frac{\partial X^{assoc}}{\partial \overline{n}}\right)_{T,V}$ for the PC-SAFT EOS) This example makes use of the results of the examples presented earlier in this section for a mixture of 1 mole of propane, 1 mole of n-hexane, and 1 mole of water at $T = 350\text{K}$

and $V = 116.38\, cm^3$. With an initial guess of:

$$(\frac{\partial \overline{X}}{\partial \overline{n}})_{short} = \begin{bmatrix} 1 & 1 & 1 \\ 1 & 1 & 1 \\ 1 & 1 & 1 \\ 1 & 1 & 1 \end{bmatrix}$$

Equation 8.41 gives;

$$\overline{\overline{F}}_{short} = \begin{bmatrix} 135.947 & 148.842 & 131.180 \\ 135.947 & 148.842 & 131.180 \\ 135.947 & 148.842 & 131.180 \\ 135.947 & 148.842 & 131.180 \end{bmatrix}$$

while by Algorithm 25:

$$(\frac{\partial \overline{X}}{\partial \overline{n}})_{long} = [1, 1, 1, 1, 1, 1, 1, 1, 1, 1, 1, 1]$$

and:

$$\begin{aligned} \overline{F}_{long} = [&135.947, 135.947, 135.947, 135.947, 148.842, \\ &148.842, 148.842, 148.842, 131.180, 131.180, 131.180, 131.180] \end{aligned}$$

Equation 8.44 and Algorithm 26 give:

$$\overline{\overline{H}}_{long} = \begin{bmatrix} 63.654, 0, 27.838, 27.838, 0, 0, 0, 0, 0, 0, 0, 0 \\ 0, 63.654, 27.838, 27.838, 0, 0, 0, 0, 0, 0, 0, 0 \\ 27.838, 27.838, 63.654, 0, 0, 0, 0, 0, 0, 0, 0, 0 \\ 27.838, 27.838, 0, 63.654, 0, 0, 0, 0, 0, 0, 0, 0 \\ 0, 0, 0, 0, 63.654, 0, 27.838, 27.838, 0, 0, 0, 0 \\ 0, 0, 0, 0, 0, 63.654, 27.838, 27.838, 0, 0, 0, 0 \\ 0, 0, 0, 0, 27.838, 27.838, 63.654, 0, 0, 0, 0, 0. \\ 0, 0, 0, 0, 27.838, 27.838, 0, 63.654, 0, 0, 0, 0 \\ 0, 0, 0, 0, 0, 0, 0, 0, 63.654, 0, 27.838, 27.838 \\ 0, 0, 0, 0, 0, 0, 0, 0, 0, 63.654, 27.838, 27.838 \\ 0, 0, 0, 0, 0, 0, 0, 0, 27.838, 27.838, 63.654, 0 \\ 0, 0, 0, 0, 0, 0, 0, 0, 27.838, 27.838, 0, 63.654 \end{bmatrix}$$

By Eq. 8.111;

$$d(\frac{\partial \overline{X}}{\partial \overline{n}})_{long} = [-1.139, -1.139, -1.139, -1.139, -1.247,$$
$$-1.247, -1.247, -1.247, -1.099, -1.099, -1.099, -1.099]$$

which yields:

$$(\frac{\partial \overline{X}}{\partial \overline{n}})_{long} = [-0.139, -0.139, -0.139, -0.139, -0.247,$$
$$-0.247, -0.247, -0.247, -0.099, -0.099, -0.099, -0.099]$$

Then by Algorithm 27:

$$(\frac{\partial \overline{X}}{\partial \overline{n}})_{short} = \begin{bmatrix} -0.139, & -0.247 & -0.099 \\ -0.139, & -0.247 & -0.099 \\ -0.139, & -0.247 & -0.099 \\ -0.139, & -0.247 & -0.099 \end{bmatrix}$$

As stated above, due to linearity only a single iteration is required to converge Eq. 8.41. The derivative of the site fractions can be used in Eq. 8.35 to calculate $(\frac{\partial^2 A^{assoc}}{\partial \overline{n}^2})$. This is shown in Example 35.

Example 35 (Calculation of $(\frac{\partial^2 A^{assoc}}{\partial \overline{n}^2})_{T,V}$ for the PC-SAFT EOS) Continuing with the examples presented in this section for a mixture of 1 mole of propane, 1 mole of n-hexane, and 1 mole of water at $T = 350\text{K}$ and $V = 116.38\,\text{cm}^3$. Based on Eq. 8.36 and Algorithm 21, we have:

$$\left(\frac{\partial^2 Q}{\partial \overline{n}^2}\right) = \begin{bmatrix} -14.286 & -25.423 & -12.447 \\ -25.423 & -45.241 & -22.122 \\ -12.447 & -22.122 & -9.569 \end{bmatrix} \left(\frac{1}{\text{mol}^1}\right)$$

Equation 8.39 and Algorithm 22 give:

$$\left(\frac{\partial^2 Q}{\partial \overline{n}\,\partial \overline{X}}\right) = \begin{bmatrix} -16.618 & -16.618 & -16.618 & -16.618 \\ -29.512 & -29.512 & -29.512 & -29.512 \\ -11.851 & -11.851 & -11.851 & -11.851 \end{bmatrix} \left(\frac{1}{\text{mol}}\right)$$

Then by Eq. 8.35

$$\left(\frac{\partial^2 A^{\text{assoc}}}{\partial \overline{n}^2}\right) = \begin{bmatrix} -14635.687 & -26142.845 & -17012.678 \\ -26142.845 & -46694.448 & -30260.102 \\ -17012.678 & -30260.102 & -14146.679 \end{bmatrix} \left(\frac{J}{\text{mol}^2}\right)$$

For reference, the hard chain term at these conditions is;

$$\left(\frac{\partial^2 A^{\text{hc}}}{\partial \overline{n}^2}\right) = \begin{bmatrix} 2050401.178 & 3612105.676 & 641934.662 \\ 3612105.676 & 6363167.338 & 1131021.081 \\ 641934.662 & 1131021.081 & 200800.840 \end{bmatrix} \left(\frac{J}{\text{mol}^2}\right)$$

while the dispersion term is:

$$\left(\frac{\partial^2 A^{\text{disp}}}{\partial \overline{n}^2}\right) = \begin{bmatrix} -700719.573 & -1268501.229 & -209792.710 \\ -1268501.229 & -2297647.709 & -378738.255 \\ -209792.710 & -378738.255 & -63294.897 \end{bmatrix} \left(\frac{J}{\text{mol}^2}\right)$$

These values give:

$$\left(\frac{\partial^2 A^{\text{res}}}{\partial \overline{n}^2}\right) = \begin{bmatrix} 1335045.917 & 2317461.602 & 415129.273 \\ 2317461.602 & 4018825.181 & 722022.724 \\ 415129.273 & 722022.724 & 123359.265 \end{bmatrix} \left(\frac{J}{\text{mol}^2}\right)$$

8.2 Chain Rule

Consider a scalar function F, which depends on the scalar variable x, having the derivative $\frac{dF}{dx}$. If x is itself a function of t, then how does F depend on t (i.e., what is $\frac{dF}{dt}$)? We begin by constructing a 1st order Taylor series (Appendix 8.19) on F:

$$F(x + dx) = F(x) + \frac{dF}{dx}dx$$

If we interpret the change in the value of x (i.e., $x + dx = x(t + dt)$) to be due to a change in t, we can gather terms to create the derivative of F with respect to t:

$$\frac{F(t + dt) - F(t)}{dt} = \frac{\frac{dF}{dx}dx}{dt}$$

We then recognize:

$$\frac{dF}{dt} = \frac{dF}{dx}\frac{dx}{dt} \tag{8.45}$$

Let's look at an example.

Example 36 (The Chain Rule) Suppose $F(x) = x^2$, and $x = 2 - t$. By Eq. 8.45, we have:

$$\frac{dF(x(t))}{dt} = 2x(-1) = -2(2-t) = -4 + 2t$$

By direct substitution, we have:

$$F(t) = (2-t)^2 = 4 - 4t + t^2$$

the derivative of which is:

$$\frac{dF}{dt} = -4 + 2t$$

The chain rule is useful for differentiating a function of one variable with respect to another variable on which the former depends. A good example is Eq. 6.28.

8.3 Computer Syntax

This text uses the following phrasing within pseudocode:

x += a ($x = x + a$)
x -= a ($x = x - a$)
x *= a ($x = a * x$)
x /= a ($x = \frac{x}{a}$)
A == B (Equivalence check between A and B, returns *True* or *False*)

8.4 δ_{ij} Operator

The δ_{ij} operator, also called the delta operator or the Kronecker delta, is a useful notation used in summations over indices. It has the following properties.

$$\delta_{ij} = \begin{cases} 1 & i = j \\ 0 & i \neq j \end{cases}$$

It finds wide use in the derivation of thermodynamic relationships, particularly in differentiating high-dimensional elements.

8.5 EOS Parameters and Constants Used in This Book

Unless otherwise noted, the following parameters and constants are used throughout this text (Tables 8.2 and 8.3).

Table 8.2 Pure component EOS parameters used in this text

Species	MW $[\frac{g}{mol}]$	T_c [K]	P_c [bar]	ω [–]
Methane	16.04	190.40	46.00	0.011
Propane	44.10	369.80	42.50	0.153
n-Butane	58.12	425.20	38.00	0.199
n-Hexane	86.18	507.50	30.10	0.299
Water	18.02	647.30	221.20	0.344

Table 8.3 Pure component PC-SAFT EOS parameters used in this text

Species	$\frac{m}{M}$ $[\frac{mol}{g}]$	σ [Å]	$\frac{\epsilon}{k}$ [K]	$\frac{\epsilon^{AB}}{k}$ [K]	κ^{AB} [–]	Donors/acceptors [–/–]
Methane	0.062	3.704	150.03	0	0	0/0
Propane	0.045	3.618	208.11	0	0	0/0
n-Butane	0.040	3.709	222.88	0	0	0/0
n-Hexane	0.035	3.798	236.77	0	0	0/0
Water	0.098	2.460	186.75	1716.	0.12	2/2

Though this text discusses EOS models with binary interaction parameters that are linear in temperature, numerically all k_{ij} in the numerical examples presented were taken to be zero.

8.6 Euler's Theory of Homogeneous Functions

Thermodynamic quantities are often characterized as *intensive* or *extensive*, which are defined in the Glossary of this text. This is a consequence of something called *homogeneity*. Suppose we have a function $F = F(\overline{a}, \overline{b})$, where $\overline{a}$ is the set of intensive variables affecting F, while $\overline{b}$ are extensive. For example, if F were the Gibbs free energy, $\overline{a}$ could[8] be T and P and $\overline{b}$ could be $\overline{n}$. A function F is said to be *homogeneous to degree D in the set of variables* $\overline{b}$ if:

$$F(\overline{a}, \lambda\overline{b}) = \lambda^D F(\overline{a}, \overline{b}) \tag{8.46}$$

To clarify, let's consider the simple example $F(k, \overline{n}) = k(n_1^2 + n_2^2)$. F is 1st order homogeneous in k because $F(\lambda k, \overline{n}) = \lambda k(n_1^2 + n_2^2)$. On the other hand, F is 2nd order homogeneous in $\overline{n}$ because $F(k, \lambda\overline{n}) = \lambda^2 k(n_1^2 + n_2^2)$.

[8] We say *could* because these are the *canoninal* variables of G, but we can write G in terms of other variables if we wish.

We will demonstrate the consequences of homogeneity of F in $\overline{b}$ by deriving a relationship between $F(\overline{a}, \overline{b})$ and $\frac{\partial F}{\partial b_i}$. Homogeneity derives from a similarity in how the function F responds to changes in the set of variables $\overline{b}$, and the fact that a constant λ can be multiplied *homogeneously* onto the set $\overline{b}$ causing a shift (λ^D) of the function's value. To derive this relationship, consider the derivative of Eq. 8.46 with respect to λ:

$$\frac{\partial F(\overline{a}, \lambda\overline{b})}{\partial \lambda} = \frac{\partial\left(\lambda^D F(\overline{a}, \overline{b})\right)}{\partial \lambda}$$

$$= \sum_r \left(\frac{\partial F(\overline{a}, \lambda\overline{b})}{\partial(\lambda b_r)} \frac{\partial(\lambda b_r)}{\partial \lambda} \right) = \sum_r \left(\frac{\partial F(\overline{a}, \lambda\overline{b})}{\partial(\lambda b_r)} b_r \right) = D\lambda^{D-1} F(\overline{a}, \overline{b})$$

Here we have used the chain rule (Appendix 8.2). We can determine the form of $\frac{\partial F(\overline{a}, \lambda\overline{b})}{\partial(\lambda b_r)}$ by differentiating equation 8.46 with respect to b_i, which again involves the chain rule:

$$\frac{\partial F(\overline{a}, \lambda\overline{b})}{\partial b_i} = \frac{\partial F(\overline{a}, \lambda\overline{b})}{\partial \lambda b_i} \frac{\partial \lambda b_i}{\partial b_i} = \frac{\partial F(\overline{a}, \lambda\overline{b})}{\partial \lambda b_i} \lambda = \lambda^D \frac{\partial F(\overline{a}, \overline{b})}{\partial b_i}$$

If we substitute this result into the previous one, the net result is:

$$\sum_r \left(\frac{\partial F(\overline{a}, \overline{b})}{\partial b_r} b_r \right) = D F(\overline{a}, \overline{b}) \tag{8.47}$$

This is significant, as it indicates that a homogeneous function of degree D in $\overline{b}$ can be constructed from its partial derivatives in $\overline{b}$. Using our example above, we therefore have:

$$k(2n_1)n_1 + k(2n_2)n_2 = 2k(n_1^2 + n_2^2)$$

There is a further consequence of homogeneity, which we can derive by differentiation of Eq. 8.47 with respect to b_i:

$$\sum_r \left(\left[\frac{\partial^2 F(\overline{a}, \overline{b})}{\partial b_i \partial b_r} \right] b_r + \frac{\partial F(\overline{a}, \overline{b})}{\partial b_r} \frac{\partial b_r}{\partial b_i} \right) = D \frac{\partial F(\overline{a}, \overline{b})}{\partial b_i}$$

With $\frac{\partial b_r}{\partial b_i} = \delta_{ir}$, we have:

$$\sum_r \left(\left[\frac{\partial^2 F(\overline{a}, \overline{b})}{\partial b_i \partial b_r} \right] b_r \right) = (D - 1) \frac{\partial F(\overline{a}, \overline{b})}{\partial b_i} \tag{8.48}$$

This result states that, if F is homogenous to degree D in $\overline{b}$, then $\frac{\partial F(\overline{a},\overline{b})}{\partial b_i}$ is homogeneous to degree $D-1$ in $\overline{b}$. Continuing with our example:

$$2kn_1 + 0n_2 = (2-1)k(2n_1)$$

We make much use of Euler's theory of homogeneous functions in this text. For example, according to Table 1.1, we can write the Helmholtz free energy of a simple systems as:

$$A(T, V, \overline{n}) = -PV + \sum_r \left[\frac{\partial A}{\partial n_r}\right]_{T,V,n_k \neq r} n_r = -PV + \sum_r \mu_r n_r \tag{8.49}$$

Note that A is a canoninical function of (but is inhomogenous in) T, and 1st order homogenous in $V, \overline{n}$. That is, in the nomenclature used here, with $A = A(T, V, \overline{n})$, $\overline{a} = T$, and $\overline{b} = V, \overline{n}$. With this understood, we can write the remaining potentials in Table 1.1 as:

$$G(T, P, \overline{n}) = \sum_r \frac{\partial G}{\partial n_r} n_r = \sum_r \mu_r n_r \tag{8.50}$$

$$\begin{aligned} H(S, P, \overline{n}) &= \left[\frac{\partial H}{\partial S}\right]_{P,\overline{n}} S + \sum_r \left[\frac{\partial H}{\partial n_r}\right]_{S,P,n_k \neq r} n_r \\ &= TS + \sum_r \mu_r n_r \end{aligned} \tag{8.51}$$

$$\begin{aligned} U(S, V, \overline{n}) &= \left[\frac{\partial U}{\partial S}\right]_{V,\overline{n}} S + \left[\frac{\partial U}{\partial V}\right]_{S,\overline{n}} V + \sum_r \left[\frac{\partial U}{\partial n_r}\right]_{S,V,n_k \neq r} n_r \\ &= TS - PV + \sum_r \mu_r n_r \end{aligned} \tag{8.52}$$

As shown, the potentials can be represented in terms of their extensive (1st order homogeneous) variables. These results are in accordance with more familiar results, such as $H = U + PV$ and $G = H - TS$. There are a number of further consequences of homogeneity relevant to computational thermodynamics. For example, differentiation on Eq. 8.50 with respect to mole numbers n_i gives:

$$\begin{aligned} [\frac{\partial G(T, P, \overline{n})}{\partial n_i}]_{T,P,n_{k \neq i}} &= \sum_r \left([\frac{\partial \mu_r}{\partial n_i}]_{T,P,n_{k \neq i}} n_r + \mu_r \delta_{ri} \right) \\ \mu_i &= \sum_r [\frac{\partial \mu_r}{\partial n_i}]_{T,P,n_{k \neq i}} n_r + \mu_i \end{aligned}$$

$$0 = \sum_r [\frac{\partial \mu_r}{\partial n_i}]_{T,P,n_{k \neq i}} n_r$$

The final form above is typically written as:

$$\sum_r [\frac{\partial \mu_r}{\partial n_i}]_{T,P,n_{k \neq i}} n_r = 0 \tag{8.53}$$

and is one form of the Gibbs–Duhem Equation. It is a very useful result that indicates the chemical potentials are related in value and cannot change independently. This is simply a consequence of homogeneity, and the Gibbs–Duhem equation can be established directly from Eq. 8.50 and the accompanying knowledge that G is 1st order homogeneous in $\overline{n}$. That is to say, from Eq. 8.48, the chemical potential is a 0th order homogeneous function of $\overline{n}$. Evidence of this comes from the understanding that doubling the total moles of a mixture without changing composition[9] will double G, but will not change any of the μ values. Equation 8.53 is much more useful when written in terms of $ln\hat{\varphi}_i$. We begin by rewriting Eq. 1.79;

$$\mu_i = g_i^o + R_g T ln\frac{P}{P_o} + R_g T ln\frac{n_i}{n} + R_g T ln\hat{\varphi}_i \tag{8.54}$$

and then differentiating:

$$\left[\frac{\partial \mu_i}{\partial n_j}\right]_{T,P,n_{k \neq j}} = R_g T\left(\frac{\delta_{ij}}{n_i} - \frac{1}{n}\right) + R_g T\left[\frac{\partial ln\hat{\varphi}_i}{\partial n_j}\right]_{T,P,n_{k \neq j}} \tag{8.55}$$

Symmetry (Appendix 8.12.3) is an important aspect of computational thermodynamics. As it is a mixed partial derivative (Appendix 8.15), $[\frac{\partial \mu_i}{\partial n_j}]_{T,P,n_{k \neq j}}$ is symmetric. The right-hand side of Eq. 8.55 is therefore symmetric, as is the δ_{ij} operator (Appendix 8.4). Therefore, $\left[\frac{\partial ln\hat{\varphi}_i}{\partial n_j}\right]_{T,P,n_{k \neq j}}$ is symmetric:

$$[\frac{\partial ln\hat{\varphi}_i}{\partial n_j}]_{T,P,n_{k \neq j}} = [\frac{\partial ln\hat{\varphi}_j}{\partial n_i}]_{T,P,n_{k \neq i}}$$

As above G is 1st order homogeneous, and μ_i is therefore 0th order, with respect to $\overline{n}$. This means that $\frac{\partial \mu_i}{\partial n_j}$ is homogeneous to degree -1 in $\overline{n}$. According to Eq. 8.54, if μ_i is 0th order homogeneous in $\overline{n}$, $ln\hat{\varphi}_i$ is too.[10] Further, since $[\frac{\partial \mu_i}{\partial n_j}]_{T,P,n_{k \neq j}}$

[9] i.e., *homogeneously* changing $\overline{n}$.

[10] The significance of this is that if $\overline{n}$ is *homogeneously* changed (e.g., all n_i doubled) at fixed T and P, there will be no change in the $ln\hat{\varphi}$ values.

is homogeneous to degree -1 in $\overline{n}$, as is $R_g T(\frac{\delta_{ij}}{n_i} - \frac{1}{n})$, then we conclude that $[\frac{\partial ln\hat{\varphi}_i}{\partial n_j}]_{T,P,n_k \neq j}$ is as well.

Continuing with the Gibbs–Duhem equation at fixed pressure, we have:

$$\sum_r [\frac{\partial(ln\frac{n_r}{n} + ln\hat{\varphi}_r)}{\partial n_i}]_{T,P,n_{k\neq i}} n_r$$
$$\sum_r (\frac{\delta_{ri}}{n_r} - \frac{1}{n}) n_r + \sum_r [\frac{\partial ln\hat{\varphi}_r}{\partial n_i}]_{T,P,n_{k\neq i}} n_r \tag{8.56}$$
$$\sum_r [\frac{\partial ln\hat{\varphi}_r}{\partial n_i}]_{T,P,n_{k\neq i}} n_r = 0$$

Equation 8.56 indicates that the $ln\hat{\varphi}_i$ values are not independent of one another. Note also that derivative in Eq. 8.56 is at fixed pressure; we can derive a similar result for the fixed volume derivatives of $ln\hat{\varphi}_i$. For this, we begin with the Helmholtz free energy in Eq. 8.49):

$$(\frac{\partial A(T, V, \overline{n})}{\partial n_i})_{T,V,n_{k\neq i}} = -(\frac{\partial P}{\partial n_i})_{T,V,n_{k\neq i}} V + \sum_r \left[(\frac{\partial \mu_r}{\partial n_i})_{T,V,n_{k\neq i}} n_r + \mu_r \delta_{ri} \right]$$
$$\mu_i = -(\frac{\partial P}{\partial n_i})_{T,V,n_{k\neq i}} V + \sum_r (\frac{\partial \mu_r}{\partial n_i})_{T,V,n_{k\neq i}} n_r + \mu_i$$
$$0 = -(\frac{\partial P}{\partial n_i})_{T,V,n_{k\neq i}} V + \sum_r (\frac{\partial \mu_r}{\partial n_i})_{T,V,n_{k\neq i}} n_r$$

The analogue to Eq. 8.53 for volume is (from Eq. 8.49):

$$\sum_r (\frac{\partial \mu_r}{\partial n_i})_{T,V,n_{k\neq i}} n_r = \left(\frac{\partial P}{\partial n_i}\right)_{T,V,n_{k\neq i}} V \tag{8.57}$$

Via Eq. 8.54 we can rewrite the summed term on left-hand side of Eq. 8.57:

$$\left(\frac{\partial \mu_i}{\partial n_j}\right)_{T,V,n_{k\neq j}} = \frac{R_g T}{P} \left(\frac{\partial P}{\partial n_j}\right)_{T,V,n_{k\neq j}} + R_g T \left(\frac{\delta_{ij}}{n_i} - \frac{1}{n}\right)$$
$$+ R_g T \left(\frac{\partial ln\hat{\varphi}_i}{\partial n_j}\right)_{T,V,n_{k\neq j}} \tag{8.58}$$

In contrast to Eq. 8.55, Eq. 8.58 indicates that $\left(\frac{\partial ln\hat{\varphi}_i}{\partial n_j}\right)_{T,V,n_{k\neq j}}$ is asymmetric due to the $\left(\frac{\partial P}{\partial n_j}\right)_{T,V,n_{k\neq j}}$ term:

$$(\frac{\partial ln\hat{\varphi}_i}{\partial n_j})_{T,V,n_{k\neq j}} \neq (\frac{\partial ln\hat{\varphi}_j}{\partial n_i})_{T,V,n_{k\neq i}} \text{ for } i \neq j$$

In terms of homogeneity, we conclude from the fact that A is 1st order homogeneous in V and $\overline{n}$, that μ_i is 0th order homogenous in these variables. The homogeneity of $ln\hat{\varphi}$ values at fixed volume requires careful attention, as these are not homogeneous with respect to $\overline{n}$, but rather with respect to the combined[11] set V *and* $\overline{n}$. This can be understood by again considering Eq. 8.54. Although we are working at fixed V and not P, we recognize by Eq. 1.14 that $-P$ is the derivative of a 1st order homogeneous function of the variable set V and $\overline{n}$ (i.e. A) with respect to one of those variables (i.e. V), and therefore by Eq. 8.48, the function P must be 0th order in the variable set V and $\overline{n}$. These remarks collectively demonstrate that $ln\hat{\varphi}$ is 0th order homogenous in V and $\overline{n}$.

Continuing with the development of the Gibbs–Duhem equation at fixed volume, we can rewrite the left side of Eq. 8.57:

$$\begin{aligned} R_g T \sum_r \left[\frac{1}{P}\left(\frac{\partial P}{\partial n_i}\right)_{T,V,n_{k\neq i}} n_r + (\frac{\delta_{ri}}{n_r} - \frac{1}{n})n_r + (\frac{\partial ln\hat{\varphi}_r}{\partial n_i})_{T,V,n_{kj\neq i}} n_r\right] \\ = \frac{nR_g T}{P}\left(\frac{\partial P}{\partial n_i}\right)_{T,V,n_{k\neq i}} + R_g T \sum_r (\frac{\partial ln\hat{\varphi}_r}{\partial n_i})_{T,V,n_{k\neq i}} n_r \end{aligned} \tag{8.59}$$

Equation 8.57 then becomes:

$$\sum_r (\frac{\partial ln\hat{\varphi}_r}{\partial n_i})_{T,V,n_{j\neq i}} n_r = \left(\frac{\partial P}{\partial n_i}\right)_{T,V,n_{i\neq j}} \left(\frac{V}{R_g T} - \frac{n}{P}\right) \tag{8.60}$$

With $P = \frac{ZnR_gT}{V}$ (Eq. 1.64);

$$(\frac{\partial P}{\partial n_i})_{T,V,n_{j\neq i}} = \left[(\frac{\partial Z}{\partial n_i})_{T,V,n_{i\neq j}} n + Z\right]\frac{R_g T}{V}$$

[11] This means that doubling $\overline{n}$ at fixed T and V will change the $ln\hat{\varphi}$ values. On the other hand, homogeneously changing V and $\overline{n}$ leaves them unchanged. This illustrates the origin and prevalence of molarity (or concentration), which is often used in chemical modeling.

Equation 8.59 becomes:

$$\sum_r \left(\frac{\partial ln\hat{\varphi}_r}{\partial n_i}\right)_{T,V,n_{j\neq i}} n_r = \left[\left(\frac{\partial Z}{\partial n_i}\right)_{T,V,n_{i\neq j}} n + Z\right]\left[\frac{Z-1}{Z}\right] \quad (8.61)$$

Comparing the developments of Eqs. 8.56 and 8.61, we see a number of interesting results:

- $\left[\frac{\partial ln\hat{\varphi}_i}{\partial n_j}\right]_{T,P,n_{k\neq i}}$ is symmetric, while $\left(\frac{\partial ln\hat{\varphi}_i}{\partial n_i}\right)_{T,V,n_{j\neq i}}$ is not. This contrasts with μ_i, of which both the constant P and V derivatives are symmetric.
- Other than for the ideal-gas state (for which $Z = 1$), in general $\sum_r \left[\frac{\partial ln\hat{\varphi}_r}{\partial n_i}\right]_{T,V,n_{j\neq i}} n_r \neq \overline{0}$, while identically $\sum_r \left[\frac{\partial ln\hat{\varphi}_r}{\partial n_i}\right]_{T,P,n_{j\neq i}} n_r = \overline{0}$.

Example 37 clarifies these concepts.

Example 37 (An examination of Symmetry and the Gibbs–Duhem Equations) Consider a mixture of methane, n-butane, and n-hexane (1 mole each) at 200 K and 80 bar, modeled with the SRK EOS. As shown in Fig. 6.2, these conditions lie within the single-phase liquid region for this mixture. The constant pressure mole number derivatives of $ln\hat{\varphi}$ are according to Eq. 5.16:

$$\left[\frac{\partial ln\hat{\varphi}_i}{\partial n_j}\right]_{T,P,n_{k\neq j}} = \begin{bmatrix} -0.078 & 0.033 & 0.044 \\ 0.033 & -0.014 & -0.019 \\ 0.044 & -0.019 & -0.025 \end{bmatrix}$$

With the usual convention that the first index of a 2D matrix corresponds to the row, and the second to the column, Eq. 8.56 instructs us to sum down the columns. By inspection, with $\overline{n} = 1, 1, 1$, it is clear that Eq. 8.56 holds. Due to the symmetry ($\left[\frac{\partial ln\hat{\varphi}_i}{\partial n_j}\right]_{T,P,n_{k\neq j}} = \left[\frac{\partial ln\hat{\varphi}_j}{\partial n_i}\right]_{T,P,n_{k\neq i}}$), we can also sum along each row. The constant volume mole number derivatives of $ln\hat{\varphi}$ are according to Eq. 3.16:

$$\left(\frac{\partial ln\hat{\varphi}_i}{\partial n_j}\right)_{T,V,n_{k\neq j}} = \begin{bmatrix} -12.497 & -25.077 & -36.110 \\ -8.835 & -17.946 & -25.837 \\ -5.734 & -11.703 & -16.849 \end{bmatrix}$$

As stated above, $\left(\frac{\partial ln\hat{\varphi}_i}{\partial n_j}\right)_{T,V,n_{k\neq j}}$ is asymmetric. Though not obvious, we can verify Eq. 8.61 by noting that $n = 3$, $Z = 0.432$ and $\left(\frac{\partial Z}{\partial n_j}\right)_{T,V,n_{k\neq j}} = 6.727, 13.749, 19.859$. The right side of Eq. 8.61 is then $-27.066, -54.727, -78.796$. These numbers are indeed produced by the mole sum down each column of $\left(\frac{\partial ln\hat{\varphi}_i}{\partial n_j}\right)_{T,V,n_{k\neq j}}$.

8.7 Gradient

Suppose we have a scalar function of multiple variables; $F = F(\overline{s})$. The *gradient* of F with respect to $\overline{s}$ is:

$$\overline{\nabla} F_i = \left[\frac{\partial F}{\partial s_i}\right]_{s_{j \neq i}} \tag{8.62}$$

Let's look at an example.

Example 38 (Gradient of a Scalar Function) Consider $F = F(x, y) = x^2 + y^2$, then the gradient of F is:

$$\overline{\nabla} F = 2x, 2y$$

As shown, the gradient of a scalar is a vector quantity, and geometrically we can interpret $\overline{\nabla} F$ as vector with direction and magnitude indicating the maximal rate of change of F. In other words, it is a vector normal to the surface of constant F. The rate of change of F along the x axis is of course $\left(\frac{\partial F}{\partial x}\right)_y = 2x$, with the corresponding term for y. For a *unit* vector $\overline{v} = a, b$, the rate of change in F (change per unit length) is:

$$\frac{dF}{dv} = \overline{\nabla} F \cdot \overline{v} = \left(\frac{\partial F}{\partial x}\right)_y a + \left(\frac{\partial F}{\partial y}\right)_x b \tag{8.63}$$

Returning to our example, the rate of change of F at a point $\overline{s} = 1, 1$ in the direction $\overline{v} = 1, 0$ (along the direction of the x axis) is $2 * 1 + 2 * 0 = 2$. The same rate is found along the direction of the y axis. A higher rate of change is observed along the direction of the unit vector $\frac{1}{\sqrt{2}}, \frac{1}{\sqrt{2}}$: $dF = 2\frac{1}{\sqrt{2}} + 2\frac{1}{\sqrt{2}} = 2\sqrt{2}$. Though F has the same value in these cases, the direction in the first and second cases is not fully aligned with the gradient vector, while in the last case, it is and therefore produces a higher rate of change in F. This is made clear in Fig. 8.1.

8.8 Index and Summation Notation

Computational thermodynamics makes heavy use of summation, with many expressions calling for summation over species, sites, and/or phases. From a notational perspective, the choice of index variable for summation is largely arbitrary. For example, the summation over squared mole numbers:

$$Sum = n_1^2 + n_2^2 + n_3^2 \ldots . + n_{N-2}^2 + n_{N-1}^2 + n_N^2$$

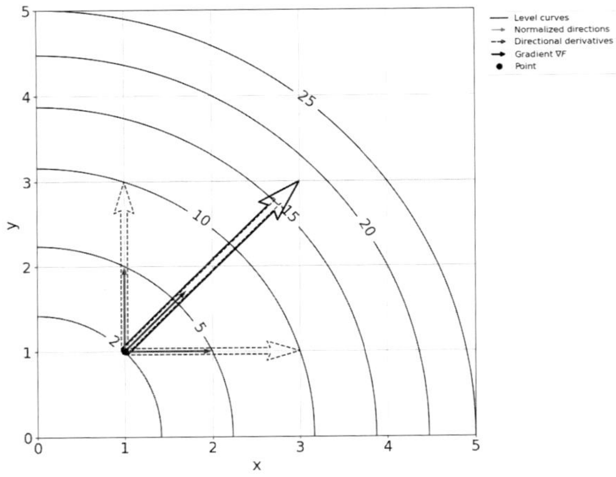

Fig. 8.1 Level curves, gradient and directional derivatives of $F = x^2 + y^2$

can be represented by either;

$$Sum = \sum_{i=1}^{N} n_i^2 \tag{8.64}$$

or:

$$Sum = \sum_{r=1}^{N} n_r^2 \tag{8.65}$$

Without wider context, there is no difference between the expressions above; i or r can be used as the summation index. However, if we are interested in computing how S depends on the mole numbers of species i (this is being the *wider context*), we cannot use i to designate both the species of interest and the summation index that runs over *all* species. Using the notation in Eq. 8.64, this would give:

$$\left(\frac{\partial Sum}{\partial n_i}\right)_? = \sum_{i=1}^{N} 2n_i \tag{8.66}$$

Not only is this derivative ambiguous, hence the "?" on the left side of Eq. 8.66, but the right-hand side is not the result we're seeking ($2n_i$), but instead gives:

$$2n_1 + 2n_2 + 2n_3 + \ldots + 2n_{N-2} + 2n_{N-1} + 2n_N$$

Using the notation in Eq. 8.65, we have:

$$\left(\frac{\partial Sum}{\partial n_i}\right)_{n_{k\neq i}} = \sum_{r=1}^{N}\left(\frac{\partial n_r^2}{\partial n_i}\right)_{n_{j\neq i}} = \sum_{r=1}^{N} 2n_r\left(\frac{\partial n_r}{\partial n_i}\right)_{n_{j\neq i}} \tag{8.67}$$

On the left side of Eq. 8.67, we have introduced (necessarily) a third index k, which is used to represent the species that are to remain constant in the differentiation. This cannot be r, because r includes i. While we are being very specific, we have avoided the ambiguity in Eq. 8.66. On the right side of Eq. 8.67, we can see clearly that we require the derivative of n_r with respect to n_i. Using the δ_{ij} notation in Appendix 8.4 (where in this example we consider the mole numbers to be independent quantities), we have:

$$\left(\frac{\partial Sum}{\partial n_i}\right)_{n_{k\neq i}} = \sum_{r=1}^{N} 2n_r\delta_{ri} = 2n_i \tag{8.68}$$

which is the correct result.

It is emphasized here is that index variables must be selected carefully and consistently, but otherwise the choice is yours. The author has tried to be consistent in the use of i for a particular species index (with summation using r or s), and j for a particular phase (with summation using p or q), though admittedly there are examples that do not follow this convention.

8.9 Integration by Parts

Due to the product rule of differential calculus:

$$d(FG) = FdG + GdF \tag{8.69}$$

it is possible to rewrite integrals according to:

$$\int FdG = \int \left[d(FG) - GdF\right] = FG - \int GdF \tag{8.70}$$

Let's look at an example.

Example 39 (Integration by Parts) If $F = lnx$, then:

$$\int lnxdx = ln(x)x - \int xdlnx$$
$$= ln(x)x - \int x\frac{1}{x}dx$$
$$= ln(x)x - \int dx = ln(x)x - x \tag{8.71}$$

Note that Example 39 has been written for an *indefinite* integral. If an integral is to be evaluated over limits, the expression for integration by parts requires a modification of Eq. 8.70;

$$\int_a^b FdG = \left[FG\right]_a^b - \int_a^b GdF \tag{8.72}$$

8.10 Lagrange Multipliers

Optimization, finding the inputs to a function that minimizes or maximizes a function is an important technique in computational thermodynamics. In a simple case, we might seek the minimum value of:

$$F = x^2$$

Via differential calculus, we seek the zero of the derivative of this function:

$$\frac{dF}{dx} = 2x = 0$$

With $x = 0$, we find that $F = 0$ We recognize this as minimum of F, because:

$$\frac{\partial^2 F}{\partial^2} = 2 > 0$$

This function has only one such point, while more important problems may have multiple. This type of problem is called an unconstrained optimization; there is no constraint on the independent variable(s). However, many important problems are constrained. For example, suppose we wish to find the extreme values of:

$$F(x, y) = x^2 + y^2$$

subject to the constraint that:

$$G(x, y) = x + y - 1 = 0$$

We might be inclined to say that F has a global minimum at $x, y = 0, 0$, but this point does not satisfy the constraint; $G(0, 0) = -1 \neq 0$. One option is to substitute the constraint into the objective function:

$$y = 1 - x$$

$$F(x) = x^2 + (1 - x)^2 = 2x^2 - 2x + 1$$

We then differentiate our now unconstrained problem:

$$\frac{dF}{dx} = 4x - 2 = 0$$

The solution is $x = 0.5$, and from the constraint we have $y = 0.5$, which gives $F = 0.5$. Because there is only one extreme value, and the derivatives of F are all non-negative, this solution is a global minimum of this constrained equilibrium. What can we do if we don't wish to, or cannot, substitute the constraint into the objective function? We use the method of Lagrange multipliers.

Consider the function $F(x, y) = x^2 + y^2$; each value of F determines a circle. At every point on such a circle F is constant, and a normal vector can be drawn that points in the direction increasing F. This is the gradient of F (see Appendix 8.7):

$$\nabla F(x, y) = 2x, 2y$$

Similar remarks can be said of the constraint $G = x + y - 1$, which defines a line, and the gradient of g is:

$$\nabla G(x, y) = 1, 1$$

These concepts are shown in Fig. 8.2. Notice that the points $x, y = 0, 1$ and $1, 0$ lie on the contours of F and G, but the gradients of F and G at these points are not aligned. Since ∇F points in the direction of maximal increase in F, the orthogonal direction corresponds to constant F. The corresponding argument can be made for G. This means that at the points $x, y = 0, 1$ and $1, 0$, it is possible to move in a direction keeping G constant while changing F.[12] Figure 8.2 also shows the point $x, y = 0.5, 0.5$, which satisfies $G = 0$ and lies on $F = 0.5$. Here the normals to the sets of constant F and G are aligned, meaning movement along constant G corresponds to constant F. Note that Fig. 8.2 makes it clear that we have not found the global minimum or maximum of F; $F = 0.25$ is shown, but such points on this curve do not satisfy the constraint.

[12] It is also possible to keep F constant while changing G, but since the latter is the constraint in our problem, it is helpful to think of it the other way around.

Alignment of the gradients of the objective function (F) and constraint (G) is mathematically expressed as:

$$\nabla F = \lambda \nabla G \tag{8.73}$$

which, along with the fact that $G = 0$, implies that the function:

$$\mathcal{L}(x, y, \lambda) = F(x, y) - \lambda G(x, y) \tag{8.74}$$

has the following properties:

$$\frac{\partial \mathcal{L}}{\partial x} = \frac{\partial F}{\partial x} - \lambda \frac{\partial G}{\partial x} = 0$$
$$\frac{\partial \mathcal{L}}{\partial y} = \frac{\partial F}{\partial y} - \lambda \frac{\partial G}{\partial y} = 0$$
$$\frac{\partial \mathcal{L}}{\partial \lambda} = -G = 0$$

This means that the gradient of $\mathcal{L}$ *with respect to the variables* x, y, λ is zero. Another way of understanding the method of Lagrange multipliers is to think of it as converting a constrained optimization into an unconstrained optimization in a new (λ added) set of variables. Using the example above, we write:

$$\mathcal{L}(x, y, \lambda) = F(x, y) - \lambda G(x, y) = x^2 + y^2 - \lambda(x + y - 1)$$

Note that, at the solution with $G(x, y) = x + y - 1 = 0$, the functions $\mathcal{L}$ and F will be numerically equal. The parameter λ is called the Lagrange multiplier. If we seek the minimum of this function, we must find the zeros of the derivatives of this function with respect to the variables x, y and λ:

$$\begin{aligned}
\frac{\partial \mathcal{L}(x, y, \lambda)}{\partial x} &= \frac{\partial F}{\partial x} - \lambda \frac{\partial G}{\partial x} = 2x - \lambda(1) = 0 \\
\frac{\partial \mathcal{L}(x, y, \lambda)}{\partial y} &= \frac{\partial F}{\partial y} - \lambda \frac{\partial G}{\partial y} = 2y - \lambda(1) = 0 \\
\frac{\partial \mathcal{L}(x, y, \lambda)}{\partial \lambda} &= -G = -(x + y - 1) = 0
\end{aligned} \tag{8.75}$$

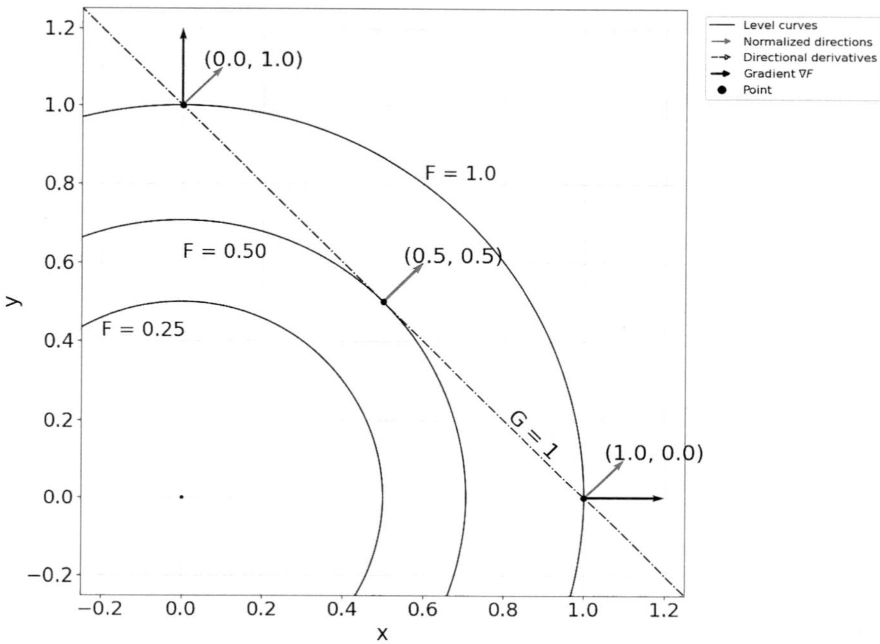

Fig. 8.2 The concept of Lagrange multipliers in solving constrained optimization of $F = x^2 + y^2$ subject to $G = x + y = 1$. Vectors representing the gradient vectors of F are distinguished from those of G by bold shading

As above, the first two equations imply that the gradients[13] are aligned while the third equation states that the constraint is satisfied. The solution to Eq. 8.75 is: $x = 0.5$, $y = 0.5$, $\lambda = 1$, from which we find $F = 0.5$, as shown in Fig. 8.2. Note that multiple constraints can be incorporated into an optimization problem. For N constraints, we have:

$$\mathcal{L}(\overline{a}, \overline{\lambda}) = F(\overline{a}) - \sum_{r=1}^{N} \lambda_r G_r(\overline{a})$$

$$\frac{\partial \mathcal{L}}{\partial a_i} = \frac{\partial F}{\partial a_i} - \sum_{r=1}^{N} \lambda_r \frac{\partial G_r}{\partial a_i} = 0$$

$$\frac{\partial \mathcal{L}}{\partial \lambda_j} = -G_j(\overline{a}) = 0$$

[13] Note that due to the constraint we do not solve for (unconstrained) minima or maxima of F: $\nabla F \neq \overline{0}$.

8.11 Linear Least Squares

Linear least squares is a fitting method by which a set of data is regressed to a linear model. That is, a straight line is fit to the data by determining the slope and intercept. Imagine you have a set of x values that produce a set of y values. Now suppose we wish to fit these data to a linear model $y = mx + b$; what should the slope (m) and intercept (b) values be? We can define these by requiring that these parameters minimize the error between the measured (y_i) and the model $(mx_i + b)$ values. We can define the error as:

$$e_i = y_i - (mx_i + b) \tag{8.76}$$

Depending on the choice of m and b, e_i could be positive, zero, or negative, while summation of the *square* of each error defines a function L:

$$L = \sum_r e_r^2 = \sum (y_r - mx_r - b)^2 \tag{8.77}$$

As above, we define the slope and intercept by requiring that these minimize L:

$$\frac{\partial L}{\partial m} = \sum_r 2(y_r - mx_r - b)(-x_r) = 0 \tag{8.78}$$

$$\frac{\partial L}{\partial b} = \sum_r 2(y_r - mx_r - b)(-1) = 0 \tag{8.79}$$

Note carefully that Eq. 8.79 simplifies to;

$$\begin{aligned} &\sum_r y_r - m\sum_r x_r - b\sum_r 1 \\ &= \sum_r y_r - m\sum_r x_r - bN \\ &= 0 \end{aligned} \tag{8.80}$$

where N is the number of points in the summation. This allows us write the intercept as:

$$b = \overline{y} - m\overline{x} \tag{8.81}$$

where $\overline{k} = \frac{\sum_r^N k_r}{N}$. We can also simplify Eq. 8.78, yielding:

$$\sum_r (y_r x_r) - m\sum_r x_r^2 - b\sum_r x_r = 0 \tag{8.82}$$

Substitution of Eq. 8.81 into Eq. 8.82 gives the following expression for the slope:

$$m = \frac{\sum_r x_r(y_r - \overline{y})}{\sum_r x_r(x_r - \overline{x})} = \frac{\overline{xy} - \overline{x}\,\overline{y}}{\overline{x^2} - \overline{x}^2} \tag{8.83}$$

8.12 Matrix Operations

Linear algebra is a very important and broad branch of mathematics, on which computational thermodynamics heavily relies. For a proper coverage of this subject, the reader is referred to the many excellent texts on this subject. For the purposes of clarifying the material presented in this text, the following details are provided.

8.12.1 Determinant

The *determinant* of a matrix provides useful information on the properties of that matrix. Determinants naturally arise in the context of solving systems of linear equation. Consider the two-by-two system:

$$\begin{aligned} ax + by &= e \\ cx + dy &= f \end{aligned} \tag{8.84}$$

One strategy for solving this system is by simple algebra, eliminating factors of y:

$$(da - bc)x + (db - bd)y = (de - bf)$$

We can then attempt to solve for x:

$$x = \frac{de - bf}{da - bc}$$

and y;

$$y = \frac{e - ax}{b} = \frac{fa - ce}{da - bc}$$

provided that:

$$da - bc \neq 0$$

To make the connection with matrix determinants, consider the linear algebra representation of Eq. 8.84:

$$\begin{bmatrix} a & b \\ c & d \end{bmatrix} \begin{bmatrix} x \\ y \end{bmatrix} = \begin{bmatrix} e \\ f \end{bmatrix} \tag{8.85}$$

We can solve Eq. 8.85 using the methods of Sect. 8.12.2:

$$\begin{bmatrix} x \\ y \end{bmatrix} = \begin{bmatrix} a & b \\ c & d \end{bmatrix}^{-1} \begin{bmatrix} e \\ f \end{bmatrix}$$

We now recognize that the combination of elements a, b, c, d as $ad - bc$ is the determinant of the matrix in Eq. 8.85;

$$det \begin{bmatrix} a & b \\ c & d \end{bmatrix} = ad - bc \tag{8.86}$$

and its value dictates whether a solution to Eq. 8.84 can be found (or more generally whether the matrix in Eq. 8.85 can be inverted). This is a very cursory look at a very deep field of mathematics, and the reader should recognize that this simple treatment of determinants only applies to two dimensional matrices. The determinant of higher dimensional matrices can be computed, but Eq. 8.86 does not apply. The reader is referred to a general text on linear algebra should more information be desired. As with matrix inversion we will not concern ourselves with manual calculation of determinants, but instead rely on built-in methods available in computer software.

8.12.1.1 Eigenvalues

If a matrix $\overline{\overline{M}}$ exhibits the property that:

$$\overline{\overline{M}}\overline{x} = \lambda \overline{x} \tag{8.87}$$

we say that λ is an eigenvalue, and $\overline{x}$ is an eigenvector, of $\overline{\overline{M}}$. We can solve Eq. 8.87 by rewriting it as:

$$\overline{\overline{M}}\overline{x} - \overline{\overline{I}}\lambda x = (\overline{\overline{M}} - \lambda \overline{\overline{I}})x = \overline{\overline{0}} \tag{8.88}$$

If $\overline{x} \neq \overline{0}$, Eq. 8.88 implies that:

$$det(\overline{\overline{M}} - \lambda \overline{\overline{I}}) = 0$$

Here we have computed the *determinant* of the matrix $\overline{\overline{M}} - \lambda \overline{\overline{I}}$. If we could invert that matrix, we would then find that $\overline{x} = \overline{0}$. Let's look at an example.

Example 40 (Eigenvalues of a Matrix) Suppose we have a matrix:

$$\overline{\overline{M}} = \begin{bmatrix} 1 & 2 \\ -1 & 4 \end{bmatrix} \tag{8.89}$$

Equation 8.88 requires:

$$det(\overline{\overline{M}} - \lambda\overline{\overline{I}}) = det\begin{bmatrix} 1-\lambda & 2 \\ -1 & 4-\lambda \end{bmatrix} = (1-\lambda)(4-\lambda) - (2)(-1) = 0 \tag{8.90}$$

Equation 8.90 yields the *characteristic polynomial* of $\overline{\overline{M}}$:

$$\lambda^2 - 5\lambda + 6 = 0 \tag{8.91}$$

The roots of the characteristic polynomial (*eigenvalues*) are:

$$\lambda = 2, 3$$

while the *eigenvectors* are solutions to:

$$\begin{bmatrix} 1-\lambda & 2 \\ -1 & 4-\lambda \end{bmatrix}\begin{bmatrix} x_1 \\ x_2 \end{bmatrix} = \begin{bmatrix} 0 \\ 0 \end{bmatrix} \tag{8.92}$$

There are no unique[14] solutions to Eq. 8.92. We can therefore arbitrarily set $x_1 = 1$ and solve x_2. The eigenvectors are

$$\begin{aligned} (1-2)(1) + 2x_2 &= 0 \\ -1 + 2x_2 &= 0 \\ x_2 &= 0.5 \\ (x_1, x_2 &= 1, 0.5) \end{aligned} \tag{8.93}$$

and:

$$\begin{aligned} (1-3)(1) + 2x_2 &= 0 \\ -2 + 2x_2 &= 0 \\ x_2 &= 1 \\ (x_1, x_2 &= 1, 1) \end{aligned} \tag{8.94}$$

[14] For a given eigenvalue, scalar multiples of $\overline{x}$ are also solutions to Eq. 8.92. Further, note that having set $det(\overline{\overline{M}} - \lambda\overline{\overline{I}}) = 0$, the rows of $(\overline{\overline{M}} - \lambda\overline{\overline{I}})$ are proportional to each other, meaning the (in this case two) equations produced by matrix multiplication are redundant.

With these solutions, (λ and $\overline{x}$ pairs), we find that Eq. 8.89 is indeed satisfied for the matrix $\overline{\overline{M}}$;

$$\begin{bmatrix} 1 & 2 \\ -1 & 4 \end{bmatrix} \begin{bmatrix} 1 \\ 0.5 \end{bmatrix} = \begin{bmatrix} 2 \\ 1 \end{bmatrix} = 2 \begin{bmatrix} 1 \\ 0.5 \end{bmatrix}$$

and:

$$\begin{bmatrix} 1 & 2 \\ -1 & 4 \end{bmatrix} \begin{bmatrix} 1 \\ 1 \end{bmatrix} = \begin{bmatrix} 3 \\ 3 \end{bmatrix} = 3 \begin{bmatrix} 1 \\ 1 \end{bmatrix}$$

8.12.2 Inverse

The need to solve sets of multidimensional coupled equations naturally leads to the methods of linear algebra, particularly matrix inversion. Just as the scalar equation;

$$A * x + B = 0$$

has the solution $x = -B/A$, provided $A \neq 0$, the vector equation;

$$\overline{\overline{A}} * \overline{x} + \overline{B} = \overline{0}$$

has the solution $\overline{x} = -\overline{\overline{A}}^{-1} * \overline{B}$, provided the inverse of $\overline{\overline{A}}$ exists. If $\overline{\overline{M}}$ is the inverse of $\overline{\overline{A}}$, then by definition:

$$\overline{\overline{M}} * \overline{\overline{A}} = \overline{\overline{I}}$$

Written for a $3x3$ case in terms of elements we have:

$$\begin{bmatrix} M_{11} & M_{12} & M_{13} \\ M_{21} & M_{22} & M_{23} \\ M_{31} & M_{32} & M_{33} \end{bmatrix} \begin{bmatrix} A_{11} & A_{12} & A_{13} \\ A_{21} & A_{22} & A_{23} \\ A_{31} & A_{32} & A_{33} \end{bmatrix} = \begin{bmatrix} 1 & 0 & 0 \\ 0 & 1 & 0 \\ 0 & 0 & 1 \end{bmatrix} \tag{8.95}$$

Reviewing briefly matrix multiplication, the above product is calculated according to:

$$\sum_r M_{ir} A_{rj} = \delta_{ij}$$

That is, the left side elements are:

$$M_{i1}A_{1j} + M_{i2}A_{2j} + \ldots. + M_{in}A_{nj}$$

while δ_{ij} is defined in Appendix 8.4. This operator will be used throughout this text. As an example of matrix inversion, consider the matrix:

$$\overline{\overline{A}} = \begin{bmatrix} 1 & 3 & 3 \\ 1 & 4 & 3 \\ 1 & 3 & 4 \end{bmatrix} \tag{8.96}$$

which has an inverse given by:

$$\overline{\overline{M}} = \overline{\overline{A}}^{-1} = \begin{bmatrix} 7 & -3 & -3 \\ -1 & 1 & 0 \\ -1 & 0 & 1 \end{bmatrix} \tag{8.97}$$

There are several algorithms for matrix inversion, most of which have been implemented in computing environments. As a result, we will not concern ourselves with manually computing the inverse of a matrix. This section is provided simply to provide a reminder of how to apply this concept.

8.12.3 Transpose

The transpose of a matrix $\overline{\overline{M}}$, written $\overline{\overline{M}}^T$, is defined by:

$$M_{ij}^T = M_{ji} \tag{8.98}$$

For example, for the matrix:

$$\overline{\overline{M}} = \begin{bmatrix} M_{11} & M_{12} & M_{13} \\ M_{21} & M_{22} & M_{23} \end{bmatrix}$$

the transpose is:

$$\overline{\overline{M}}^T = \begin{bmatrix} M_{11} & M_{21} \\ M_{12} & M_{22} \\ M_{13} & M_{23} \end{bmatrix}$$

Note that, as shown above, a matrix need not be square in order to have a transpose, while a square matrix that is equal to its own transpose is symmetric. Also note that one can consider the transpose of a matrix with dimension greater than 2.

8.13 Newton's Method

Computational thermodynamics can be thought of as an exercise in solving coupled nonlinear algebraic equations. As a result, Newton's method is indispensable to our purposes. Here we review the basic ideas of this technique. Let us first consider the simple scalar case. Suppose we seek the value of x that produces $F(x) = 0$. For linear equations, this is trivial. For example, with $F(x) = mx + b = 0$, the solution is known to be $x = -\frac{b}{m}$. Analytical solutions also exist for low-order polynomials (e.g., the quadratic equation, and for cubics, which apply to cubic EOS). However, analytical solutions may not exist for an arbitrary problem, and we must therefore use numerical methods to find the roots. Newton's method is based on a Taylor series (Sect. 8.19) representation of a function, based on a "guess value" $(x_1, F(x_1))$:

$$F(x) = F(x_1) + \frac{dF(x_1)}{dx}(x - x_1) + O(x - x_1)^2 \tag{8.99}$$

Here the symbol O represents order, meaning that remaining terms are second order or higher in $(x - x_1)$. If we truncate the approximation to 1st order and seek the root ($F(x) = 0$), we have:

$$0 \approx F(x_1) + \frac{dF(x_1)}{dx}(x - x_1) \tag{8.100}$$

We can then solve for the update to x:

$$\Delta x = (x - x_1) = -\frac{F(x_1)}{\frac{dF(x_1)}{dx}} \tag{8.101}$$

allowing us to update the value of x:

$$x = x_1 + \Delta x \tag{8.102}$$

These iterations are continued until the change in x is acceptably low. To implement a solver using Newton's method, we also require a stop condition. Typically, the absolute fractional changes in the independent variable(s) are checked against a prescribed tolerance ($|\Delta x/x| \leq Tol$), which can be set in the range of 10^-6 or some other criteria based on the physical problem. Of course, such an exit relies on $x \neq 0$. For such cases, a different exit condition is needed. It is also good practice (and sometimes necessary) to incorporate stepsize control and limits on the number of iterations. Algorithm 28 outlines an approach for solving $F(x) = 0$ by Newton's method, while Example 41 demonstrates these concepts.

Example 41 (Newton's Method to Find the Roots of a Scalar Function) The roots of $F(x) = x^2 - 3x + 2 = 0$ are known to be $x = 1$ and $x = 2$, but for

Algorithm 28 The use of Newton's method for solving $F(x) = 0$

Initialize $Iter = 0$, $MaxIter = 100$, $Tol = 10^{-6}$, $LI = 0$, $MLI = 10$, $Flag = True$
Initialize $x = x_{guess}$ (This value is based on insight into the problem to be solved)
Calculate $F(x)$
while $(Flag == True)$ **do**
  Calculate $H(x) = \frac{dF}{dx}$
  Calculate $\Delta x = -\frac{F}{H}$ (This is the essence of Newton's method)
  Compute $Err = |\frac{\Delta x}{x}|$ and update $Err+ = 1$
  Stepsize Controller
  if $|\Delta x/x| > 0.1$ **then**
    $\Delta x = sign(\Delta x) * |x| * 0.1$
  end if
  $x+ = \Delta x$ (Step)
  Exit condition
  if $Iter \geq MaxIter$ or $Err \leq Tol$ **then**
    Flag = False
  else
    Calculate $F(x)$
  end if
end while
Check convergence
if $Iter \geq MaxIter$ **then**
  Failed to converge: return error message
else
  Return x
end if

illustrative purposes, consider an initial guess $x_0 = 0$. With $H(x) = \frac{dF(x)}{dx} = 2x - 3$, Table 8.4 and Fig. 8.3 summarize the solution.

These iterations are shown in Fig. 8.3.

Notice that the next iteration is derived by following the tangent to the function to $F = 0$. The value of x is then used as the next guess. With a guess of $x = 3$, we also produce the expected result ($x = 2$) in a few iterations.

However useful, Newton's method is not foolproof. Suppose we wish to find both (*all*) the roots of $F(x) = x^2 - 3x + 2 = 0$. After our initial guess of $x = 0$ yields $x = 1.0$, if we then guess, say, $x = 1.4$ (you will not always have a simple graph of the function, especially in higher dimensions), Newton's method will simply converge back to $x = 1.0$. This is shown in Fig. 8.4.

Table 8.4 Newton iterations to solve $F = x^2 - 3x + 2 = 0$ initialized at $x = 0$

Iter	x	F	H	Δx
1	0	2	-3	0.667
2	0.667	0.444	-1.667	0.267
3	0.933	7.111E-2	-1.133	6.275E-2
4	0.996	3.937E-2	-1.008	3.960E-2
5	≈ 1	1.526E-5	-1.000	1.526E-5

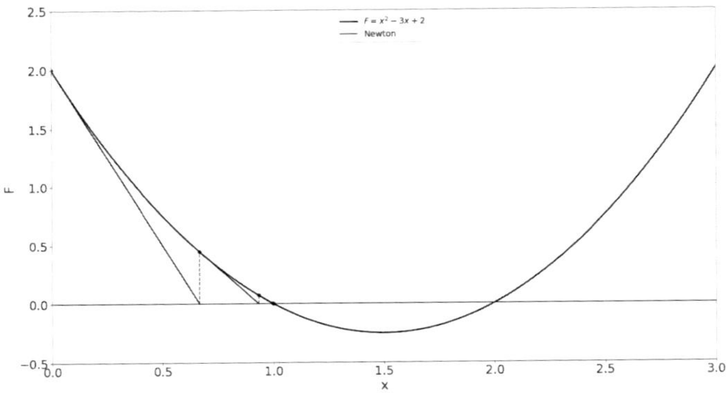

Fig. 8.3 Use of Newton's method to solve $F = x^2 - 3x + 2$ with initial guess of $x = 0$

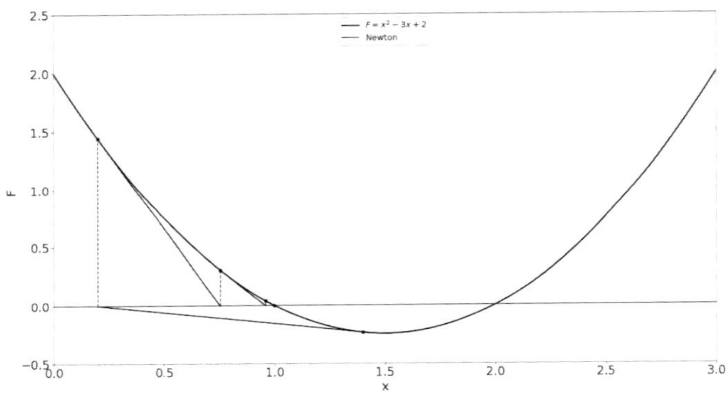

Fig. 8.4 Use of Newton's method to solve $F = x^2 - 3x + 2$ with initial guess of $x = 1.4$

The reason for this can be seen in Eq. 8.101. For this example, $\frac{dF(x)}{dx} = 2x - 3$ is negative on the interval $x < 1.5$, where the sign of dx is controlled by the sign of $F(x)$. When $F(x) > 0$, we have $\Delta x > 0$, and vice versa. This results in convergence toward to $x = 1.0$ for any guess $x_0 < 1.5$. A change in sign of $\frac{dF(x)}{dx}$ is required to "get away" from the solution at $x = 1$. We can see this if we guess $x = 1.6$, $F(x_0) = -0.24$, and $\frac{dF(x)}{dx} = 0.2$, meaning $dx = 1.2$. While this step does initially overshoot the solution[15] at $x = 2$, direct calculation shows that this will indeed converge correctly. In the design of a root solver, we will include steps that allow us to "get away" from roots already determined. Finally, we must avoid guesses close

[15] this problem is forgiving, but other problems require stepsize control for this reason.

to $\frac{dF(x)}{dx} = 0$, which produce extremely large step sizes. For the current example, a guess of $x_1 = 1.5$ produces:

$$F(1.5) = -1/4$$
$$\frac{dF(1.5)}{dx} = 0$$
$$dx = ?!$$

Strategies for dealing with these issues are dealt with in Chap. 4 and other parts of this text. These observations underscore that we cannot "blindly" input guesses into Newton's method and expect to find all the roots, particularly since we cannot in the general case plot the function whose roots we are seeking. Newton's method can also be used for multivariate problems. The development proceeds as in the 1D case, with a Taylor series expansion around an initial guess:

$$\overline{F}(\overline{x}) = \overline{F}(\overline{x}_0) + \frac{\partial \overline{F}(\overline{x}_0)}{\partial \overline{x}} * (\overline{x} - \overline{x}_0) + O(\overline{x} - \overline{x}_0)^2 \tag{8.103}$$

Equation 8.103 is a vector equation, and the multiplication shown represents the product of the matrix $\frac{\partial \overline{F}(\overline{x}_0)}{\partial \overline{x}} * (\overline{x} - \overline{x}_0)$ and the vector $(\overline{x} - \overline{x}_0)$. If we truncate the Taylor series to 1st order, we have:

$$\overline{F}(\overline{x}) \approx \overline{F}(\overline{x}_0) + \frac{\partial \overline{F}(\overline{x}_0)}{\partial \overline{x}} * (\overline{x} - \overline{x}_0) \tag{8.104}$$

In this text, the matrix $\frac{\partial \overline{F}}{\partial \overline{x}}$ is given the symbol $\overline{\overline{H}}$

$$H_{ij} = \frac{\partial F_i}{\partial x_j} \tag{8.105}$$

We can then solve for the update to $\overline{x}$ by attempting to make $F = 0$;

$$\Delta\overline{x} = (\overline{x} - \overline{x}_0) = -\overline{\overline{H}}^{-1} * \overline{F}(\overline{x_0}) \tag{8.106}$$

allowing us to update the value of $\overline{x}$:

$$\overline{x}_1 = \overline{x}_0 + \Delta\overline{x} \tag{8.107}$$

Example 42 solidifies these concepts.

Example 42 (Newton's Method to Find the Roots of a Vector Function) Let us attempt to solve for the roots of $x^2 + y^2 - 8 = 0$ and $x^2 - 2y = 0$. By algebra we can easily show that $x = \pm 2$ and $y = 2$, but using the approach of Eqs. 8.104 through 8.107, we have:

$$\overline{F} = \begin{bmatrix} x^2 + y^2 - 8 \\ x^2 - 2y \end{bmatrix} \tag{8.108}$$

$$\overline{\overline{H}} = \begin{bmatrix} 2x & 2y \\ 2x & -2 \end{bmatrix} \tag{8.109}$$

At an initial guess of $x_0, y_0 = 1, 1$ the iterations are summarized in Table 8.5. As shown, we converge to $x, y = 2, 2$ is reached. Similarly, if the initial guess is $x_0, y_0 = -1, 1$, we converge to $x, y = -2, 2$.

Throughout this text, the following notation is used for representing Newton's method:

Scalar case:

$$dx = -\frac{F}{H} \tag{8.110}$$

Vector Case:

$$d\overline{x} = -\overline{\overline{H}}^{-1} * \overline{F} \tag{8.111}$$

In both cases, it is understood that F and H are evaluated at the previous iteration (x), for which the update dx is generated.

Table 8.5 Newton iterations to solve $F = x^2 - 3x + 2 = 0$ initialized at $x = 0$

Iter	x, y	$\overline{F}$	H	$\Delta x, \Delta y$
1	1,1	−6,1	2,2 2,−2	1.75, 1.25
2	2.75 2.25	4.625 3.0625	5.5 4.5 5.5 −2	−0.644, −0.240
3	2.106, 2.010	0.473, 0.415	4.212, 4.019 4.212, −2	−0.103, −9.600E-3
4	2.003, 2.000	0.011, 0.011	4.005, 4.000 4.005, −2	−2.662E-3, −1.536E-5
5	2.000, 2.000	7.086E-6, 7.085E-6	4.000, 4. 4.000, −2	−1.771E-6, −3.932E-11

8.13.1 Newton's Method in Multiphase TPn Flash Calculations

Multiphase TP flash calculations can be converged using Newton's method, though care must be taken to ensure that the correct number of phases has been identified. It is therefore advisable to apply the following methods only after stability analysis (Chap. 6.4) has indicated phase separation, and more robust solution methods have been used to approach the correct solution. One strategy is to apply successive substitution combined with acceleration (Appendix 8.18.1) and then apply Newton's method after two or three successive accelerations (analogous to the approach in 6.4.1).

In the following development, we will consider C-components ($i = 1 \ldots C$) and Π-phases ($j \neq R$). The set of equations to be solved is analogous to Eq. 6.121:

$$F_i^j = \begin{cases} EQ_i^j = lnk_i^j + ln\hat{\varphi}_i^j - ln\hat{\varphi}_i^R = 0 \; C(\Pi - 1) \text{ equations} \\ RR^j = \sum_r \frac{z_r(k_r^j - 1)}{1 + \sum_{s \neq R} f^s(k_i^s - 1)} = 0 \quad (\Pi - 1) \text{ equations} \end{cases} \tag{8.112}$$

Formally considering the unknowns to be the lnk and f values, we have:

$$\alpha_i^j = \begin{cases} lnk_i^j \; C(\Pi - 1) \text{ equations} \\ f^j \quad (\Pi - 1) \text{ equations} \end{cases} \tag{8.113}$$

F and α are two-dimensional, with size $C + 1$ by $(N - 1)$. The Jacobian takes the form:

$$H_{iq}^{jt} = \frac{\partial F_i^j}{\partial \alpha_q^t} = \begin{cases} \frac{\partial EQ_i^j}{\partial lnk_q^t} \; \frac{\partial EQ_i^j}{\partial f^t} \\ \frac{\partial RR^j}{\partial lnk_q^t} \; \frac{\partial RR^j}{\partial f^t} \end{cases} \tag{8.114}$$

Note that all derivatives in this section are at fixed T and P, but for clarity this has been dropped from the notation. We now develop the first structure within H_{iq}^{jt}; $\frac{\partial EQ_i^j}{\partial lnk_q^t}$

$$\begin{aligned} \frac{\partial EQ_i^j}{\partial lnk_q^t} &= \frac{\partial(lnk_i^j + ln\hat{\varphi}_i^j - ln\hat{\varphi}_i^R)}{\partial lnk_q^t} \\ &= \delta_{iq}\delta_{jt} + \frac{\partial ln\hat{\varphi}_i^j}{\partial lnk_q^t} - \frac{\partial ln\hat{\varphi}_i^R}{\partial lnk_q^t} \end{aligned} \tag{8.115}$$

Now we must recognize that $ln\hat{\varphi}_i^j$ depends only[16] Therefore:

$$\frac{\partial ln\hat{\varphi}_i^j}{\partial lnk_q^t} = \sum_r \frac{\partial ln\hat{\varphi}_i^j}{\partial n_r^j} \frac{\partial n_r^j}{\partial lnk_q^t} \tag{8.116}$$

Here we have chosen to regard $ln\hat{\varphi}_i^j$ as a function of n_i^j defined[17] by Eq. 1.71. However, the variables of our present context are lnk and f, which are intensive variables. We therefore appeal to Appendix 8.6, recognizing that $\frac{\partial ln\hat{\varphi}_i^j}{\partial n_r^j}$ is homogeneous to degree -1:

$$\frac{\partial ln\hat{\varphi}_i^j}{\partial n_r^j} = \frac{1}{n^j} \frac{\partial ln\hat{\varphi}_i^j}{\partial c_r^j} \tag{8.117}$$

With the mole numbers in phase j given by:

$$n_i^j = nf^j c_i^j = n^j c_i^j \tag{8.118}$$

At fixed total moles (n) and phase fractions $\overline{f}$, plus the fact that c_i^j only depends on the lnk values of species i (see Eq. 8.159), we interpret $\frac{\partial n_r^j}{\partial lnk_q^t}$ as:

$$\frac{\partial n_r^j}{\partial lnk_q^t} = \delta_{rq} n^j \frac{\partial c_q^j}{\partial lnk_q^t}$$

We can therefore write Eq. 8.116 as:

$$\frac{\partial ln\hat{\varphi}_i^j}{\partial lnk_q^t} = \frac{\partial ln\hat{\varphi}_i^j}{\partial c_q^j} \frac{\partial c_q^j}{\partial lnk_q^t} \tag{8.119}$$

One cannot differentiate with respect to mole fractions. In the evaluation of $\frac{\partial ln\hat{\varphi}_i^j}{\partial c_q^j}$ we simply call our function $\frac{\partial ln\hat{\varphi}_i^j}{\partial n_q^j}$ using the output from Eq. 8.159 as "mole numbers." Equation 8.119 also applies to the reference phase. Therefore we have:

$$\frac{\partial EQ_i^j}{\partial lnk_q^t} = \delta_{iq}\delta_{jt} + \frac{\partial ln\hat{\varphi}_i^j}{\partial c_q^j} \frac{\partial c_q^j}{\partial lnk_q^t} - \frac{\partial ln\hat{\varphi}_i^R}{\partial c_q^R} \frac{\partial c_q^R}{\partial lnk_q^t} \tag{8.120}$$

[16] Do not be confused here. True, lnk values in any phase affect the composition in that phase, which in turn relate to those in another phase, but this is handled by the Rachford–Rice equation. Formally, $ln\hat{\varphi}_i^j$ only depends on (all of the) mole numbers in phase j.

[17] Equation 1.71 does not have a phase index. We are simply calculating the properties of a particular phase.

We can develop the derivative of EQ^j with respect to phase fractions as above:

$$\frac{\partial EQ_i^j}{\partial f^t} = \frac{\partial(lnk_i^j + ln\hat{\varphi}_i^j - ln\hat{\varphi}_i^R)}{\partial f^t}$$
$$= \frac{\partial ln\hat{\varphi}_i^j}{\partial f^t} - \frac{\partial ln\hat{\varphi}_i^R}{\partial f^t} \tag{8.121}$$

The derivative of the $ln\hat{\varphi}$ terms proceeds as above:

$$\frac{\partial ln\hat{\varphi}_i^j}{\partial f^t} = \sum_r \frac{\partial ln\hat{\varphi}_i^j}{\partial n_r^j}\frac{\partial n_r^j}{\partial f^t} \tag{8.122}$$

As before we make use of Eq. 8.117 and then regard the mole numbers in phase j as:

$$n_i^j = nf^j c_i^j \tag{8.123}$$

the derivative of which with respect to f^j is:

$$\frac{\partial n_i^j}{\partial f^t} = n\delta_{jt}c_i^j + nf^j\frac{\partial c_i^j}{\partial f^t} \tag{8.124}$$

With $\frac{\partial c_i^j}{\partial f^t}$ given by Eq. 8.164, we can substitute Eq. 8.124 into Eq. 8.122, giving:

$$\frac{\partial ln\hat{\varphi}_i^j}{\partial f^t} = \sum_r \frac{\partial ln\hat{\varphi}_i^j}{\partial c_r^j}\frac{1}{n^j}(n\delta_{jt}c_i^j + nf^j\frac{\partial c_i^j}{\partial f^t})$$
$$= \frac{\delta_{jt}}{f^j}\sum_r \frac{\partial ln\hat{\varphi}_i^j}{\partial c_r^j}c_i^j + \sum_r \frac{\partial ln\hat{\varphi}_i^j}{\partial c_r^j}\frac{\partial c_i^j}{\partial f^t} \tag{8.125}$$

The first term in Eq. 8.125 is zero, not because of δ_{jt}, but through Eq. 8.56. Therefore we have:

$$\frac{\partial ln\hat{\varphi}_i^j}{\partial f^t} = \sum_r \frac{\partial ln\hat{\varphi}_i^j}{\partial c_r^j}\frac{\partial c_i^j}{\partial f^t} \tag{8.126}$$

Equation 8.126 also applies to the reference phase. The remaining terms in the Jacobian (Eq. 8.114) have been defined in Sect. 8.16.2 (Eqs. 8.166 and 8.167).

As indicated above, the various elements of our target function F (Eq. 8.112), our unknowns α (Eq. 8.113), and Jacobian H (Eq. 8.114) have multidimensional structures. As written, for a C-component π-phase mixture, F and α are $(C + 1)$ by

$(\pi - 1)$, giving H the dimensions $(C + 1)$ by $(\pi - 1)$ by $(C + 1)$ by $(\pi - 1)$. In order to use Newton's method, we have to restructure F and α into 1D vectors and then restructure H into a corresponding 2D matrix.[18]

The main idea is to create a mapping between the 2D and 1D forms of α and F, which we will call here the *short* and the *long* forms. These are shown for a 3-component, 3-(only 2 are independent)phase system below:

$$\overline{\overline{\alpha}}_{short} = \begin{bmatrix} lnk_1^1 & lnk_1^2 \\ lnk_2^1 & lnk_2^2 \\ lnk_3^1 & lnk_3^2 \\ f^1 & f^2 \end{bmatrix} \tag{8.127}$$

$$\overline{\overline{F}}_{short} = \begin{bmatrix} EQ_1^1 & EQ_1^2 \\ EQ_2^1 & EQ_2^2 \\ EQ_3^1 & EQ_3^2 \\ RR^1 & RR^2 \end{bmatrix} \tag{8.128}$$

A suitable mapping is shown below:

$$Map = \begin{bmatrix} 1 & 5 \\ 2 & 6 \\ 3 & 7 \\ 4 & 8 \end{bmatrix} \tag{8.129}$$

The number of rows in $\overline{\overline{F}}_{short}$ is one more than the number of components in the mixture, while the number of columns is one less than the number of phases. Algorithm 29 demonstrates how to populate the Map;

Algorithm 29 Mapping for Newton's method in multiphase TPn flash

```
Initialize idx
for each row do
    for each column do
        Map[row, col] = idx
        idx += 1
    end for
end for
```

[18] This results from the way we have structured the elements of F and α (e.g., giving independent indices to components and phases within lnk_i^j and $ln\hat{\varphi}_i^j$). This is of course intuitive, but creates difficulties here because we cannot invert H in the usual way. An alternative might be to structure lnk and $ln\hat{\varphi}$ as 1D vectors (note that they will have different lengths) from which we can extract the relevant values of a particular phase by knowing how many components are in the mixture. This is the basis of the technique that we will use to restructure F, α and H.

Algorithm 30 demonstrates how to use the Map to populate $\overline{\alpha}_{long}$.

Algorithm 30 Population of α_{long} used for Newton's method in multiphase TPn flash

for each row **do**
  for each column **do**
    $\alpha_{long}[Map[row, col]] = \alpha_{short}[row, col]$
  end for
end for

$$\overline{\alpha}_{long} = \begin{bmatrix} lnk_1^1 \\ lnk_2^1 \\ lnk_3^1 \\ f^1 \\ lnk_1^2 \\ lnk_2^2 \\ lnk_3^2 \\ f^2 \end{bmatrix} \tag{8.130}$$

The same mapping provides a definition for $\overline{F}_{long}$:

$$\overline{F}_{long} = \begin{bmatrix} EQ_1^1 \\ EQ_2^1 \\ EQ_3^1 \\ RR^1 \\ EQ_1^2 \\ EQ_2^2 \\ EQ_3^2 \\ RR^2 \end{bmatrix} \tag{8.131}$$

This mapping also defines a long form structure for $\overline{\overline{H}}_{long}$, which uses the Map according to Algorithm 31.

Algorithm 31 Population of $\overline{\overline{H}}_{long}$ used for Newton's method in multiphase TPn flash

$\frac{\partial EQ_i^j}{\partial lnk_q^t}$
for i (comp) **do**
  for j (phase) **do**
    for q (comp) **do**
      for t (phase) **do**
        $H_{long}[Map[i, j], Map[q, t]] = \frac{\partial EQ_i^j}{\partial lnk_q^t}$
      end for
    end for
  end for
end for

$\frac{\partial RRi^j}{\partial lnk_q^t}$
Initialize row
for j (phase) **do**
  $row+ = C$
  for q (comp) **do**
    for t (phase) **do**
      col = Map[q,t]
      $H_{long}[row, col] = \frac{\partial RRi^j}{\partial lnk_q^t}$
    end for
  end for
end for

$\frac{\partial EQ_i^j}{\partial f^t}$
Initialize col
for t (phase) **do**
  $col+ = C$
  for i (comp) **do**
    for j (phase) **do**
      $row = Map[i, j]$
      $H_{long}[row, col] = \frac{\partial EQ_i^j}{\partial f^t}$
    end for
  end for
end for

$\frac{\partial RR^j}{\partial f^t}$
Initialize row
for $j(phase)$ **do**
  $row+ = C$
  Initialize col
  for t phase **do**
    $col+ = C$
    $H_{long}[row, col] = \frac{\partial RR^j}{\partial f^t}$
  end for
end for

The result is:

$$H_{long} = \begin{bmatrix} \frac{\partial EQ_1^1}{\partial lnk_1^1} & \frac{\partial EQ_1^1}{\partial lnk_2^1} & \frac{\partial EQ_1^1}{\partial lnk_3^1} & \frac{\partial EQ_1^1}{\partial f^1} & \frac{\partial EQ_1^1}{\partial lnk_1^2} & \frac{\partial EQ_1^1}{\partial lnk_2^2} & \frac{\partial EQ_1^1}{\partial lnk_3^2} & \frac{\partial EQ_1^1}{\partial f^2} \\ \frac{\partial EQ_2^1}{\partial lnk_1^1} & \frac{\partial EQ_2^1}{\partial lnk_2^1} & \frac{\partial EQ_2^1}{\partial lnk_3^1} & \frac{\partial EQ_2^1}{\partial f^1} & \frac{\partial EQ_2^1}{\partial lnk_1^2} & \frac{\partial EQ_2^1}{\partial lnk_2^2} & \frac{\partial EQ_2^1}{\partial lnk_3^2} & \frac{\partial EQ_2^1}{\partial f^2} \\ \frac{\partial EQ_3^1}{\partial lnk_1^1} & \frac{\partial EQ_3^1}{\partial lnk_2^1} & \frac{\partial EQ_3^1}{\partial lnk_3^1} & \frac{\partial EQ_3^1}{\partial f^1} & \frac{\partial EQ_3^1}{\partial lnk_1^2} & \frac{\partial EQ_3^1}{\partial lnk_2^2} & \frac{\partial EQ_3^1}{\partial lnk_3^2} & \frac{\partial EQ_3^1}{\partial f^2} \\ \frac{\partial RR^1}{\partial lnk_1^1} & \frac{\partial RR^1}{\partial lnk_2^1} & \frac{\partial RR^1}{\partial lnk_3^1} & \frac{\partial RR^1}{\partial f^1} & \frac{\partial RR^1}{\partial lnk_1^2} & \frac{\partial RR^1}{\partial lnk_2^2} & \frac{\partial RR^1}{\partial lnk_3^2} & \frac{\partial RR^1}{\partial f^2} \\ \frac{\partial EQ_1^2}{\partial lnk_1^1} & \frac{\partial EQ_1^2}{\partial lnk_2^1} & \frac{\partial EQ_1^2}{\partial lnk_3^1} & \frac{\partial EQ_1^2}{\partial f^1} & \frac{\partial EQ_1^2}{\partial lnk_1^2} & \frac{\partial EQ_1^2}{\partial lnk_2^2} & \frac{\partial EQ_1^2}{\partial lnk_3^2} & \frac{\partial EQ_1^2}{\partial f^2} \\ \frac{\partial EQ_2^2}{\partial lnk_1^1} & \frac{\partial EQ_2^2}{\partial lnk_2^1} & \frac{\partial EQ_2^2}{\partial lnk_3^1} & \frac{\partial EQ_2^2}{\partial f^1} & \frac{\partial EQ_2^2}{\partial lnk_1^2} & \frac{\partial EQ_2^2}{\partial lnk_2^2} & \frac{\partial EQ_2^1}{\partial lnk_3^2} & \frac{\partial EQ_2^2}{\partial f^2} \\ \frac{\partial EQ_3^2}{\partial lnk_1^1} & \frac{\partial EQ_3^2}{\partial lnk_2^1} & \frac{\partial EQ_3^2}{\partial lnk_3^1} & \frac{\partial EQ_3^2}{\partial f^1} & \frac{\partial EQ_3^2}{\partial lnk_1^2} & \frac{\partial EQ_3^2}{\partial lnk_2^2} & \frac{\partial EQ_3^1}{\partial lnk_3^2} & \frac{\partial EQ_3^2}{\partial f^2} \\ \frac{\partial RR^2}{\partial lnk_1^1} & \frac{\partial RR^2}{\partial lnk_2^1} & \frac{\partial RR^2}{\partial lnk_3^1} & \frac{\partial RR^2}{\partial f^1} & \frac{\partial RR^2}{\partial lnk_1^2} & \frac{\partial RR^2}{\partial lnk_2^2} & \frac{\partial RR^2}{\partial lnk_3^2} & \frac{\partial RR^2}{\partial f^2} \end{bmatrix} \tag{8.132}$$

This allows an update to α_{long} via the usual application of Newton's method:

$$d\overline{\alpha}_{long} = -\overline{\overline{H}}_{long}^{-1} * \overline{F}_{long} \tag{8.133}$$

We can then reverse the mapping to extract the short form of the updated α_{long} via Algorithm 32;

Algorithm 32 Reverse mapping for Newton's method in multiphase TPn flash

---

Initialize idx
for each col **do**
  for each row **do**
    $\alpha_{short}[row, col] = \alpha_{long}[idx]$
    $idx += 1$
  end for
end for

---

which provides all the elements required to recalculate $\overline{\overline{F}}_{short}$. This is clarified in the following example.

Example 43 (Solving a Three-Phase Flash Using Newton's Method) Consider a flash calculation of an equimolar mixture of nitrogen, n-hexane, and water at $T = 478$K and $P = 131$ bar, using the SRK EOS. Suppose that a successive substitution flash has, following successful accelerations, yielded the following compositions and phase fractions:

$$\overline{\overline{c}} = \begin{bmatrix} 9.857E-4 & 0.252 & 0.492 \\ 4.270E-5 & 0.439 & 0.266 \\ 0.999 & 0.309 & 0.242 \end{bmatrix}$$

$$\overline{f} = \begin{bmatrix} 0.507 & 0.417 \end{bmatrix}$$

The volumes of the three phases are found to be;

$$\overline{v} = \begin{bmatrix} 28.703 & 162.518 & 244.568 \end{bmatrix} \quad (\frac{cm^3}{mol})$$

which can be used to establish that the phases are correctly sorted by mass density, meaning the lnk values are:

$$ln\overline{\overline{k}} = \begin{bmatrix} 5.544 & 6.214 \\ 9.237 & 8.736 \\ -1.173 & -1.418 \end{bmatrix}$$

The $ln\hat{\varphi}$ values are found to be:

$$ln\overline{\overline{\hat{\varphi}}} = \begin{bmatrix} 6.574 & 1.0310 & 0.361 \\ 7.684 & -1.554 & -1.053 \\ -2.010 & -0.837 & -0.591 \end{bmatrix}$$

We can then compute the elements of F as in Eq. 8.112:

$$\overline{\overline{EQ}} = \begin{bmatrix} -1.250E-6 & 1.215E-6 \\ 1.208E-6 & -1.631E-6 \\ 3.893E-6 & -6.379E-7 \end{bmatrix}$$

$$\overline{RR} = \begin{bmatrix} 3.489E-8 & 5.484E-8 \end{bmatrix}$$

making the short (Eq. 8.112) and long (Eq. 8.131 forms of F:

$$\overline{\overline{F}}_{short} = \begin{bmatrix} -1.250E-6 & 1.215E-6 \\ 1.208E-6 & -1.631E-6 \\ 3.893E-7 & -6.349E-7 \\ 3.489E-8 & 5.484E-8 \end{bmatrix}$$

$$\overline{F}_{long} = [-1.250E-6 \quad 1.215E-6 \quad 1.208E-6 \quad -1.631E-6 \quad 3.893E-7 \quad -6.349E-7 \quad 3.489E-8 \quad 5.484E-8]$$

The short (Eq. 8.113) and long (Eq. 8.130 forms of α are:

$$\overline{\overline{\alpha}}_{short} = \begin{bmatrix} 5.544 & 6.214 \\ 9.237 & 8.736 \\ -1.173 & -1.418 \\ 0.507 & 0.417 \end{bmatrix}$$

$$\overline{\alpha}_{long} = \begin{bmatrix} 5.544 & 9.237 & -1.173 & 0.507 & 6.214 & 8.736 & -1.418 & 0.417 \end{bmatrix}$$

The long form of H (Eq. 8.132) is;

$$\overline{\overline{H}}_{long} = \begin{bmatrix} 0.616 & 0.122 & 0.137 & 0.425 & 0.368 & -0.124 & -0.068 & 1.094 \\ 0.114 & 0.953 & -0.020 & -0.308 & -0.159 & 0.041 & 0.041 & -0.585 \\ 0.122 & -0.039 & 0.958 & -0.159 & -0.122 & 0.039 & 0.024 & -0.379 \\ 0.156 & 0.146 & 0.633 & -2.194 & -0.155 & -0.146 & 0.209 & -2.286 \\ 0.146 & -0.197 & -0.040 & 0.075 & 0.838 & 0.195 & 0.077 & 0.521 \\ -0.226 & 0.265 & 0.090 & -0.319 & 0.181 & 0.729 & -0.083 & -0.989 \\ -0.081 & 0.104 & 0.026 & -0.068 & 0.081 & -0.104 & 0.962 & -0.311 \\ -0.188 & -0.177 & 0.356 & -2.286 & 0.189 & 0.177 & 0.471 & -2.655 \end{bmatrix}$$

which makes the update to $\overline{\alpha}_{long}$:

$$d\overline{\alpha}_{long} = -\overline{\overline{H}}_{long}^{-1}\overline{F}_{long} =$$

$$4.035E-6 - 2.146E-6 - 1.286E-6 - 9.715E-6$$

$$-1.224E-6\ 1.678E-5\ 6.481E-6\ 9.468E-6$$

Extraction from the updated $\overline{\alpha}_{long}$ (using the mapping) and recalculation of F_{short} yields;

$$\overline{\overline{F}}_{short} = \begin{bmatrix} 5.067E-8 & 3.125E-8 \\ -2.272E-8 & -4.942E-8 \\ -1.692E-8 & -1.733E-8 \\ 3.499E-8 & -5.468E-8 \end{bmatrix}$$

indicating a significant reduction in the magnitudes of the target functions.

8.14 Numerical Derivatives

Key thermodynamic concepts are generally based on analytical derivatives of one quantity with respect to another. However, there do arise circumstances where analytical derivatives are either impossible or impractical to compute, and in these scenarios we rely on numerical estimates. The simplest numerical derivative is the so-called two-point method:

$$\frac{dF}{dx} \approx \frac{F(x+\Delta x) - F(x)}{\Delta x} \tag{8.134}$$

For a linear function of x, Eq. 8.134 is exact, while for nonlinear functions, its accuracy is dependent on the topology of F (i.e., where we are applying this approximation), and on the stepsize (Δx) used. Let's look at an example.

Example 44 (Numerical Differentiation) Consider $F(x) = x^2$, for which $\frac{dF}{dx} = 2x$. Numerical differentiation at $x = 1$ with $\Delta x = 0.01$ gives (according to Eq. 8.134;

$$\frac{dF}{dx} \approx \frac{(1+0.01)^2 - (1)^2}{0.01} = \frac{1.0201 - 1}{0.01} = 2.01$$

while a stepsize of $\Delta x = 1E - 6$ gives:

$$\frac{dF}{dx} \approx \frac{(1+1E-6)^2 - (1)^2}{1E-6} = \frac{1.0201 - 1}{1E-6} = 2.000001$$

As indicated, numerical differentiation gives only an approximation of the analytical derivative, which at $x = 1$ is 2.0.

Better accuracy can be achieved with a higher-order approximation, such as the so-called five-point method, which is derived as follows. Consider a Taylor series of F around the point x:

$$F(x+\Delta x) = F(x) + \frac{dF}{dx}[\Delta x] + \frac{1}{2}\frac{d^2F}{dx^2}[\Delta x]^2 + \frac{1}{6}\frac{d^3F}{dx^3}[\Delta x]^3 \ldots$$

This series is a *forward* step. Now consider the *reverse* step:

$$F(x-\Delta x) = F(x) + \frac{dF}{dx}[-\Delta x] + \frac{1}{2}\frac{d^2F}{dx^2}[\Delta x]^2 + \frac{1}{6}\frac{d^3F}{dx^3}[-\Delta x]^3 \ldots$$

Taking the difference, the even-order powers in Δx cancel:

$$F(x+\Delta x) - F(x-\Delta x) = 2\frac{dF}{dx}[\Delta x] + \frac{1}{3}\frac{d^3F}{dx^3}[\Delta x]^3 \ldots \tag{8.135}$$

This difference uses the two points on each side of x. We can also calculate this term for the next two points:

$$F(x+2\Delta x) - F(x-2\Delta x) = 2\frac{dF}{dx}[2\Delta x] + \frac{1}{3}\frac{d^3F}{dx^3}[2\Delta x]^3 \ldots$$
$$= 4\frac{dF}{dx}[\Delta x] + \frac{8}{3}\frac{d^3F}{dx^3}[\Delta x]^3 \ldots \quad (8.136)$$

The 3rd order term can be canceled by substitution of Eq. 8.135 into Eq. 8.136:

$$\frac{dF}{dx} \approx \frac{-F(x+2\Delta x) + 8F(x+\Delta x) - 8F(x-\Delta x) + F(x-2\Delta x)}{12\Delta x} \quad (8.137)$$

Using Eq. 8.137, the numerical derivative of $F(x) = x^2$ at $x = 1$ with $\Delta x = 1E{-}6$ is 2.00000000001, a significant improvement on the two-point method (Eq. 8.134), yet still an approximation.

Finally, though illustrated here for scalar, univariate functions, these same approximations can be used for computing the partial derivatives of multivariate, multidimensional functions.

8.15 Partial Derivatives

Given a function $F = F(\overline{s})$, the *partial derivative* of F with respect to s_i is:

$$[\frac{\partial F}{\partial s_i}]_{s_{j \neq i}}$$

That is, one simply differentiates the function with respect to s_i, leaving all other variables unchanged. For example, given $F = 2 + xlny$, the relevant partial derivatives are:

$$[\frac{\partial F}{\partial x}]_y = lny$$
$$[\frac{\partial F}{\partial y}]_x = \frac{x}{y}$$

A very useful property of partial derivatives is that the order of differentiation in mixed partial derivatives is immaterial. For example, the mixed partial derivatives of the above form of F are:

$$[\frac{\partial^2 F}{\partial x \partial y}] = [\frac{\partial [\frac{\partial F}{\partial y}]_x}{\partial x}]_y = [\frac{\partial (\frac{x}{y})}{\partial x}]_y = \frac{1}{y}$$
$$[\frac{\partial^2 F}{\partial y \partial x}] = [\frac{\partial [\frac{\partial F}{\partial x}]_y}{\partial y}]_x = [\frac{\partial lny}{\partial y}]_x = \frac{1}{y}$$

This result is very useful in thermodynamic modeling.

8.16 The Rachford–Rice Equation

The Rachford–Rice equation is a very useful form of material balance, used in phase boundary and flash calculations. In this section we develop the main equations related to the Rachford–Rice equation, for two as well as multiphase cases.

8.16.1 Two-phase Rachford–Rice

Consider n^F moles of a mixture of overall composition $\overline{z}$, separated into a liquid and a vapor, with mole numbers n^ℓ and n^ν, and composition $\overline{x}$ and $\overline{y}$. A material balance on the amount of each component i:

$$\begin{aligned} n^F z_i &= n^\ell x_i + n^\nu y_i \\ z_i &= f^\ell x_i + f^\nu y_i \\ z_i &= x_i + f(y_i - x_i) \end{aligned} \tag{8.138}$$

In Eq. 8.138, f^ℓ and f^ν represent the phase *fractions* (by moles) of the liquid and vapor, respectively. We therefore recognize that only one of these two phase fractions is independent. Choosing the liquid phase as our *reference*, f represents the vapor phase fraction.[19] Equation 8.138 thus indicates that x_i and y_i are not independent of each other. We can therefore solve Eq. 8.138 for the liquid (our reference) phase composition via:

$$x_i = \frac{z_i}{1 + f(\frac{y_i}{x_i} - 1)} = \frac{z_i}{1 + f(k_i - 1)} \tag{8.139}$$

Equation 8.139 has introduced the notion of k values (or k factors), which allow us to express the composition of the vapor as:

$$y_i = k_i x_i = \frac{k_i z_i}{1 + f(k_i - 1)} \tag{8.140}$$

If we consider $\overline{z}$ to be normalized and take the component sum over Eq. 8.138, we have:

$$\sum_r z_r = \sum_r x_r + f(\sum_r y_r - \sum_r x_r) = 1 \tag{8.141}$$

[19] f is referred to as the *quality* in the context of steam generation.

This demonstrates that if *either* $\overline{x}$ or $\overline{y}$ is normalized, other will be too. We can phrase this normalization requirement through an equation called the *Rachford–Rice* equation:

$$RR = \sum_r y_r - \sum_r x_r = \sum_r \frac{z_r(k_r - 1)}{1 + f(k_r - 1)} = 0 \tag{8.142}$$

For many applications, derivatives of the Rachford–Rice equation are required. One context relates to the computation of VLE boundaries (Sect. 6.3). If we consider the *either* the temperature or the pressure to be given, along with the feed composition and phase fraction, we can estimate the temperature (or the pressure) and k values using the Wilson approximation (Sect. 8.20). The target function is a combination of Eqs. 8.142 and 8.198.

$$F(T, P) = \sum_r \frac{z_r(k_r{}^W(T, P) - 1)}{1 + f(k_r{}^W(T, P) - 1)} = 0 \tag{8.143}$$

While the RR equation is not formally a function of T and P (it is a function of f, $\overline{k}$ and $\overline{z}$, or it can be regarded as a function of composition), the Wilson approximation provides an approximate relationship between T,P, and $\overline{k}$. The T and P derivative of Eq. 8.143 become:

$$\begin{aligned} \left[\frac{\partial F}{\partial T}\right]_P &= \sum_r \left[\frac{z_r \left[\frac{\partial k_r{}^W}{\partial T}\right]_P}{1 + f(k_r{}^W - 1)} - \frac{z_r(k_r - 1) f \left[\frac{\partial k_r{}^W}{\partial T}\right]_P}{[1 + f(k_r{}^W - 1)]^2} \right] \\ &= \sum_r \left[\frac{z_r \left[\frac{\partial k_r{}^W}{\partial T}\right]_P}{[1 + f(k_r{}^W - 1)]^2} \right] \end{aligned} \tag{8.144}$$

and:

$$\left[\frac{\partial F}{\partial P}\right]_T = \sum_r \left[\frac{z_r \left[\frac{\partial k_r{}^W}{\partial P}\right]_T}{[1 + f(k_r{}^W - 1)]^2} \right] \tag{8.145}$$

Using the results in Appendix 8.20, Eq. 8.143 is monotonic in T and can therefore easily be solved for the temperature by Newton's method, as outlined in Algorithm 33 and in Example 11 in Sect. 6.3. It is also straightforward to solve for P.

Figure 8.5 summarizes many important points from this section, including the monotonic behavior of the Rachford-Rice equation, the tendency for the Rachford-Rice equation (Eq. 8.143) to take on very large magnitude values at low T, and how

Algorithm 33 Solving the Rachford–Rice equation with Wilson k values for temperature

Initialize $Iter = 0$, $MaxIter = 100$, $Tol = 10^{-6}$, $LI = 0$, $MLI = 10$, $Flag = True$
Calculate $T_{sat\,i}$ for each component via Wilson (Eq. 8.199)
Initialize T_{sat} for the mixture via $T \approx \sum_r z_r T_{sat\,r}$
Calculate F (Eq. 8.143) and increase T until F is finite
while $(Flag == True)$ **do**
  Calculate H (Eq. 8.144)
  Update T via Newton: $dT = -\frac{F}{H}$
  Compute $Err = |\frac{dT}{T}|$ and update $Err += 1$
  Compute $T_{New} = T + dT$ and control stepsize
  if $(T_{New} \leq 0)$ or $(|\frac{dT}{T}| \geq 0.1)$ **then**
    $Step = True$, $LI = 0$
    while $(Step == True)$ **do**
      $dT / = 2$
      $T_{New} = T + dT$
      $LI + = 1$
      if $((LI \geq MLI)$ or $(T_{New} > 0)$ and $(|\frac{dT}{T}| < 0.1))$ **then**
        $Step = False$
      end if
    end while
  end if
  Step: $T = T_{New}$
  if $((Iter \geq MaxIter) or (Err < Tol))$ **then**
    $Flag = False$
  else
    Calculate F (Eq. 8.143)
  end if
end while
Check convergence
if $Iter \geq MaxIter$ **then**
  Failed to converge: return error message
else
  Return T
end if

the choice of phase fraction ($f = 0, 0.5, 1.0$) affects the value of T_{sat} (the point where $RR = 0$). This is also shown in Fig. 6.2, as the *width* of a phase diagram for a mixture. Algorithm 33 can be modified to solve for the estimated saturation pressure for a given T, $\overline{z}$ and f. The stepsize controller avoids negative pressure and limits the change in P, and the derivative is now given by Eq. 8.145.

In other parts of this text, derivatives of Eqs. 8.139 and 8.140 with respect to k values are required. As shown, x_i and y_i only depend on their respective k_i values. That is:

$$\left[\frac{\partial x_i}{\partial k_j}\right] = \delta_{ij}\left[\frac{\partial x_i}{\partial k_i}\right] = -\delta_{ij}\frac{x_i^2}{z_i}f \tag{8.146}$$

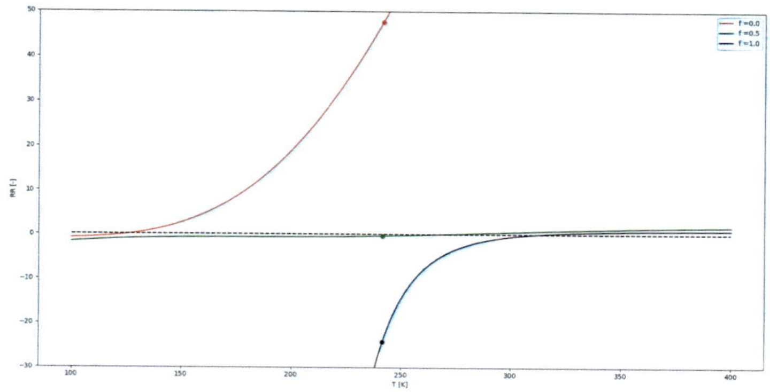

Fig. 8.5 Rachford–Rice Equation using Wilson k values (Eq. 8.143) as a function of temperature with varying phase fraction (f) for an equimolar mixture of methane, n-butane and n-hexane at 1 bar. The initial estimate of T_{sat} is shown on each curve, and is indepenent of f

and:

$$\left[\frac{\partial y_i}{\partial k_j}\right] = \delta_{ij}\left[\frac{\partial y_i}{\partial k_i}\right] = \delta_{ij}\left(x_i - \frac{x_i y_i}{z_i} f\right) = \delta_{ij} x_i \left(1 - \frac{y_i}{z_i} f\right) \tag{8.147}$$

Consistent with the above, we can directly calculate from Eq. 8.142:

$$\begin{aligned}\frac{\partial RR}{\partial k_j} &= \sum_r \left(\frac{z_r \delta_{rj}}{1 + f(k_r - 1)} - \frac{z_r (k_r - 1) f \delta_{rj}}{[1 + f(k_r - 1)]^2}\right) \\ &= \sum_r z_r \delta_{rj} \frac{1 + f(k_r - 1) - (k_r - 1) f}{[1 + f(k_r - 1)]^2} \\ &= \frac{z_j}{[1 + f(k_j - 1)]^2} \\ &= \frac{x_j^2}{z_j}\end{aligned} \tag{8.148}$$

We also require derivatives of the Rachford–Rice equation with respect to the phase fraction, using the derivatives of the compositions;

$$\left[\frac{\partial x_i}{\partial f}\right] = -\frac{x_i^2}{z_i}(k_i - 1) = -\frac{x_i (y_i - x_i)}{z_i} \tag{8.149}$$

and:

$$\left[\frac{\partial y_i}{\partial f}\right] = k_i \left[\frac{\partial x_i}{\partial f}\right] = -\frac{y_i(y_i - x_i)}{z_i} \tag{8.150}$$

The need for these derivatives is primarily in solving the Rachford–Rice equation for the phase fraction in flash calculations via successive substitution. The target function is:

$$F = RR = \sum_r (y_r - x_r) = \sum_r \frac{z_r(k_r - 1)}{1 + f(k_r - 1)} = 0 \tag{8.151}$$

while the derivative is:

$$\begin{aligned} H &= \frac{\partial F}{\partial f} = \frac{\partial RR}{\partial f} = \sum_r \left(\frac{\partial y_i}{\partial f} - \frac{\partial x_i}{\partial f}\right) \\ &= -\sum_r \frac{(y_r - x_r)^2}{z_r} = -\sum_r \frac{z_r(k_r - 1)^2}{(1 + f(k_r - 1))^2} \end{aligned} \tag{8.152}$$

As we can see, the Rachford–Rice equation is monotonic[20] with respect to several variables considered here. This makes solving the Rachford–Rice equation for these variables quite straightforward, with the initialization being the more important consideration. In the case of phase fraction for two phases, we can narrow the search range f considerably. As demonstrated by Michelsen and Mollerup [Michelsen book], for a species with $k_i > 1$, very small values of f may yield $y_i > 1$ (see Eq. 8.140). We investigate this by solving the following inequality for $k_i > 1$;

$$\frac{k_i z_i}{1 + f(k_i - 1)} < 1$$

yielding f values:

$$f > \frac{k_i z_i - 1}{k_i - 1}$$

We then select:

$$f_{min} = max\left(\frac{k_i z_i - 1}{k_i - 1}\right) \tag{8.153}$$

[20] H does not change sign.

Table 8.6 Initialization and convergence of two-phase Rachford–Rice. Conditions excerpted from Example 24

Iteration	f	F Eq. 8.151	H Eq. 8.152	$df = -\frac{F}{H}$
1	0.536	0.048	−1.355	0.036
2	0.572	1.674E-4	−1.348	1.241E-4
3	0.572	6.951E-12	−1.348	5.155E-12

If we find $f_{min} > 0$, this value can be used to set a lower limit.[21] Otherwise, we set $f_{min} = 0$. In the same way, $k_i < 1$ raises the possibility that $x_i > 1$ (see Eq. 8.139). Accordingly we have:

$$\frac{z_i}{1 + f(k_i - 1)} < 1$$

yielding f values:[22]

$$f < \frac{z_i - 1}{k_i - 1}$$

We accordingly select:

$$f_{max} = min(\frac{z_i - 1}{k_i - 1}) \tag{8.154}$$

Analogous to the above, if we find $f_{max} < 1$, this value can be used to set an upper limit. Otherwise, we set $f_{max} = 1$. With the two limits set, we then initialize our search for the phase fraction via:

$$f_{init} = \frac{f_{min} + f_{max}}{2} \tag{8.155}$$

Let's look at an example of using these limits and solving the two-phase Rachford–Rice equation.

Example 45 (Solving the Two-Phase Rachford–Rice Equation) Suppose we have a two-component mixture with $z = 0.5, 0.5$ for which $lnk = 1.494, -1.197$ (see Example 24). Using the limits found in this section, we have $f_{min} = 0.355$ (Eq. 8.153) and $f_{max} = 0.716$ (Eq. 8.154), generating the initial guess of $f = 0.536$. We can then solve the Rachford–Rice equation via Newton's method with Eqs. 8.151 and 8.152. As shown in Table 8.6 the initialization and monotonic behavior leads in this case to rapid convergence.

[21] Alternatively, you may wish to solve what is called a Negative Flash (a flash with $f < 0$ or $f > 1$), which is not dealt with in this text.

[22] Recall that $k_i < 1$, and the effect of negative values on an inequality.

8.16.2 Multiphase Rachford–Rice

The Rachford–Rice equation can be written for more than two phases. We begin by rewriting Eq. 8.138 for N phases:

$$z_i = f^1 c_i^1 + f^2 c_i^2 + \ldots + f^N c_i^N \tag{8.156}$$

We now select a reference phase, which throughout this text and often in computational modeling is taken to be the phase with the highest (mass) density. We then rewrite Eq. 8.156 as:

$$\begin{aligned} z_i &= f^R c_i^R + \sum_{s \neq R} f^s c_i^R \\ &= (1 - \sum_{s \neq R}) c_i^R + \sum_{s \neq R} f^s c_i^s \\ &= c_i^R + \sum_{s \neq R} f^s (c_i^s - c_i^R) \end{aligned} \tag{8.157}$$

We develop the analogy to Eq. 8.139 by solving Eq. 8.156 for the reference phase composition:

$$\begin{aligned} c_i^R &= \frac{z_i}{1 + \sum_{s \neq R} f^s (\frac{c_i^s}{c_i^R} - 1)} \\ &= \frac{z_i}{1 + \sum_{s \neq R} f^s (k_i^s - 1)} \end{aligned} \tag{8.158}$$

Equation 8.158 implies a definition of the k value of species i in phase s ($s \neq R$), allowing the composition in any other phase[23] to be written as:

$$\begin{aligned} c_i^j &= k_i^j c_i^R \\ &= \frac{k_i^j z_i}{1 + \sum_{s \neq R} f^s (k_i^s - 1)} \end{aligned} \tag{8.159}$$

[23] Note that we cannot use index s to refer to an arbitrary phase; we have used this index for summation over the non-reference phases.

The multiphase Rachford–Rice equation is analogous to Eq. 8.142

$$RR^j = \sum_r c_r^j - \sum_r c_r^R = \sum_r \frac{z_r(k_r^j - 1)}{1 + \sum_{s \neq R} f^s(k_r^s - 1)} = 0 \tag{8.160}$$

The multiphase Rachford–Rice equation (and the composition expressions) is formally dependent on k and f values. Derivatives of these equations with respect to these variables are therefore necessary. It should be noted carefully that the composition of species i in any phase depends on the k values in *all* of the phases, but *only* on those of species i; the derivatives with unequal composition indices are zero. The derivatives of the compositions with respect to lnk values are then, beginning with Eq. 8.158:

$$\begin{aligned}
\left[\frac{\partial c_i^R}{\partial lnk_i^t}\right] &= -\frac{(c_i^R)^2}{z_i} \sum_{s \neq R} f^s \frac{\partial k_i^s}{\partial k_i^t} k_i^t \\
&= -\frac{(c_i^R)^2}{z_i} \sum_{s \neq R} f^s \delta_{st} k_i^t \\
&= -\frac{(c_i^R)^2}{z_i} f^t k_i^t \\
&= -\frac{c_i^R c_i^t}{z_i} f^t
\end{aligned} \tag{8.161}$$

Equation 8.161 differs in appearance from its analogue Eq. 8.146, as we are now using j to represent the phase index, and we are differentiating with respect to lnk. This aside, the agreement between the two equations can be clearly seen. Now with Eq. 8.159, we have:

$$\begin{aligned}
\left[\frac{\partial c_i^j}{\partial lnk_i^t}\right] &= \left[\frac{\partial (k_i^j c_i^R)}{lnk_i^t}\right] \\
&= \delta_{jt} k_i^t c_i^R + k_i^j \left[\frac{\partial c_i^R}{lnk_i^t}\right] \\
&= \delta_{jt} c_i^j - \frac{c_i^j c_i^t}{z_i} f^t \\
&= c_i^j \left(\delta_{jt} - \frac{c_i^t}{z_i} f^t\right)
\end{aligned} \tag{8.162}$$

The derivatives of the compositions with respect to phase fractions are:

$$\left[\frac{\partial c_i^R}{\partial f^t}\right] = -\frac{(c_i^R)^2}{z_i}\sum_{s\neq R}\frac{\partial f^s}{\partial f^t}(k_i^s - 1)$$
$$= -\frac{(c_i^R)^2}{z_i}\sum_{s\neq R}\delta_{st}(k_i^s - 1)$$
$$= -\frac{(c_i^R)^2(k_i^t - 1)}{z_i}$$
$$= -c_i^R\frac{(c_i^t - c_i^R)}{z_i} \tag{8.163}$$

and

$$\left[\frac{\partial c_i^j}{\partial f^t}\right] = \left[\frac{\partial k_i^j c_i^R}{\partial f^t}\right]$$
$$= -k_i^j c_i^R\frac{(c_i^t - c_i^R)}{z_i}$$
$$= -c_i^j\frac{(c_i^t - c_i^R)}{z_i} \tag{8.164}$$

With these results, we can write the derivatives of the multiphase Rachford–Rice equation with respect to the lnk and f values:

$$\left[\frac{\partial RR^j}{\partial lnk_q^t}\right] = \sum_r\left(\frac{\partial c_r^j}{\partial lnk_q^t} - \frac{\partial c_r^R}{\partial lnk_q^t}\right) \tag{8.165}$$

As stated in the developments in Eqs. 8.161 and 8.162:

$$\left[\frac{\partial c_i}{\partial lnk_q}\right] = \delta_{iq}\left[\frac{\partial c_q}{\partial lnk_q}\right]$$

Therefore:

$$\left[\frac{\partial RR^j}{\partial lnk_q^t}\right] = \frac{\partial c_q^j}{\partial lnk_q^t} - \frac{\partial c_q^R}{\partial lnk_q^t} \tag{8.166}$$

The derivatives of the Rachford–Rice equation with respect to phase fraction is:

$$\left[\frac{\partial RR^j}{\partial f^t}\right] = \sum_r \left(\frac{\partial c_r^j}{\partial f^t} - \frac{\partial c_r^R}{\partial f^t}\right) \tag{8.167}$$

This last derivative is used in solving the multiphase Rachford–Rice equation for the phase fractions.

Example 46 (Solving the Multiphase Rachford–Rice equation) Suppose we have a three-component mixture with $z = \frac{1}{3}, \frac{1}{3}, \frac{1}{3}$ for which:

$$ln\overline{\overline{k}} = \begin{bmatrix} 10.779 & 12.284 \\ 19.801 & 18.601 \\ -2.831 & -2.585 \end{bmatrix}$$

and the independent phase fractions are initialized as:

$$\overline{f} = \begin{bmatrix} 0 \; 0.721 \end{bmatrix}$$

(see Example 25).

We can then solve the Rachford–Rice equation via Newton's method with Eqs. 8.160 and 8.167. As shown in Table 8.7 the initialization and monotonic behavior leads in this case to rapid convergence.

It is worth noting that there are only two independent phase fractions, and the phase with initially zero phase fraction was introduced as the incipient phase from a stability analysis.

8.17 Spinodal Volumes vs. the Spinodal Boundary

The term spinodal *volume* is used in root solving and refers to the condition:

$$\frac{\partial P}{\partial V} = 0$$

In contrast, the term spinodal *boundary* describes the condition where a mixture is inherently unstable with respect to phase separation, with:

$$det\left(\overline{\overline{\frac{\partial^2 A}{\partial n^2}}}\right) = 0.$$

Table 8.7 Initialization and convergence of multiphase Rachford–Rice. Conditions excerpted from Example 25

Iteration	$\overline{f}$	$\overline{\overline{c}}$ Eqs. 8.158 and 8.159	$\overline{F}$ Eq. 8.160	$\overline{\overline{H}}$ Eq. 8.167	$d\overline{f}$ Eq. 8.111
1	0, 0.721	2.139E-6, 0.103, 0.462 3.862E-9, 1.535, 0.462 0.999, 0.059, 0.075	0.696, −2.547E-13	−9.754, −4.881 −4.881, −3.847	0.196, −0.248
2	0.196, 0.473	2.987E-6, 0.143, 0.646 2.481E-9, 0.986, 0.297 0.880, 0.052, 0.066	0.301, 0.129	−5.037, −3.179 −3.179, −3.501	0.0857, −0.041
3	0.281, 0.432	3.119E-6, 0.150, 0.674 2.039E-9, 0.810, 0.244 0.992, 0.059, 0.075	0.027, 1.026E-3	−4.652, −3.465 −3.465, −4.066	0.015, −0.013
4	0.296, 0.419	3.179E-6, 0.153, 0.687 1.985E-9, 0.789, 0.238 0.999, 0.059, 0.075	5.945E-4, 3.6715E-4	−4.591, −3.485 −3.485, −4.148	1.7201E-4, −5.601E-5
5	0.297, 0.419	3.179E-6, 0.153, 0.687 1.984E-9, 0.788, 0.237 0.999, 0.059, 0.075	3.863E-9, −6.785E-8	−4.591, −3.487 −3.487, −4.150	3.662E-8, −4.712E-8

These two definitions are confusing, as they are not in general consistent with each other (one does not in general imply the other). In this appendix, we examine the connection between these two formalisms. We start by recognizing that pressure and chemical potential are each 0th order homogeneous functions of V and $\overline{n}$ (see Sect. 8.6):

$$\frac{\partial P}{\partial V}V + \sum_r \frac{\partial P}{\partial n_r}n_r = 0 \tag{8.168}$$

$$\frac{\partial \mu_i}{\partial V}V + \sum_r \frac{\partial \mu_i}{\partial n_r}n_r = 0 \tag{8.169}$$

The relationship between pressure and chemical potential is through mixed partial derivatives of the Helmholtz free energy:

$$\frac{\partial P}{\partial n_i} = -\frac{\partial^2 A}{\partial V \partial n_i} = -\frac{\partial^2 A}{\partial n_i \partial V} = -\frac{\partial \mu_i}{\partial V} \tag{8.170}$$

Rewriting Eq. 8.169 via Eq. 8.170, we have:

$$-V\sum_s \frac{\partial P}{\partial n_s}n_s + \sum_s \sum_r \frac{\partial \mu_s}{\partial n_r}n_r n_s = 0 \tag{8.171}$$

Adding Eq. 8.168 and 8.171, noting that a single index of summation is arbitrary, and the symmetry of mixed partial derivatives, we have:

$$\frac{\partial P}{\partial V}V^2 + \sum_r \sum_s \frac{\partial \mu_r}{\partial n_s}n_r n_s = 0 \tag{8.172}$$

Equation 8.172 is generally true, but for a pure substance ($r = s = 1$), it can be written as:

$$\frac{\partial P}{\partial V}V^2 + \frac{\partial \mu}{\partial n}n^2 = 0 \tag{8.173}$$

This result states that, for a pure substance, at a point where $\frac{\partial P}{\partial V} = 0$ we have that $\frac{\partial \mu}{\partial n} = 0$. This is (trivally) equivalent to $det\frac{\partial \mu}{\partial n} = 0$. This result, which is not true for mixtures, is the reason why, in the context of root solving, points where $\frac{\partial P}{\partial V} = 0$ are (misleadingly called) *spinodal volumes*.

8.18 Successive Substitution

Successive substitution is a simple method for solving nonlinear equations of the form:

$$F(x) = 0 \tag{8.174}$$

The idea is to recast F into:

$$x^{(i+1)} = G(x^{(i)}) \tag{8.175}$$

where i is the iteration number. Let's look at a simple one-dimensional example.

Example 47 (Successive Substitution) Suppose we wish to solve:

$$F(x) = x - \sqrt{x} = 0.$$

Instead of using Newton's method, which requires computation of the derivative of G, we can recast the problem into:

$$x = G(x) = \sqrt{x} \tag{8.176}$$

We then make an initial guess based on physical insights into the problem or from previous results. In this example, suppose we choose $x = 3$. Via Eq. 8.176, the updates to x are (1.732, 1.316, 1.472, 1.071. . .). The process is continued until the change in x (or G) is sufficiently close to 0. Figure 8.6 shows the convergence to the root ($F = 0$) at $x = 1$.

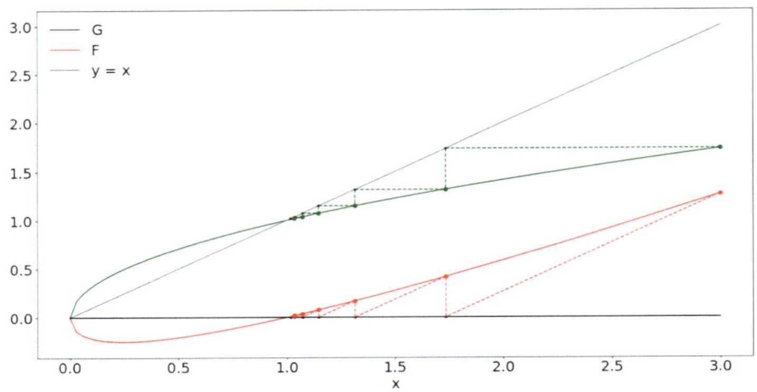

Fig. 8.6 Root finding for $F = x - \sqrt{(}x)$ by successive substitution, recasting into $G = x = \sqrt{(}x)$, with initial guess $x = 3$. The dashed lines shown the respective updates to F and G

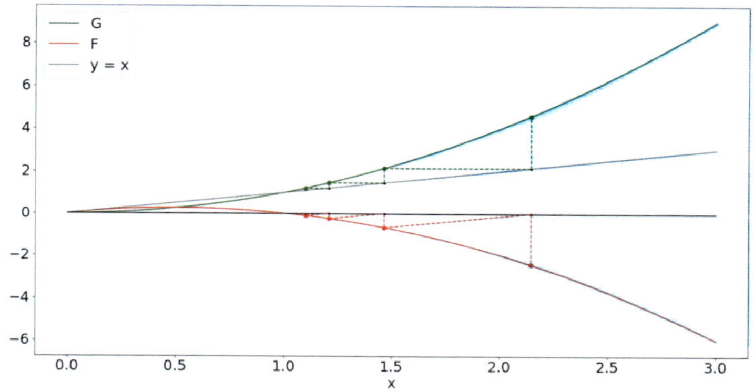

Fig. 8.7 Root finding for $F = x^2 - x$ by successive substitution, recasting into $G = x = x^2$, with initial guess $x = 1.1$. The dashed lines shown the respective updates to F and G

Figure 8.6 also shows the essential feature of Successive Substitution; instead of paying the computational cost of calculating the derivative, a slope of one[24] is assumed. Successive substitution, like any nonlinear solver, is not foolproof. We can see by eye in Fig. 8.6 that an initial guess $x < 1$ will also converge to the root at $x = 1$, but we cannot use successive substitution to find the root at $x = 0$. The issue is that a slope of one is a reasonable approximation to the tangent to F for $x > 0.25$, but is of incorrect sign for $x < 0.25$. This limitation is avoided by applying Newton's method to F; initial guesses with $x < 0.25$ converge to the root at $x = 0$ (provided stepsize control is used), while initial guesses with $0.25 < x < 1.0$ converge to $x = 1.0$. If we repeat this analysis with $F(x) = x - x^2$, we fail to converge to the root $x = 1$, even starting at an initial guess of $x = 1.1$. This is shown in Fig. 8.7, which emphasizes that simply rewriting an arbitrary function F as $x + (F - x)$, and using Successive Substitution on $G = -(F - x)$, does not in the general case lead to convergence.

By eye we can see in Fig. 8.7 that, for initial guesses of $x < 1$, Successive Substitution will converge to the root at $x = 0$ (never to the root $x = 1$), while Newton's method can converge to either of these, depending on the initial guess.

Successive substitution can also be used to solve multidimensional (non-scalar) problems. As shown in this text, stability and flash calculations are important examples. Equation 6.73 serves to illustrate why:

$$F_i = lnW_i + ln\hat{\varphi}_i(\overline{W}) - d_i = 0$$

[24] The successive substitution update is $x^{k+1} = G(x^k)$. With $F(x) = x - F(x)$, this update is written as $x^{k+1} = x^k - F(x^k)$. In terms of a Newton update to x in solving $F = 0$, this implies, according to Eq. 8.102, $\frac{dF}{dx} = 1$.

For cases where the fugacity coefficients are weak functions of composition;

$$[\frac{\partial ln\hat{\varphi}_i}{\partial W_j}]_{T,P,W_{k\neq j}} \approx 0$$

the derivative of the above equation (refer to Eq. 6.78) yields:

$$H_{ij} = \frac{\partial F_i}{\partial ln W_j} = \delta_{ij} + \left[\frac{\partial ln\hat{\varphi}_i}{\partial W_j}\right]_{T,P,W_{k\neq j}} W_j \approx \delta_{ij}$$

This is the multidimensional analogue of the scalar case articulated above; successive substitution is reliable when the Jacobian is well approximated by the identity matrix. Example 48 illustrates this further.

Example 48 (Successive Substitution at Fixed Pressure vs. Volume) Let us revisit Example 17. At $T = 350$K, $P = 5$ bar, we seek a liquid-like phase $\overline{W} = 0.083, 1.920$, we have (via Eq. 6.78):

$$\overline{\overline{H}} = \begin{bmatrix} 0.997 & 3.182E-3 \\ 1.384E-4 & 0.999\ldots \end{bmatrix}$$

The above result indicates that the condition for successive substitution, a Jacobian well-approximated by the identity matrix, is fulfilled for this example.[25] This allows us to solve multicomponent problems with reliable convergence without needing to compute Jacobians As has been emphasized in this subsection, successive substitution is not applicable. This can be illustrated by revisiting Example 18 and considering Eq. 6.89;

$$F_i = ln(W_i) + \frac{1}{R_g T}\left(\frac{\partial A^{Res}}{\partial W_i}\right)_{T,V,W_{k\neq i}} - b_i$$

the derivative of which is given by Eq. 6.91:

$$H_{ij} = \delta ij + \frac{1}{R_g T}\left(\frac{\partial^2 A^{Res}}{\partial W_i \partial W_j}\right) W_j$$

As shown in Table 6.9 At $T = 350K$, $P = 5bar$ and $\overline{W} = 1.648, 37.894$, with a volume given by the solver as $V = 6223.64cm^3$, we have (via Eq. 6.78):

$$\overline{\overline{H}} = \begin{bmatrix} 1.165 & 6.084 \\ 0.265 & 10.315 \end{bmatrix}$$

[25] Michelsen and Mollerup [Michelsen book]discuss the convergence of successive substitution in further detail.

For this fixed volume problem, the conditions required for successive substitution are not fulfilled (the Jacobian differs significantly from the identity matrix), and as a result, stability and flash calculations at fixed TVn will not reliably converge by this method.

8.18.1 Accelerated Successive Substitution (Dominant Eigenvalue Method)

Successive substitution is a simple and robust method of converging stability and flash calculations at fixed TPn. Its primary advantage is the avoidance of fugacity coefficient derivatives, while its primary disadvantage the slow rate of convergence near critical states. This can be understood by recognizing that successive substitution replaces the Jacobian with the identity matrix (see Sect. 8.18). Suppose we are performing a stability analysis at fixed T, P and $\overline{z}$ (Sect. 6.4.1), attempting to solve Eq. 6.73:

$$[lnW_i - lnz_i] + [ln\hat{\varphi}_i(\overline{W}) - ln\hat{\varphi}_i(\overline{z})] = 0.$$

The update from successive substitution becomes:

$$[lnW_i - lnz_i] = -[ln\hat{\varphi}_r(\overline{W}) - ln\hat{\varphi}_r(\overline{z})]$$

This form may seem strange, but we need to cast the problem into the form of successive substitution; $\overline{x} = \overline{F}(\overline{x})$. The factoring we've used allows us to recognize that, near the critical point, $[ln\hat{\varphi}_r(\overline{W}) - ln\hat{\varphi}_r(\overline{z})] \rightarrow 0$, leading to very small updates to the quantity $[lnW_i - lnz_i]$. This implies small updates to lnW_i.

While a second-order (Newton) approach can address this, it requires the derivatives of $ln\hat{\varphi}_r(\overline{W})$. Accelerated successive substitution offers us to the chance to retain convergence behavior of successive substitution without the burden of computing these derivatives. If we consider how Newton's method would update[26] $\overline{x}$ in an equation $\overline{x} + \overline{F}(\overline{x}) \rightarrow \overline{0}$ we would have:

$$d\overline{x} = -\overline{\overline{H}}(\overline{x}^k)^{-1} * \overline{F}(\overline{x}^k)$$

The next iteration becomes:

$$\overline{x}^{k+1} = \overline{x}^k - \overline{\overline{H}}(\overline{x}^k)^{-1} * \overline{F}(\overline{x}^k)$$

[26] For stability analysis $\overline{x} = ln\overline{W}$ and $\overline{F} = [ln\hat{\varphi}_r(\overline{W}) - ln\hat{\varphi}_r(\overline{z})]$.

The offset between an iteration and the ultimate solution (i.e., $\overline{x}^{\infty}$) is the error:

$$e^k = \overline{x}^k - \overline{x}^{\infty} \tag{8.177}$$

This gives:

$$\overline{e}^{k+1} = \overline{e}^k - \overline{\overline{H}}(\overline{x}^k)^{-1} * \overline{F}(\overline{x}^k) \tag{8.178}$$

A Taylor series of $\overline{F}(\overline{x}^k)$ in the vicinity of the solution (where $\overline{F}(\overline{x}^{\infty}) = \overline{0}$) gives:

$$\begin{aligned} \overline{F}(\overline{x}^k) &= \overline{F}(\overline{x}^{\infty}) + \overline{\overline{H}}(x^{\infty}) * (\overline{x}^k - \overline{x}^{\infty}) \\ &= \overline{\overline{H}}(x^{\infty}) * (\overline{x}^k - \overline{x}^{\infty}) \\ &= \overline{\overline{H}}(\overline{x}^{\infty}) * e^k \end{aligned} \tag{8.179}$$

Substitution of Eq. 8.179 into Eq. 8.178 gives:

$$\begin{aligned} \overline{e}^{k+1} &= \overline{e}^k - \overline{\overline{H}}(\overline{x}^k)^{-1} * \overline{\overline{H}}(\overline{x}^{\infty}) * \overline{e}^k \\ &= \overline{\overline{H}}(\overline{x}^k)^{-1} * \overline{\overline{H}}(x^k) * \overline{e}^k - \overline{\overline{H}}(\overline{x}^k)^{-1} * \overline{\overline{H}}(\overline{x}^{\infty}) * \overline{e}^k \\ &= \overline{\overline{H}}(\overline{x}^k)^{-1} * \left[\overline{\overline{H}}(x^k) - \overline{\overline{H}}(\overline{x}^{\infty})\right] * \overline{e}^k \\ &= \overline{\overline{M}}(\overline{x}^k) * \overline{e}^k \end{aligned} \tag{8.180}$$

In the final form of Eq. 8.180, we have shown that sequential values of the error ($\overline{e}$) are related by a matrix $\overline{\overline{M}}$. As we approach the critical point, and the rate of successive substitution convergence slows, the difference in subsequent $\overline{x}$ will be small, allowing us to regard the elements of $\overline{\overline{M}}$ to be relatively constant. As discussed by Michelsen and Mollerup [Michelsen book], we can write the error vectors in terms of normalized[27] eigenvectors of $\overline{\overline{M}}$, $\overline{u}$. Using their notation:

$$\overline{e}^k = \sum_j c_r \overline{u}_j \tag{8.181}$$

[27] See Sect. 8.12.1.1, particularly the comment that the eigenvalues of a matrix are not unique.

Substitution of Eq. 8.181 in Eq. 8.180 yields:

$$\begin{aligned}
\overline{e}^{k+1} &= \overline{\overline{M}}\overline{e}^{k} \\
&= \overline{\overline{M}}\sum_j c_j \overline{u}_j \\
&= \sum_j c_j \overline{\overline{M}}\overline{u}_j \\
&= \sum_j c_j \lambda_j \overline{u}_j
\end{aligned} \tag{8.182}$$

For an arbitrary iteration, we have:

$$\overline{e}^{k+m} = \sum_j c_j \lambda_j^m \overline{u}_j \tag{8.183}$$

As pointed out by Michelsen and Mollerup [Michelsen book], convergence requires that the eigenvalues must all be smaller than one in magnitude. The *dominant* eigenvalue method (DEM) considers the largest magnitude eigenvalue by truncating Eq. 8.183:

$$\overline{e}^{k+m} \approx c_1 \lambda_1^m \overline{u}_1 \tag{8.184}$$

where index 1 is that of the dominant eigenvalue.[28]

The next iteration becomes:

$$\overline{e}^{k+m+1} \approx c\lambda^{m+1}\overline{u} = \lambda c \lambda^m \overline{u} = \lambda \overline{e}^{k+m} \tag{8.185}$$

Convergence is then linearly dependent on the dominant eigenvalue, where the subscript 1 has been dropped. Continuing with the development of Michelsen and Mollerup [Michelsen book], (but using the index I to be distinct from the previous use of k and m), we now consider the case where several iterations of successive substitution have been performed, yielding an error at iteration I:

$$\overline{e}^{I} \approx c\overline{u}$$

The next two iterations yield (via Eq. 8.185):

$$\begin{aligned}
\overline{e}^{I+1} &\approx \lambda c \overline{u} \\
\overline{e}^{I+2} &\approx \lambda^2 c \overline{u}
\end{aligned}$$

[28] For clarity, this algorithm does not call for all the eigenvalues to be determined; we are simply saying that we can approximate by considering only the eigenvalue that is largest in magnitude.

Computing differences in subsequent errors, recognizing by Eq. 8.177 that a difference in errors is a difference in $\overline{x}$, we have:

$$\begin{aligned}\overline{\Delta}^{I+1} &= \overline{e}^{I+1} - \overline{e}^{I} = \overline{x}^{I+1} - \overline{x}^{I} \approx (\lambda - 1)c\overline{u} \\ \overline{\Delta}^{I+2} &= \overline{e}^{I+2} - \overline{e}^{I+1} = \overline{x}^{I+2} - \overline{x}^{I+1} \approx \lambda(\lambda - 1)c\overline{u}\end{aligned} \tag{8.186}$$

These differences (i.e., $\overline{\Delta}$ values are known from differences in $\overline{x}$) provide sufficient information to compute λ and $c\overline{u}$:

$$\begin{aligned}(\overline{\Delta}^{I+2})^T * \overline{\Delta}^{I+1} &\approx \lambda(\lambda - 1)^2 c^2 (\overline{u}^T * \overline{u}) = \lambda(\lambda - 1)^2 c^2 \\ (\overline{\Delta}^{I+1})^T * \overline{\Delta}^{I+1} &\approx (\lambda - 1)^2 c^2 (\overline{u}^T * \overline{u}) = (\lambda - 1)^2 c^2\end{aligned} \tag{8.187}$$

Note that the superscript T symbolizes the transpose operator (Appendix 8.12.3). If you prefer not to use this, consider the multiplication of the vectors in Eq. 8.187 as dot products. Also, since $\overline{u}_1$ is normalized, $(\overline{u}_1^T * \overline{u}_1) = 1$.

By division we have:

$$\lambda \approx \frac{(\overline{\Delta}^{I+2})^T * \overline{\Delta}^{I+1}}{(\overline{\Delta}^{I+1})^T * \overline{\Delta}^{I+1}} \tag{8.188}$$

Substitution of λ into Eq. 8.186 gives:

$$c\overline{u} \approx \frac{\overline{\Delta}^{I+1}}{(\lambda - 1)} \tag{8.189}$$

With $\overline{e}^{I+2} = \overline{x}^{k+2} - \overline{x}^{\infty} \approx \lambda^2 c\overline{u}$, we solve:

$$\overline{x}^{\infty} \approx \overline{x}^{I+2} - \lambda^2 c\overline{u} \tag{8.190}$$

As discussed by Michelsen and Mollerup [Michelsen book], the denominator term $(\lambda - 1)$ has a large effect on our estimate of $\overline{x}^{\infty}$. Therefore it is advisable to limit the use of acceleration when $\lambda << 1$.

Let's clarify the acceleration by DEM with an example of its application to stability analysis, where we will use $ln\overline{W}$ as $\overline{x}$.

Example 49 (Converging a Stability Analysis at Fixed T, P, and $\overline{n}$ Using the Dominant Eigenvalue Method) Consider the conditions in Fig. 6.7; 1 mole of propane and 1 mole of n-hexane at $T = 463.55K$, $P = 45.60$ bar bar. As stated in the discussion of Fig. 6.7, these conditions are very close to the mixture's critical point. Table 8.8 summarizes the result for a liquid-like initialization, where acceleration is applied after the 10th successive subsitution.

Table 8.8 Iterations prior to application of acceleration (DEM) for equimolar propane / n-hexane at $T = 463$K and $P = 45.5$ bar (near critical) using SRK EOS. Err is calculated as the maximum magnitude term in Eq. 6.73

I	lnW^I	$lnErr$	$\Delta^I * 10^3$	λ	$cu * 10^2$	lnW^∞
8	−0.754, −0.636	−5.205				
9	−0.750, −0.639	−5.431	4.380, −3.869			
10	−0.746, −0.643	−5.630	3.589, −3.200	0.823	−2.472, 2.183	−0.729, −0.659
11	−0.729, −0.659	−6.721				

The effect of acceleration by DEM is shown in Fig. 6.7, and the result of the stability analysis is $W = 0.514, 0.486$, with $t_m = -2.373E - 6$.

incorporating acceleration by DEM into a stability analysis based on successive substitution is quite straightforward, as indicated Algorithm 34.

Acceleration (DEM) can also be applied to flash calculations. In a two-phase flash calculation, we seek to solve Eq. 6.121.

$$\overline{F} = \begin{cases} \overline{EQ} = ln\overline{k} + ln\hat{\varphi}^v(T, P, \overline{y}) - ln\hat{\varphi}^\ell(T, P, \overline{x}) = \overline{0} \\ RR = \sum_r \frac{z_r(k_r - 1)}{1 + f(k_r - 1)} = 0 \end{cases}$$

As explained in Sect. 6.7.1, solving a flash via successive substitution involves alternating between applying successive substitution to the equilibrium (EQ) relationship and numerically (i.e., not via successive substitution, but using the Jacobian) solving the Rachford–Rice (RR). This is a distinction from how successive substitution is applied to stability analysis. Also, in stability analysis the amount of the incipient phase (i.e., $\overline{W}$) has a zero phase fraction, meaning that during the stability analysis, there is no variation in the Gibbs free energy of the *actual* mixture; there *could* be if a negative tangent plane distance is detected and an additional phase is formed. However, in TP *flash* calculations, we seek to reduce the Gibbs free energy of the multiphase mixture while respecting the material balance:

$$\begin{aligned} G^{Mixture} &= \sum_j^{phases} \sum_i^{species} \mu_i^j n_i^j \\ &= \sum_j^{phases} \sum_i^{species} \mu_i^j c_i^j n^j \\ &= \sum_j^{phases} \sum_i^{species} \mu_i^j c_i^j f^j n \end{aligned} \tag{8.191}$$

Algorithm 34 Incorporating DEM into Successive Substitution TPn stability calculations

```
Calculate d (Eq. 6.54)
Set Iter = 1, MaxIter = 100, Tol = 1E − 8, Flag = True
Set A_Count = 1, A_Thresh = 10
Initialize lnW̄^(I+1), lnW̄^(I+2) (as NANs, for example)
while (Flag == True) do
    W̄ = exp(lnW̄) (Extract W̄)
    Calculate lnφ̂(W̄) for the desired root
    Iter+ = 1
    Err = max(abs(lnW̄ + lnφ̂(W̄) − d̄))

    Update lnW̄ = d̄ − lnφ̂(W̄) (Successive substitution)
    A_Count+ = 1
    lnW̄^(I+2) = lnW̄
    if A_Count ≥ A_Thresh then
        A_Count = 0 (Reset the Acceleration Count)
        D1 = lnW^(I+1) − lnW^I
        D2 = lnW^(I+2) − lnW^(I+1)
        Calculate λ from Eq. 8.188
        Calculate cū from Eq. 8.189
        Accelerate: lnW̄ = lnW^(I+2) − (λ^2)cū
    else
        Pass
    end if
    Cycle: lnW̄^I = lnW̄^(I+1), lnW̄^(I+1) = lnW̄^(I+2)
    if (Iter ≥ MaxIter) or (Err < Tol) then
        Flag = False
    end if
end while
Check convergence
if (Iter ≥ MaxIter) then
    Return Error Message
else
    Return W̄
end if
```

We can rewrite Eq. 8.191 as the molar Gibbs free energy of the multiphase mixture:[29]

$$g^{Mixture} = \sum_{j}^{phases} \sum_{i}^{species} \mu_i^j c_i^j f^j \tag{8.192}$$

[29] G is a 1st order homogeneous function of T, P, and $\overline{n}$, as shown in Appendix 8.6.

We do not want to retain chemical potentials, due to their embedded reference states. We replace them via fugacity, specifically[30] using Eq. 1.80:

$$\mu_i^j = g_i + R_g T ln c_i^j + R_g T ln \hat{\varphi}_i^j - R_g T ln \varphi_i$$

For clarity we have not used the superscript *pure*, and we indicate the phase of interest using the index j. Substituting this form of Eq. 1.80 into Eq. 8.192, we have:

$$\begin{aligned} g^{Mixture} &= \sum_j^{phases} \sum_i^{species} (g_i + R_g T ln c_i^j + R_g T ln \hat{\varphi}_i^j - R_g T ln \varphi_i) c_i^j f^j \\ &= \sum_j^{phases} \sum_i^{species} g_i c_i^j f^j + \sum_j^{phases} \sum_i^{species} (R_g T ln c_i^j + R_g T ln \hat{\varphi}_i^j - R_g T ln \varphi_i) c_i^j f^j \\ &= \sum_i^{species} g_i \sum_j^{phases} c_i^j f^j + \sum_j^{phases} \sum_i^{species} (R_g T ln c_i^j + R_g T ln \hat{\varphi}_i^j - R_g T ln \varphi_i) c_i^j f^j \\ &= \sum_i^{species} g_i z_i + \sum_j^{phases} \sum_i^{species} (R_g T ln c_i^j + R_g T ln \hat{\varphi}_i^j - R_g T ln \varphi_i) c_i^j f^j \end{aligned} \tag{8.193}$$

The first term in the last line of Eq. 8.193 results from the material balance (Eq. 8.156) and represents the molar Gibbs free energy of the pure (separated) components. We can therefore rewrite Eq. 8.193 as the molar Gibbs free energy of mixing:[31]

$$\begin{aligned} \frac{g^{Mix}}{R_g T} &= \frac{g^{Mixture} - \sum_i^{species} g_i z_i}{R_g T} \\ &= \sum_j^{phases} \sum_i^{species} (ln c_i^j + ln \hat{\varphi}_i^j - ln \varphi_i) c_i^j f^j \end{aligned} \tag{8.194}$$

Here, since we are concerned with g^{Mix} at a particular T, P, and $\overline{n}$, we have factored out the constant $R_g T$. We must be cautious in applying acceleration to flash calculations, ensuring that the appropriate potential (Gibbs for TPn) is in fact reduced. Otherwise, the application of acceleration to flash calculations presents no

[30] Very often the ideal gas referenced fugacity expression (Eq. 1.80) is used, but here the reference to the pure component fugacities is more convenient.

[31] We developed the *single* phase *two-component* version of this equation in our study of tangent plane analysis (Eq. 6.50).

significant challenge; the variable $\overline{x}$ in Eq. 8.175 is now $ln\overline{k}$. Let's revisit the flash calculation in Example 24 at the near critical conditions given in Example 49.

Example 50 (Converging a Two-Phase Flash at Fixed T, P and $\overline{n}$ Using the Dominant Eigenvalue Method) Using the results of Example 49, we see that two[32] phases will coexist at equilibrium.

As is required for the calculation of g^{Mix}, the pure component fugacities are:

$$ln\overline{\varphi} = -0.172, -1.169$$

The initial compositions are:

$$\overline{\overline{c}} = \begin{bmatrix} 0.5 \; 0.514 \\ 0.5 \; 0.486 \end{bmatrix}$$

The phase fractions are:

$$\overline{f} = \begin{bmatrix} 1.0 \; 0.0 \end{bmatrix}$$

The volumes are:

$$\overline{v} = \begin{bmatrix} 328.068 & 351.678 \end{bmatrix} \quad \left(\frac{cm^3}{mol}\right)$$

With the mass densities (via Eq. 1.81) being;

$$\overline{\rho} = \begin{bmatrix} 0.199 & 0.183 \end{bmatrix} \quad \left(\frac{g}{cm^3}\right)$$

the phases are already correctly sorted. The fugacities coefficients are then:

$$ln\overline{\overline{\hat{\varphi}}} = \begin{bmatrix} 0.046 & 0.018 \\ -0.980 & -0.951 \end{bmatrix}$$

[32] According to the Gibbs Phase Rule (Eq. 1.63), a two-phase binary mixture has two degrees of freedom, which for a TPn flash are taken by T and P. While it is possible for a binary mixture to form three phases, there is only one degree of freedom. Having specified the temperature, there are no degrees of freedom remaining, and the three phase compositions and pressure can be determined by calculation. If the pressure we have specified happens to correspond exactly to the three-phase pressure, we will have two phases coexisting, unless our composition is exactly one of the three phase compositions. In this case, we will have one phase with all the material in it, with the other two phases being incipient (analogous to saturated liquid, with a steam quality of zero).

By generating $l n \bar{\bar{k}}$ values from the compositions, we can calculate the value of $\overline{\overline{EQ}}$ in Eq. 6.121;

$$\overline{EQ} = \begin{bmatrix} -2.356E-6 \\ -2.357E-6 \end{bmatrix}$$

from which the error (Err) is taken to be maximum of the absolute values of $\overline{EQ}$:

$$Err = 2.357E-6$$

This value should be compared to t_m in Example 49. We can then estimate the update to the $l n \bar{\bar{k}}$ values via Eq. 6.122:

$$\overline{lnk} = \begin{bmatrix} 0.028 \\ -0.029 \end{bmatrix}$$

We can now solve[33] the Rachford–Rice equation (Appendix 8.16.1), which gives:

$$f = 2.863E-3$$

The corresponding compositions are:

$$\bar{\bar{c}} = \begin{bmatrix} 0.49996 \; 0.5143 \\ 0.50004 \; 0.4857 \end{bmatrix}$$

This completes one successive substitution, which we can now continue through multiple iterations. The left side of Fig. 8.8 shows the convergence by successive substitution (SS), which for this near critical state is very slow, requiring more than 700 iterations.

Also interesting is that the left side of the figure indicates that early iterations are characterized by an *increase* in the error, though the right side of the figure shows that the Gibbs free energy decreases monotonically. It is also worth noting that the change in Gibbs energy between the feed and the two-phase state is very small, owing to the proximity to the critical point. The figure also shows that if we use DEM acceleration and Newton's method, we can significantly improve the performance. We continue the present example with a demonstration of this.

[33] If we attempted to solve the Rachford–Rice equation without first updating the lnk values (i.e., just using the lnk values from the compositions estimated from stability), we would of course produce phase fractions of 1, 0.

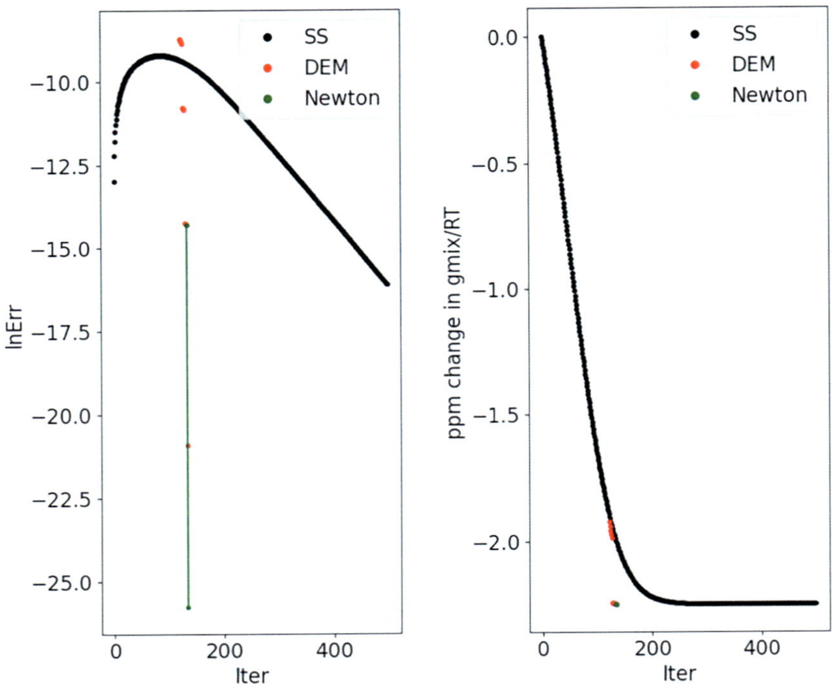

Fig. 8.8 Converging a two-phase flash of 1 mole of propane and 1 mole of n-hexane at $T = 463$K and $P = 45.5$ bar, using the SRK EOS. Acceleration attempted after 3 repeated decreases in error with $\lambda < 0.999$. Newton's method applied after 4 consecutive successful DEM accelerations

The acceleration terms leading to the 119th iteration are:

$$\overline{\Delta}_1 = ln\overline{k}_1 - ln\overline{k}_0 = \begin{bmatrix} 3.717E-2, -3.708E-2 \\ -3.772E-2 - -3.763E-2 \end{bmatrix} = \begin{bmatrix} 8.747E-5 \\ -8.583E-5 \end{bmatrix}$$

$$\overline{\Delta}_2 = ln\overline{k}_2 - ln\overline{k}_1 = \begin{bmatrix} 3.726E-2 - 3.717E-2 \\ -3.780E-2 - -3.772E-2 \end{bmatrix} = \begin{bmatrix} 8.689E-5 \\ -8.526E-5 \end{bmatrix}$$

$$\lambda = \frac{\overline{\Delta}_2 \overline{\Delta}_1}{\overline{\Delta}_1 \overline{\Delta}_1} = 0.993$$

$$c\overline{u} = \frac{\overline{\Delta}_1}{(\lambda - 1)} = \begin{bmatrix} -1.307E-2 \\ 1.282E-2 \end{bmatrix}$$

$$ln\overline{k}^{\infty} = ln\overline{k}^{2} - \lambda^2 c\overline{u} = \begin{bmatrix} 5.015E-2 \\ -5.046E-2 \end{bmatrix}$$

Table 8.9 Iterations leading to DEM acceleration for a two-phase flash of equimolar propane and n-hexane at $T = 463.55\text{K}$ and $P = 45.60\,\text{bar}$, using the SRK EOS

I	$\overline{\overline{c}}$	$\overline{v}$ (cm^3/mol)	$ln\overline{\overline{\hat{\varphi}}}$	$ln\overline{k}$	f
117	0.4944, 0.5130	310.845, 351.522	5.726E-2, 2.017E-2	3.708E-2	0.302
	0.5056, 0.4870		−0.9913, −0.95437	−3.763E-2	
118	0.4944, 0.5130	311.0489, 351.430	5.734E-2, 2.017E-2	3.717E-2	0.304
	0.5056, 0.4870		−0.991, −0.954	−3.772E-2	
119	0.4943, 0.5130	311.2403, 351.3454	5.742E-2, 2.017E-2	3.726E-2	0.305
	0.5057, 0.4870		−0.991, −0.954	−3.780E-2	

$$f = 0.439$$

$$\overline{\overline{c}} = \begin{bmatrix} 0.489, 0.514 \\ 0.511, 0.486 \end{bmatrix}$$

$$\overline{v} = \begin{bmatrix} 311.420 \; 351.267 \end{bmatrix}$$

$$ln\overline{\overline{\hat{\varphi}}} = \begin{bmatrix} 0.068 & 0.018 \\ -0.001 & -0.952 \end{bmatrix}$$

As stated above, it is important to ensure that the acceleration reduces[34] the Gibbs free energy:

$$\frac{{g^{Mix}}^{119}}{R_g T} = -0.489833672$$

$$\frac{{g^{Mix}}^{Accel}}{R_g T} = -0.489833679$$

Comparing the compositions and phase fractions in Table 8.9 to those resulting from the acceleration indicates that the effect is significant. After the first acceleration, the Err actually increases, but the rate of convergence is dramatically improved. Subsequent accelerations are successful and bring about more rapid convergence, as shown in Fig. 8.8. Figure 8.8 also shows the rapid convergence that is possible with Newton's method (see Example 43 in Appendix 8.13.1). As discussed throughout this text, convergence with Newton's method cannot always be assumed, and here it is applied after 4 successful accelerations.

[34] In this example, earlier attempts to accelerate yield *increased* $\frac{g^{Mix}}{R_g T}$ are therefore discarded.

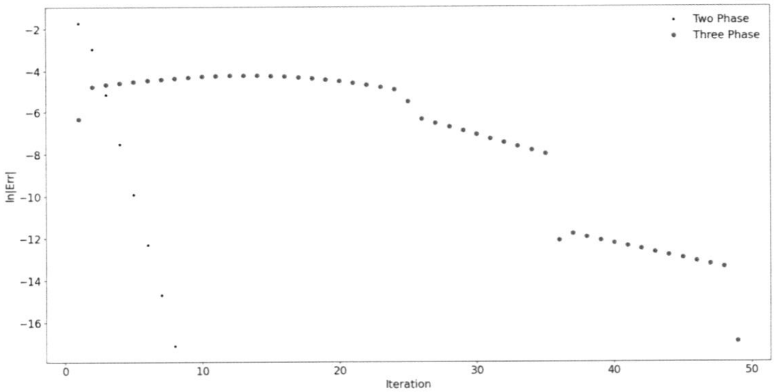

Fig. 8.9 Application of DEM acceleration in a three-phase flash for equimolar nitrogen, n-hexane, and water at $T = 478\text{K}$ and $P = 131\,\text{bar}$, using the SRK EOS. The two-phase flash converges rapidly by successive substitution, while the subsequent three-phase flash benefits from acceleration. The λ values for the three-phase acceleration at iteration 26 are 0.870 and 0.861

8.18.2 Acceleration in Multiphase TPn Flash Calculations

Acceleration in multiphase TPn flash calculations presents no major challenge beyond the two-phase case. Regardless of how many phases are identified, recall that stability analysis only considers whether one phase[35] can reduce its free energy by forming another, and therefore acceleration of TPn stability always takes the form of Example 49. For a multiphase TPn flash calculation, the acceleration terms applied to phase j can be structured as follows:

$$\overline{\Delta}_1^j = ln\overline{k}_1^j - ln\overline{k}_0^j$$

$$\overline{\Delta}_2^j = ln\overline{k}_2^j - ln\overline{k}_1^j$$

$$\lambda^j = \frac{\overline{\Delta}_2^j \overline{\Delta}_1^j}{\overline{\Delta}_1^j \overline{\Delta}_1^j}$$

$$(c\overline{u})^j = \frac{\overline{\Delta}_1^j}{(\lambda^j - 1)} \tag{8.195}$$

This is examined in Fig. 8.9, which shows that a two-phase flash of equimolar nitrogen, n-hexane, and water at $T = 478\text{K}$ and $P = 131$ bar yields:

$$\overline{f} = 0.081, 0.919$$

[35] This applies whether one is dealing with a single phase or multiple phases have already been identified. See the commentary in Example 24.

and

$$\bar{\bar{c}} = \begin{bmatrix} 9.882E-4 & 0.363 \\ 4.278E-5 & 0.363 \\ 0.999 & 0.274 \end{bmatrix}$$

This mixture is unstable ($t_m = -1.710E-3$), with an incipient phase composition of $\overline{w} = 0.494, 0.265, 0.241$. The three-phase flash stabilizes at:

$$\overline{f} = [0.076, 0.507, 0.417]$$

and

$$\bar{\bar{c}} = \begin{bmatrix} 9.857E-4 & 0.252 & 0.492 \\ 4.270E-5 & 0.439 & 0.266 \\ 0.999 & 0.309 & 0.242 \end{bmatrix}$$

To aid in understanding these results, one can consider that the two-phase system results from unfavorable mixing between the water and the other two species. Given the formal choice of only two phases, the water concentrates in one phase, while the nitrogen and n-hexane concentrates are almost equally in the other. The requirement for each chemical potential to be equal across the two phases (plus the material balance) yields the two-phase composition shown. However, when the possibility of forming a third phase is presented, unfavorable mixing between nitrogen and n-hexane can be expressed, yielding the three-phase solution shown. Though this type of visualization may be helpful, it is important to avoid thinking that, after the two-phase solution is found, *one* of the phases then separates; this is false. Though the water-rich[36] phase does not appear to change significantly between the two- and three-phase results, it is important to understand that all phases have adjusted between the two- and the three-phase solutions.

Michelsen and Mollerup [Michelsen book]discuss the limitations of applying acceleration to mixtures consisting of more than two phases.

8.19 Taylor Series

A Taylor series is an approximation of a function, based on a local value of the function and its derivatives at that point. We can represent a scalar function $F(x)$

[36] This is likely an aqueous liquid phase.

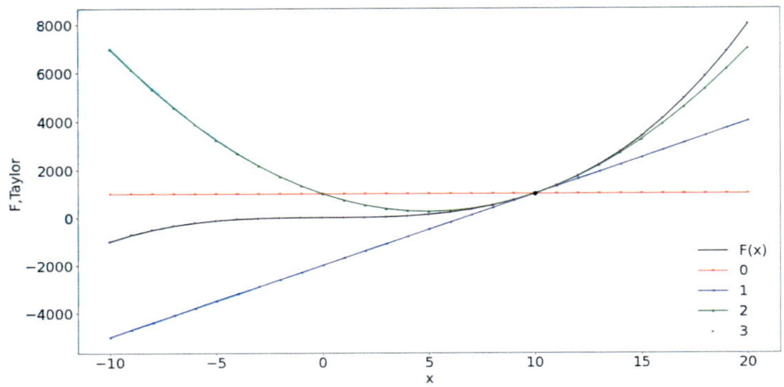

Fig. 8.10 Approximation of $F(x) = x^3$ about $x = 10$ by 0th through 3rd Taylor series (Eq. 8.197)

around the point a as:

$$F(x) = F(a) + \left[\frac{dF}{dx}\right]_{x=a}(x-a) + \frac{1}{2}\left[\frac{d^2F}{dx^2}\right]_{x=a}(x-a)^2 + \frac{1}{6}\left[\frac{d^3F}{dx^3}\right]_{x=a}(x-a)^3 + \ldots + \frac{1}{n!}\left[\frac{d^nF}{dx^n}\right]_{x=a}(x-a)^n \tag{8.196}$$

Example 51 provides further clarity.

Example 51 (Taylor Series) Suppose we have a function $F(x) = x^3$. With $\frac{dF}{dx} = 3x^2$, $\frac{d^2F}{dx^2} = 6x$, $\frac{d^3F}{dx^3} = 6$, and $\frac{d^{N>3}F}{dx^{N>3}} = 0$, Eq. 8.196 gives:

$$F(x) = a^3 + (3a^2)(x-a) + \frac{1}{2}(6a)(x-a)^2 + \frac{1}{6}(6)(x-a)^3 \tag{8.197}$$

We can truncate Eq. 8.197 to between 0th and 3rd order, as shown in Fig. 8.10.

As shown in Fig. 8.10, the approximations improve in accuracy as the order increases. In this example, 3rd order is exact for $F(x) = x^3$.

8.20 The Wilson Approximation

In many thermodynamic routines, k factors are considered independent variables, which must be determined numerically (e.g., by successive substitution or Newton's method). In such cases, initial guesses for the k factors are required. Wilson's k factor approximation [Michelsen book] provides initial estimates for vapor–liquid equilibria:

$$lnk_i{}^W = ln\frac{P_{ci}}{P} + 5.373(1+\omega_i)(1-\frac{T_{ci}}{T}) \tag{8.198}$$

T_c, P_c and ω are the critical temperature, critical pressure, and the acentric factor of the species of interest. This serves as an indication that, even if you are using an EOS that is not parameterized on these quantities (i.e., PC-SAFT), your code will still need access to these values.

For pure substances it also provides an estimate of saturation behavior. Setting $k_i^W = 0$ in Eq. 8.198 one has:

$$T_{sat\,i} \approx \frac{T_{ci}}{1 - \frac{ln\frac{P_{sat}}{P_{ci}}}{5.373(1+\omega_i)}} \tag{8.199}$$

$$P_{sat\,i} \approx P_{ci}exp\left(5.373(1+\omega_i)(1-\frac{T_{ci}}{T})\right) \tag{8.200}$$

Some applications require derivatives[37] of the lnK_i^W;

$$\left[\frac{\partial lnk_i^W}{\partial T}\right]_P = 5.373(1+\omega_i)\frac{T_{ci}}{T^2} \tag{8.201}$$

$$\left[\frac{\partial lnk_i^W}{\partial P}\right]_T = -\frac{1}{P} \tag{8.202}$$

An important observation on Eqs. 8.201 and 8.202 indicates that $lnk_i{}^W$ is monotonic in T and P; the signs of the respective derivatives do not change.

[37] Note that $\frac{dlnQ}{ds} = \frac{1}{Q}\frac{dQ}{ds}$. With $T, P, 1+\omega, k \geq 0$, the signs of $\left[\frac{\partial lnk_i^W}{\partial T}\right]_P$ and $\left[\frac{\partial lnk_i^W}{\partial P}\right]_T$ are positive and negative, respectively.

References

Michelsen, M. L.; Mollerup, J. M. *Thermodynamic Models: Fundamentals and Computational Aspects*; Tie-Line Publications, 2007.

Soave, G. Equilibrium Constants from a Modified Redlich–Kwong Equation of State. *Chem. Eng. Sci.* **1972**, *27*(6), 1197–1203.

Koningsveld, R.; Stockmayer, W. H.; Nies, E. *Polymer Phase Diagrams: A Textbook*; Oxford University Press: Oxford, U.K., 2001.

Gross, J.; Sadowski, G. Perturbed-Chain SAFT: An Equation of State Based on a Perturbation Theory for Chain Molecules. *Ind. Eng. Chem. Res.* **2001**, *40* (4), 1244–1260.

Mikyška, J.; Firoozabadi, A. Investigation of Mixture Stability at Given Volume, Temperature, and Number of Moles. *Fluid Phase Equilib.* **2012**, *321*, 1–9.

Heidemann, R. A.; Khalil, A. M. The Calculation of Critical Points. *AIChE J.* **1980**, *26* (5), 769–779.

Michelsen, M. L.; Hendriks, E. M. Physical Properties from Association Models. *Fluid Phase Equilib.* **2001**, *180*, 165–174.

Marshall, B. D. A PC-SAFT Model for Hydrocarbons IV:Water-Hydrocarbon Phase Behavior Including Petroleum Pseudo-Components. *Fluid Phase Equilib.* **2019**, *497*, 79–86.

B. R. Greenhalgh, *An Introduction to Computational Thermodynamics*,
https://doi.org/10.1007/978-3-032-30317-2

Index

B. R. Greenhalgh, *An Introduction to Computational Thermodynamics*,
https://doi.org/10.1007/978-3-032-30317-2

Kolibri 360 GmbH
Ferdinand-Jühlke-Straße 7
99095 Erfurt, Deutschland
produktsicherheit@kolibri360.de